GRADUATE STUDIES IN MATHEMATICS **204**

Hochschild Cohomology for Algebras

Sarah J. Witherspoon

AMERICAN MATHEMATICAL SOCIETY

Providence, Rhode Island

2010 *Mathematics Subject Classification.* Primary 16E40, 16E05, 16E30.

For additional information and updates on this book, visit
www.ams.org/bookpages/gsm-204

Library of Congress Cataloging-in-Publication Data

Cataloging-in-Publication Data has been applied for by the AMS.
See http://www.loc.gov/publish/cip/.
Softcover ISBN: 978-1-4704-6286-4

Contents

Introduction

Homological techniques first arose in topology, in work of Poincaré [174], at the end of the 19th century. They appeared in algebra several decades later in the 1940s, when Eilenberg and Mac Lane [59–61] introduced homology and cohomology of groups and Hochschild [114] introduced homology and cohomology of algebras. Since that time, both Hochschild cohomology and group cohomology, as they came to be called, have become indispensable in algebra, algebraic topology, representation theory, and other fields. They remain active areas of research, with frequent discoveries of new applications. There are excellent books on group cohomology such as [2, 21, 22, 35, 47, 76]. These are good references for those working in the field and are also important resources for those learning group cohomology in order to begin using it in their research. There are fewer such resources for Hochschild cohomology, notwithstanding some informative chapters in the books [146, 223]. This book aims to begin filling the gap by providing an introduction to the basic theory of Hochschild cohomology for algebras and some of its current uses in algebra and representation theory.

Hochschild cohomology records meaningful information about rings and algebras. It is used to understand their structure and deformations, and to identify essential information about their representations. This book takes a concrete approach with many early examples that reappear later in various settings.

We begin in Chapter 1 with Hochschild's own definitions from [114], only slightly rephrased in modern terminology and notation, and then connect to definitions based on arbitrary resolutions under suitable conditions. We present some of the important contributions of Gerstenhaber [82] beginning in the 1960s that lead us now to think of a Hochschild cohomology ring as a Gerstenhaber algebra, that is, it has both an associative product and a nonassociative Lie bracket. Many properties of Hochschild cohomology rings that are essential in today's applications can be seen in these classical definitions of Hochschild and Gerstenhaber. In Chapter 2 we give detailed descriptions of many equivalent definitions of the associative product (cup product) on Hochschild cohomology. In Chapter 3 we examine several different types of examples: smooth commutative algebras, Koszul algebras, algebras defined by quivers and relations, and algebras built from others such as skew group algebras and (twisted) tensor product algebras. We present the seminal Hochschild-Kostant-Rosenberg (HKR) Theorem on Hochschild homology and cohomology of smooth finitely generated commutative algebras.

Current algebraic applications and developments in the algebraic theory of Hochschild cohomology include the following, explored in detail in the rest of the book.

Some classical geometric notions such as smoothness may be viewed as essentially homological properties of commutative function algebras, allowing interpretations of them in noncommutative settings via Hochschild cohomology. We present these and related ideas in Chapter 4, including Hochschild dimension, smoothness, noncommutative differential forms, Van den Bergh duality, Calabi-Yau algebras, the Connes differential, and Batalin-Vilkovisky structures.

Understanding how some algebras may be viewed as deformations of others calls on Hochschild cohomology, as explained in Chapter 5. There we discuss formal deformations, rigidity of algebras, the Maurer-Cartan equation, Poisson brackets, and deformation quantization. We present the fundamental Poincaré-Birkhoff-Witt (PBW) Theorem as a consequence of a more general theorem on deformations of Koszul algebras. In algebraic deformation theory, the Lie structure on Hochschild cohomology arises naturally; we spend some additional time studying this important structure in detail in Chapter 6. Further probing the associative and Lie algebra structures on Hochschild cohomology and related complexes uncovers infinity algebras. There, binary operations are layered with n-ary operations which in turn have important implications for the original algebra structure. We give a brief introduction to infinity structures and their applications to Hochschild cohomology in Chapter 7.

In representation theory, support varieties may sometimes be defined in terms of Hochschild cohomology; these are geometric spaces assigned to modules that encode representation-theoretic information. Support varieties for finite-dimensional algebras are introduced and explored in Chapter 8. This theory began in the parallel setting of finite group cohomology. There are strong connections between Hochschild cohomology and group cohomology that we analyze more generally for Hopf algebras in Chapter 9. Hopf algebras are those algebras whose categories of modules are tensor categories, and include many examples of interest such as group algebras, universal enveloping algebras of Lie algebras, and quantum groups. Relationships between Hochschild cohomology and Hopf algebra cohomology lead to better understanding of both and of all their applications. Inspecting these relationships, we connect the two first appearances of homological techniques in algebra in the form of group cohomology [59–61] and Hochschild cohomology [114].

We include an appendix with needed background material from homological algebra. The appendix is largely self-contained, however, proofs are omitted, and instead the reader is referred to standard homological algebra textbooks such as [48, 112, 151, 168, 187, 223] for proofs and more details.

This introductory text is not intended to be a comprehensive treatment of the whole subject of Hochschild cohomology, which long ago expanded well beyond the reach of a single book. Necessarily many important topics are left out. For example, we do not treat Tate-Hochschild cohomology, relative Hochschild cohomology, Hochschild cohomology of presheaves and schemes, connections to cyclic homology and K-theory, Hochschild cohomology of abelian categories, topological Hochschild cohomology, Hochschild cohomology of differential graded and A_∞-algebras and categories, nor operads. Hochschild homology is an important subject in its own right, and we spend only a little time on it in this book. Also, here we will almost exclusively work with algebras over a field, both for simplicity of presentation in this introductory text and to take advantage of a great array of good properties and current applications for algebras over a field.

We provide a few references for the reader looking for details on some of the topics that are not in this book. This list is not meant to be complete, but rather a beginning, and further references may be found in each of these: more on Hochschild homology can be found in the standard references [146, 223]. Tate-Hochschild cohomology, stable Hochschild cohomology, and singular Hochschild cohomology are $\mathbb{Z}$-graded theories while Hochschild cohomology itself is $\mathbb{N}$-graded; see, for example, [29, 74]. Relative Hochschild cohomology and secondary Hochschild cohomology are designed for a ring and subring pair; see, for example, [106, 115, 205]. There is a version of Hochschild cohomology for coalgebras and bicomodules [57]. Hochschild cohomology is defined for presheaves of algebras and schemes, and used in algebraic geometry; see, for example, [85, 86, 132, 213]. Topological Hochschild homology and cohomology are related theories in algebraic topology; see, for example, [173]. Hochschild cohomology is used in functional analysis, with connections to properties of Banach algebras, von Neumann algebras, and locally compact groups; see, for example, [**122**, **198**]. Many important applications of the theory of Hochschild cohomology involve its connections to cyclic homology and cohomology and algebraic K-theory; see, for example, [**146**, **223**]. Hochschild homology and cohomology can be defined for some types of categories; see, for example, [**150**, **161**]. Some operads underlie much of the structure of Hochschild cohomology, a hint of which appears in the infinity structures of Chapter 7 here; see, for example, [**152**, **153**]. Formality and Deligne's Conjecture are barely touched in Chapter 7 here, and more details may be found in the references given in Section 7.6 and in [**153**]. Hochschild cohomology may be realized as the Lie algebra of the derived Picard group of an algebra; see, for example, [**128**]. Hochschild cohomology of differential graded and A_∞-algebras and categories, for example, are in [**127**, **130**].

This book is written for graduate students and working mathematicians interested in learning about Hochschild cohomology. It can serve as a reference for many facts that are currently only found in research papers, and as a bridge to some more advanced topics that are not included here. The main prerequisite for students is a graduate course in algebra. It would also be helpful to have taken further introductory courses in homological algebra or algebraic topology and in representation theory, or else to have done some reading in these subjects. However, all of the required homological algebra background is summarized in the appendix, with references, and a motivated reader might rely solely on this as homological algebra background. Beyond the first three chapters of this book, the remaining chapters are largely independent of each other, and so there are many options for basing a one-semester graduate course on this book. A one-semester course could start with a treatment of Chapter 1 and selected sections from Chapters 2 and 3, possibly including material from the appendix depending on the background of the students. Then the course could focus on a subset of the remaining chapters: a course with a focus on noncommutative geometry could continue with Chapter 4; a course with a focus on algebraic deformation theory and related structures could instead continue with Chapter 5 and the related Chapters 6 and 7, as time allowed; a course with a focus on Hopf algebras, group algebras, or support varieties could instead continue with Chapters 8 and/or 9. A full-year course might include most of the book and time for a complete introduction to or review of homological algebra based on the appendix.

This book came into being as an aftereffect of some lecture series that I gave and through interactions with many people. I first thank Universidad de Buenos Aires, and especially Andrea Solotar and her students, postdocs, and colleagues, for hosting me for several weeks in 2010. During that time I gave a short course on Hopf algebra cohomology that led to an early version of Chapter 9 on which they gave me valuable feedback. I thank the Morningside Center in Beijing and the organizers and students of a workshop there in 2011 for the opportunity to give lectures on support varieties that expanded into the current Chapter 8. I thank the Mathematisches Forschungsinstitut Oberwolfach for its hospitality during several workshops where the idea for this book began in discussions with Karin Erdmann and Henning Krause.

Most of this book was written during the academic year 2016–17 that I spent at the University of Toronto visiting Ragnar-Olaf Buchweitz and his research group. It was with deep sadness that I learned of his death the following fall. His legacy lives on and continues to grow through the mathematical writings that are still being completed by his many collaborators, as well as others he influenced. He was a great friend and mentor to so many of us.

I am grateful to the University of Toronto for hosting me in 2016–17 and to Buchweitz and his team for the many helpful conversations and stimulating seminar talks. This interaction significantly influenced some of my choices of topics for the book. I particularly thank Buchweitz and his students Benjamin Briggs and Vincent Gélinas for many pointers on smooth algebras and infinity algebras that helped me prepare Chapters 4 and 7. I also had fruitful discussions with Cris Negron and Yury Volkov in relation to the material in Chapter 6. Special thanks go to my long-time collaborator Anne Shepler for many conversations and joint projects over the years that led to my current point of view on Chapter 5.

I thank Chelsea Walton and her Spring 2017 homological algebra class at Temple University for trying out beta versions of several chapters of this book, and for their very valuable feedback. I thank my home institution Texas A&M University and especially the mathematics department for an immensely supportive work environment. I thank my husband Frank Sottile and children Maria and Samuel for their ongoing patience and support. I thank the National Science Foundation for its support through grants DMS-1401016 and DMS-1665286.

Chapter 1

Historical Definitions and Basic Properties

We begin with Hochschild's historical definition of homology and cohomology for algebras [**114**], working in the general setting of algebras over commutative rings as in the book of Cartan and Eilenberg [**48**]. We then present some additional structure found by Gerstenhaber [**82**] under which we now say that Hochschild cohomology is a Gerstenhaber algebra. These early developments were largely based on one choice of chain complex known as the bar complex, and we focus in this chapter particularly on the many properties and structural results that can be derived from this complex. For example, we discuss here the meaning of Hochschild homology and cohomology in low degrees where connections to derivations and deformations appear. We present the cap product that pairs Hochschild homology and cohomology, and the shuffle product on Hochschild homology of a commutative algebra. We discuss Harrison cohomology and the Hodge decomposition arising from a symmetric group action on the bar complex. Other work invokes other complexes, depending on the setting, and we begin to include in this chapter some examples and discussion of the structure of Hochschild cohomology that takes advantage of other such chain complexes. We expand on this discussion of other complexes in later chapters.

1.1. Definitions of Hochschild homology and cohomology

For now, let k be a commutative associative ring (with 1), and let A be a k-algebra. That is, A is an associative ring (with multiplicative identity) that is also a k-module for which multiplication is a k-bilinear map. Denote the multiplicative identity of A also by 1, identified with the multiplicative identity of k via the unit map $k \to A$ given by $c \mapsto c \cdot 1$ for all $c \in k$. Denote by A^{op} the *opposite algebra* of A; this is A as a module over k, with multiplication $a \cdot_{\mathrm{op}} b = ba$ for all $a, b \in A$. Tensor products will often be taken over k, and unless otherwise indicated, $\otimes = \otimes_k$. Let $A^e = A \otimes A^{\mathrm{op}}$, with the tensor product multiplication:

$$(a_1 \otimes b_1) \cdot (a_2 \otimes b_2) = a_1 a_2 \otimes b_2 b_1$$

for all $a_1, a_2, b_1, b_2 \in A$. (Technically, we are really taking b_1, b_2 to be elements of A^{op}, but since the underlying k-modules are the same, we write $b_1, b_2 \in A$ where convenient.) We call A^e the *enveloping algebra* of A.

By an A-*bimodule*, we mean a k-module M that is both a left and a right A-module for which $(a_1 m)a_2 = a_1(m a_2)$ for all $a_1, a_2 \in A$ and $m \in M$, and the left and right actions of k induced by the unit map $k \to A$ agree. Thus an A-bimodule M is equivalent to a left A^e-module, where we define

$$(a \otimes b) \cdot m = amb$$

for all $a, b \in A$ and $m \in M$. It is also equivalent to a right A^e-module where the action is defined by $m \cdot (a \otimes b) = bma$ for all $a, b \in A$ and $m \in M$. We will use both structures in the sequel, but for simplicity, when we refer to a module we generally mean a left module unless otherwise specified.

Note that the algebra A is itself an A^e-module (equivalently, an A-bimodule) under left and right multiplication: $(a \otimes b) \cdot c = acb$ for all $a, b, c \in A$. More generally, let $A^{\otimes n} = A \otimes \cdots \otimes A$ (n factors of A), where $n \geq 1$. This tensor power of A is an A^e-module (equivalently, an A-bimodule) by letting

$$(a \otimes b) \cdot (c_1 \otimes c_2 \otimes \cdots \otimes c_{n-1} \otimes c_n) = ac_1 \otimes c_2 \otimes \cdots \otimes c_{n-1} \otimes c_n b$$

for all $a, b, c_1, \ldots, c_n \in A$.

Consider the following sequence of A-bimodules:

$$(1.1.1) \qquad \cdots \xrightarrow{d_3} A^{\otimes 4} \xrightarrow{d_2} A^{\otimes 3} \xrightarrow{d_1} A \otimes A \xrightarrow{\pi} A \to 0,$$

where π is the multiplication map, that is, $\pi(a \otimes b) = ab$ for all $a, b \in A$, and $d_1(a \otimes b \otimes c) = ab \otimes c - a \otimes bc$ for all $a, b, c \in A$, and in general

$$(1.1.2) \quad d_n(a_0 \otimes a_1 \otimes \cdots \otimes a_{n+1}) = \sum_{i=0}^{n} (-1)^i a_0 \otimes a_1 \otimes \cdots \otimes a_i a_{i+1} \otimes \cdots \otimes a_{n+1}$$

for all $a_0, \ldots, a_{n+1} \in A$. One may check directly that $(1.1.1)$ is a complex, that is, $d_{n-1} d_n = 0$ for all n (see Section A.1). Moreover, it is exact, as a consequence of the existence of the following contracting homotopy (see Section A.1). Let s_n be the k-linear map defined by

$$(1.1.3) \qquad s_n(a_0 \otimes \cdots \otimes a_{n+1}) = 1 \otimes a_0 \otimes \cdots \otimes a_{n+1}$$

for all $n \geq -1$ and all $a_0, \ldots, a_{n+1} \in A$. A calculation shows that indeed $s_{\bullet}$ is a contracting homotopy, and so the complex (1.1.1) is exact. We write $B_n(A) = A^{\otimes(n+2)}$ for $n \geq 0$ and often consider the truncated complex associated to (1.1.1):

$$(1.1.4) \qquad B(A): \quad \cdots \xrightarrow{d_3} A^{\otimes 4} \xrightarrow{d_2} A^{\otimes 3} \xrightarrow{d_1} A \otimes A \to 0.$$

This is the *bar complex* of the A^e-module A. As a complex, its homology is concentrated in degree 0, where it is simply A, as a consequence of exactness of (1.1.1). Sometimes we use the isomorphism of left A^e-modules

$$(1.1.5) \qquad A^{\otimes(n+2)} \cong A^e \otimes A^{\otimes n}$$

given by $a_0 \otimes a_1 \otimes \cdots \otimes a_n \otimes a_{n+1} \mapsto (a_0 \otimes a_{n+1}) \otimes (a_1 \otimes \cdots \otimes a_n)$ for all $a_0, \ldots, a_{n+1} \in A$. (The action of A^e on $A^e \otimes A^{\otimes n}$ is multiplication on the leftmost factor A^e.) If A is free as a k-module, we see in this way that the terms in the bar complex (1.1.4) are free A^e-modules: $A^e \otimes A^{\otimes n} \cong \bigoplus_{i \in I} A^e(1 \otimes 1 \otimes \alpha_i)$, where $\{\alpha_i \mid i \in I\}$, for an indexing set I, is a basis of $A^{\otimes n}$ as a free k-module. In this case, $B(A)$ is a free resolution of the A^e-module A, also called the *bar resolution* (or *standard resolution*). See [**48**, Section IX.6].

Remark 1.1.6. The term *bar* complex arose historically due to an abbreviation of a tensor product $a_1 \otimes \cdots \otimes a_n$ as $[a_1 | \cdots | a_n]$. This convention was begun by Eilenberg and Mac Lane.

Let M be an A-bimodule. Take the tensor product of the bar complex (1.1.4) with M, writing

$$(1.1.7) \qquad C_*(A, M) = \bigoplus_{n \geq 0} M \otimes_{A^e} B_n(A),$$

a complex of k-modules with differentials $1_M \otimes d_n$, where d_n is defined by equation (1.1.2) and 1_M is the identity map on M. Call $C_*(A, M)$ the k-module of *Hochschild chains with coefficients in M*. There is a k-module isomorphism

$$(1.1.8) \qquad M \otimes_{A^e} B_n(A) \xrightarrow{\sim} M \otimes A^{\otimes n}$$

given by

$$m \otimes_{A^e} (a_0 \otimes \cdots \otimes a_{n+1}) \mapsto a_{n+1} m a_0 \otimes a_1 \otimes \cdots \otimes a_n$$

for all $m \in M$ and $a_0, \ldots, a_{n+1} \in A$. (The inverse isomorphism is given by $m \otimes a_1 \otimes \cdots \otimes a_n \mapsto m \otimes_{A^e} (1 \otimes a_1 \otimes \cdots \otimes a_n \otimes 1)$; recall the left action of A^e on $A^{\otimes(n+2)}$ involves the first and last tensor factors only, and the right action of A^e on M involves both the left and right actions of A on M.) Therefore there is a k-module isomorphism

$$C_n(A, M) \cong M \otimes A^{\otimes n}$$

for each n. Combined with the isomorphism (1.1.5), we find that the induced differential on the complex $M \otimes A^{\otimes *}$, corresponding to the map $1_M \otimes d_n$ on $M \otimes_{A^e} A^{\otimes (n+2)}$ for each $n > 0$, is the map $(d_n)_* : M \otimes A^{\otimes n} \to M \otimes A^{\otimes (n-1)}$ given by

$$(d_n)_*(m \otimes a_1 \otimes a_2 \otimes \cdots \otimes a_n)$$

$$= ma_1 \otimes a_2 \otimes \cdots \otimes a_n + \sum_{i=1}^{n-1} (-1)^i m \otimes a_1 \otimes a_2 \otimes \cdots \otimes a_i a_{i+1} \otimes \cdots \otimes a_n$$

$$+ (-1)^n a_n m \otimes a_1 \otimes a_2 \otimes \cdots \otimes a_{n-1}$$

for all $m \in M$ and $a_1, \ldots, a_n \in A$. We define Hochschild homology to be the homology of this complex.

Definition 1.1.9. The *Hochschild homology* $\mathrm{HH}_*(A, M)$ of A with coefficients in an A-bimodule M is the homology of the complex (1.1.7), equivalently

$$\mathrm{HH}_n(A, M) = \mathrm{H}_n(M \otimes A^{\otimes *}),$$

that is, $\mathrm{HH}_n(A, M) = \mathrm{Ker}((d_n)_*)/\mathrm{Im}((d_{n+1})_*)$ for all $n \geq 0$, taking $(d_0)_*$ to be the zero map, and differentials $(d_n)_*$ are as given above for $n > 0$. Elements in $\mathrm{Ker}((d_n)_*)$ are *Hochschild n-cycles* and those in $\mathrm{Im}((d_{n+1})_*)$ are *Hochschild n-boundaries*. Let

$$\mathrm{HH}_*(A, M) = \bigoplus_{n \geq 0} \mathrm{HH}_n(A, M),$$

an $\mathbb{N}$-graded k-module.

We have chosen the common notation HH to denote Hochschild homology (and cohomology below) so as to distinguish it from other versions of cohomology that we will use later. It is instead denoted with a single letter H in some of the literature.

Next we will apply $\mathrm{Hom}_{A^e}(-, M)$ to the bar complex (1.1.4): Let

$$(1.1.10) \qquad C^*(A, M) = \bigoplus_{n \geq 0} \mathrm{Hom}_{A^e}(B_n(A), M),$$

a complex of k-modules with differentials d_n^*, where $d_n^*(f) = f d_n$ for all functions f in $\mathrm{Hom}_{A^e}(A^{\otimes (n+1)}, M)$. (For simplicity here and throughout the book, we denote such a function composition by the concatenation $f d_n$ rather than $f \circ d_n$.) Call $C^*(A, M)$ the k-module of *Hochschild cochains with coefficients in M*. There is a k-module isomorphism

$$(1.1.11) \qquad \mathrm{Hom}_{A^e}(B_n(A), M) \xrightarrow{\sim} \mathrm{Hom}_k(A^{\otimes n}, M)$$

given by $g \mapsto (a_1 \otimes \cdots \otimes a_n \mapsto g(1 \otimes a_1 \otimes \cdots \otimes a_n \otimes 1))$ for all $g \in \mathrm{Hom}_{A^e}(B(A)_n, M)$ and $a_1, \ldots, a_n \in A$. (If $n = 0$ this is $g \mapsto (1 \mapsto g(1 \otimes 1))$.)

The inverse isomorphism is $g' \mapsto (a_0 \otimes \cdots \otimes a_{n+1} \mapsto a_0 g'(a_1 \otimes \cdots \otimes a_n) a_{n+1})$. We thus have an isomorphism of complexes,

$$(1.1.12) \qquad C^*(A, M) \cong \bigoplus_{n \geq 0} \operatorname{Hom}_k(A^{\otimes n}, M),$$

with differential $d_n^* : \operatorname{Hom}_k(A^{\otimes(n-1)}, M) \to \operatorname{Hom}_k(A^{\otimes n}, M)$ for each $n > 0$ given by

$$\begin{aligned} d_n^*(h)(a_1 \otimes \cdots \otimes a_n) \;=\; & a_1 h(a_2 \otimes \cdots \otimes a_n) \\ & + \sum_{i=1}^{n-1} (-1)^i h(a_1 \otimes \cdots \otimes a_i a_{i+1} \otimes \cdots \otimes a_n) \\ & + (-1)^n h(a_1 \otimes \cdots \otimes a_{n-1}) a_n \end{aligned}$$

for all $h \in \operatorname{Hom}_k(A^{\otimes(n-1)}, M)$ and $a_1, \ldots, a_n \in A$. In this expression and others, we interpret an empty tensor product to be the element 1 in k. We define Hochschild cohomology to be the cohomology of this complex.

Definition 1.1.13. The *Hochschild cohomology* $\operatorname{HH}^*(A, M)$ of A with coefficients in an A-bimodule M is the cohomology of the complex (1.1.10), equivalently

$$\operatorname{HH}^n(A, M) = \operatorname{H}^n(\operatorname{Hom}_k(A^{\otimes *}, M)),$$

that is, $\operatorname{HH}^n(A, M) = \operatorname{Ker}(d_{n+1}^*)/\operatorname{Im}(d_n^*)$ for all $n \geq 0$, where d_0^* is taken to be the zero map, and differentials d_n^* are as given above for $n > 0$. Elements in $\operatorname{Ker}(d_{n+1}^*)$ are *Hochschild n-cocycles* and those in $\operatorname{Im}(d_n^*)$ are *Hochschild n-coboundaries*. Let

$$\operatorname{HH}^*(A, M) = \bigoplus_{n \geq 0} \operatorname{HH}^n(A, M),$$

an $\mathbb{N}$-graded k-module.

As a special case, consider $M = A$ to be an A-bimodule under left and right multiplication. The resulting Hochschild homology and cohomology k-modules are sometimes abbreviated

$$\operatorname{HH}_n(A) = \operatorname{HH}_n(A, A) \quad \text{and} \quad \operatorname{HH}^n(A) = \operatorname{HH}^n(A, A),$$

and so we write $\operatorname{HH}_*(A) = \bigoplus_{n \geq 0} \operatorname{HH}_n(A)$ and $\operatorname{HH}^*(A) = \bigoplus_{n \geq 0} \operatorname{HH}^n(A)$. A disadvantage of this abbreviated notation in the case of cohomology is that it appears to indicate a functor, however HH^* is not a functor: we will see next that $\operatorname{HH}^*(A, M)$ can alternatively be defined via the bifunctor $\operatorname{Ext}_{A^e}^*(-, -)$, and this bifunctor is covariant in the second argument and contravariant in the first argument. Another disadvantage is that some authors use this notation to refer to Hochschild homology and cohomology with coefficients in a dual module to A, in which case these are functors related to cyclic homology and cohomology [146, §1.5.5]. For simplicity, we adopt the common abbreviated notation above in spite of these disadvantages.

Remarks 1.1.14. (i) In the case that k is a field (the case of focus for most of this book), A is free as a k-module and so each A^e-module $A^{\otimes(n+2)}$ ($n \geq 0$) is free, as noted earlier. It follows that the bar complex $B(A)$ given by (1.1.4) is a free left A^e-module resolution of A, called the *bar resolution*. Thus

$$(1.1.15) \quad \mathrm{HH}_n(A, M) \cong \mathrm{Tor}_n^{A^e}(M, A) \quad \text{and} \quad \mathrm{HH}^n(A, M) \cong \mathrm{Ext}_{A^e}^n(A, M)$$

for all $n \geq 0$. (See Section A.3.) More generally, as long as A is flat over the commutative ring k, the first isomorphism holds, and as long as A is projective over k, the second holds.

(ii) Even more generally, it follows from the definitions that Hochschild homology and cohomology may be realized as relative Tor and relative Ext [223, Lemma 9.1.3]. See also [115]. We will not use relative Tor or relative Ext in this book.

We will use the equivalent definitions of Hochschild homology and cohomology given by the isomorphisms (1.1.15) in the case that k is a field. An advantage is that we may thus choose any flat (respectively, projective) resolution of A as an A^e-module to define Hochschild homology (respectively, cohomology). Depending on the algebra A, there may be more convenient resolutions than the bar resolution; the latter is quite large and not conducive to explicit computation. The bar resolution may also obscure important information that stands out in other resolutions which are tailored more closely to the shapes of specific algebras. However, the bar resolution is very useful theoretically, as we will see.

Also useful is the following variant of the bar resolution. Identify k with the k-submodule $k \cdot 1_A$ of the k-algebra A, and write $\overline{A} = A/k$, the quotient k-module. For each nonnegative integer n, let

$$(1.1.16) \qquad\qquad \overline{B}_n(A) = A \otimes \overline{A}^{\otimes n} \otimes A.$$

Let $p_n : B_n(A) \to \overline{B}_n(A)$ be the corresponding quotient map. A calculation shows that the kernels of the maps p_n form a subcomplex of $B(A)$, and thus the quotients $\overline{B}_n(A)$ constitute a complex $\overline{B}(A)$. The contracting homotopy (1.1.3) can be shown to factor through this quotient, implying that $\overline{B}(A)$ is a free resolution of the A^e-module A in case A is free as a k-module. We give this resolution a name.

Definition 1.1.17. Assume that A is free as a k-module. The *reduced bar resolution* (or *normalized bar resolution*) is $\overline{B}(A)$ given in each degree by equation (1.1.16), a free resolution of A as an A^e-module.

See also [**85**, §13.5] for more details on the reduced bar resolution.

Hochschild cohomology modules hold substantial information about the algebra A and its modules, some of which we will find in this chapter by harnessing the bar complex, and some in later chapters via a wider range of complexes. Hochschild cohomology is invariant under some standard equivalences on rings: in case k is a field, invariance under Morita equivalence is automatic since this is by definition an equivalence of module categories and Hochschild cohomology is given by Ext in this case (see Remark 1.1.14(i)). For more details and related and more general statements, see, e.g., [**22**, Theorem 2.11.1], [**146**, Theorem 1.2.7], or the articles [**37**, **56**, **169**]. For Morita invariance of Hochschild homology, see, e.g., [**223**, §9.5]. For tilting and derived category equivalence, see, e.g., [**105**, Theorem 4.2] and [**9**, **185**, **220**], and for derived invariance of higher structures on the Hochschild complex itself, see [**127**].

We end this section with some examples that apply Remark 1.1.14(i), taking advantage of resolutions smaller than the bar resolution.

Example 1.1.18. Let k be a field and $A = k[x]$. Consider the following sequence:

$$(1.1.19) \qquad 0 \longrightarrow k[x] \otimes k[x] \xrightarrow{(x \otimes 1 - 1 \otimes x)\cdot} k[x] \otimes k[x] \xrightarrow{\pi} k[x] \longrightarrow 0,$$

where π is multiplication and the map $(x \otimes 1 - 1 \otimes x)\cdot$ is multiplication by the element $x \otimes 1 - 1 \otimes x$. Since $k[x]$ is commutative, this sequence is a complex. It is in fact exact, as can be shown directly via a calculation. Alternatively, exactness can be shown by exhibiting a contracting homotopy (see Section A.1): Let $s_{-1}(x^i) = x^i \otimes 1$ and

$$s_0(x^i \otimes x^j) = -\sum_{l=1}^{j} x^{i+j-l} \otimes x^{l-1}$$

for all i, j. (We interpret an empty sum to be 0, thus $s_0(x^i \otimes 1) = 0$ for all i.) A calculation shows that $s_{\bullet}$ is a contracting homotopy for the above sequence. Note that for each i, the map s_i is left (but not right) $k[x]$-linear. The terms in nonnegative degrees are visibly free as A^e-modules, and so the sequence (1.1.19) is a free resolution of the A^e-module A. Now apply $\mathrm{Hom}_{k[x]^e}(-, k[x])$ to the truncation of sequence (1.1.19) given by deleting the term $k[x]$. Identify $\mathrm{Hom}_{k[x]^e}(k[x] \otimes k[x], k[x])$ with $k[x]$ under the isomorphism in which a function f is sent to $f(1 \otimes 1)$. The resulting complex, with arrows reversed, becomes

$$(1.1.20) \qquad\qquad 0 \longleftarrow k[x] \longleftarrow k[x] \longleftarrow 0.$$

There is only one map to compute, namely composition with $(x \otimes 1 - 1 \otimes x)\cdot$. Let $a \in k[x]$, identified with the function f_a in $\mathrm{Hom}_{k[x]^e}(k[x] \otimes k[x], k[x])$ that takes $1 \otimes 1$ to a. Composing with the differential, since f_a is a $k[x]^e$-module homomorphism,

$$f_a((x \otimes 1 - 1 \otimes x) \cdot (1 \otimes 1)) = x f_a(1 \otimes 1) - f_a(1 \otimes 1)x = xa - ax = 0,$$

since $k[x]$ is commutative. Therefore all maps in complex (1.1.20) are 0, and the homology of the complex in each degree is just the term in the complex. We thus find that $\mathrm{HH}^0(k[x]) \cong k[x]$, that $\mathrm{HH}^1(k[x]) \cong k[x]$, and $\mathrm{HH}^n(k[x]) = 0$ for $n \geq 2$. A similar argument yields Hochschild homology $\mathrm{HH}_n(k[x])$ by first applying $k[x] \otimes_{k[x]^e} -$ to the truncation of the sequence (1.1.19) and identifying $k[x] \otimes_{k[x]^e} (k[x] \otimes k[x])$ with $k[x]$.

Example 1.1.21. Let k be a field, $n \geq 2$, and $A = k[x]/(x^n)$, called a *truncated polynomial ring*. Consider the following sequence:

$$(1.1.22) \qquad \cdots \xrightarrow{v\cdot} A^e \xrightarrow{u\cdot} A^e \xrightarrow{v\cdot} A^e \xrightarrow{u\cdot} A^e \xrightarrow{\pi} A \longrightarrow 0,$$

where $u = x \otimes 1 - 1 \otimes x$, $v = x^{n-1} \otimes 1 + x^{n-2} \otimes x + \cdots + 1 \otimes x^{n-1}$, and π is multiplication. This sequence is exact, as can be shown directly. Alternatively, the following is a contracting homotopy (see Section A.1). For each i, define a left A-linear map s_i by $s_{-1}(1) = 1 \otimes 1$ and for all $m \geq 0$,

$$s_{2m}(1 \otimes x^j) = -\sum_{l=1}^{j} x^{j-l} \otimes x^{l-1} \quad \text{and} \quad s_{2m+1}(1 \otimes x^j) = \delta_{j,n-1} \otimes 1$$

for all j, where $\delta_{j,n-1}$ is the Kronecker delta (that is, $\delta_{j,n-1} = 1$ if $j = n-1$ and $\delta_{j,n-1} = 0$ otherwise). The terms in nonnegative degrees are visibly free as A^e-modules, and so the sequence (1.1.22) is a free resolution of the A^e-module A.

Apply $\mathrm{Hom}_{A^e}(-, A)$ to (1.1.22) after truncating by deleting A, and identify each term $\mathrm{Hom}_{A^e}(A \otimes A, A)$ with A. The resulting sequence may be viewed as:

$$\cdots \xleftarrow{nx^{n-1}\cdot} A \xleftarrow{0} A \xleftarrow{nx^{n-1}\cdot} A \xleftarrow{0} A \longleftarrow 0.$$

Thus we see that if n is divisible by the characteristic of k, then $\mathrm{HH}^i(A) \cong A$ for all i. If n is not divisible by the characteristic of k, then $\mathrm{HH}^0(A) \cong A$, $\mathrm{HH}^{2m+1}(A) \cong (x)$ (the ideal generated by x) for all $m \geq 0$, and $\mathrm{HH}^{2m}(A) \cong A/(x^{n-1})$ for all $m \geq 1$.

Exercise 1.1.23. Verify the claims that formulas given for the following differentials are induced by the differential (1.1.2) on the bar complex (1.1.4):

    (a) $(d_n)_*$ right before Definition 1.1.9.

    (b) d_n^* right before Definition 1.1.13.

Exercise 1.1.24. Show that the following maps constitute contracting homotopies as claimed:

(a) s_n defined in (1.1.3).

(b) s_{-1}, s_0 defined in Example 1.1.18.

(c) s_i defined in Example 1.1.21.

Exercise 1.1.25. Finish Example 1.1.18 by finding Hochschild homology $\mathrm{HH}_n(k[x])$ for each n.

Exercise 1.1.26. Assume that k is a field and let $0 \to U \to V \to W \to 0$ be an exact sequence of A-bimodules. Use Remark 1.1.14(i) and Theorem A.4.7 (second long exact sequence for Tor) to show that there is a long exact sequence for Hochschild homology,

$$\cdots \to \mathrm{HH}_n(A, U) \to \mathrm{HH}_n(A, V) \to \mathrm{HH}_n(A, W) \to \mathrm{HH}_{n-1}(A, U) \to \cdots .$$

Derive a similar long exact sequence for Hochschild cohomology using Theorem A.4.4 (first long exact sequence for Ext).

Exercise 1.1.27. Assume that A is free as a k-module. Check that $\overline{B}(A)$ as in Definition 1.1.17 is indeed a free resolution of A as an A^e-module by verifying the claim that the kernels of the maps p_n form a subcomplex of $B(A)$ and that the contracting homotopy (1.1.3) for the bar complex (1.1.4) factors through $\overline{B}(A)$.

1.2. Interpretation in low degrees

The historical Definitions 1.1.9 and 1.1.13 of Hochschild homology $\mathrm{HH}_n(A)$ and cohomology $\mathrm{HH}^n(A)$ lead directly to specific information encoded by these k-modules when n is small. For example, in the nth Hochschild cohomology k-module $\mathrm{HH}^n(A)$ we can see the center of A in degree $n = 0$, derivations on A in degree 1, and infinitesimal deformations together with obstructions to lifting these to formal deformations in degrees 2 and 3. We make some of these observations in this section, including more general statements for A-bimodules M as well as analogous statements for Hochschild homology. More detailed discussion of deformations and obstructions is in Chapters 5 and 7.

In what follows, we will frequently identify spaces of chains under isomorphism (1.1.8) and spaces of cochains under isomorphism (1.1.11). We will abuse notation, using the same for differentials on such spaces when they have been so identified. Thus, for example, we write d_n^* for the map from $\operatorname{Hom}_{A^e}(A^{\otimes(n-1)}, M)$ to $\operatorname{Hom}_{A^e}(A^{\otimes n}, M)$ given by

$$
\begin{aligned}
d_n^*&(h)(a_0 \otimes a_1 \otimes \cdots \otimes a_n \otimes a_{n+1})\\
&= \sum_{i=0}^{n}(-1)^i h(a_0 \otimes \cdots \otimes a_i a_{i+1} \otimes \cdots \otimes a_{n+1})\\
&= a_0 a_1 h(1 \otimes a_2 \otimes \cdots \otimes a_n \otimes 1)a_{n+1}\\
&\quad + \sum_{i=1}^{n-1}(-1)^i a_0 h(1 \otimes a_1 \otimes \cdots \otimes a_i a_{i+1} \otimes \cdots \otimes a_n \otimes 1)a_{n+1}\\
&\quad + (-1)^n a_0 h(1 \otimes a_1 \otimes \cdots \otimes a_{n-1} \otimes 1)a_n a_{n+1}
\end{aligned}
$$

for all $h \in \operatorname{Hom}_{A^e}(A^{\otimes(n-1)}, M)$ and $a_0, \dots, a_{n+1} \in A$.

Degree 0. Let M be an A-bimodule. In the notation of Definition 1.1.13, $\operatorname{HH}^0(A, M) = \operatorname{Ker}(d_1^*)$. We determine necessary and sufficient conditions for a function $f \in \operatorname{Hom}_{A^e}(A \otimes A, M)$ to be in $\operatorname{Ker}(d_1^*)$. First assume that $d_1^*(f) = 0$, that is, for all $a \in A$,

$$
\begin{aligned}
0 \;=\; d_1^*(f)(1 \otimes a \otimes 1) \;&=\; f(d_1(1 \otimes a \otimes 1))\\
&=\; f(a \otimes 1 - 1 \otimes a) \;=\; af(1 \otimes 1) - f(1 \otimes 1)a.
\end{aligned}
$$

Then $f(1 \otimes 1)$ is equal to an element m of M for which $am = ma$ for all $a \in A$, and f is in fact determined by this element m:

$$
f(b \otimes c) = bf(1 \otimes 1)c = bmc
$$

for all $b, c \in A$. Conversely, any such element of M defines a function in $\operatorname{Ker}(d_1^*)$, that is, given $m \in M$ for which $am = ma$ for all $a \in A$, let $f_m \in \operatorname{Hom}_{A^e}(A \otimes A, M)$ be the function given by $f_m(b \otimes c) = bmc$ for all $b, c \in A$. Then $d_1^*(f_m) = 0$. Thus as a k-module,

$$
\operatorname{HH}^0(A, M) \cong \{m \in M \mid am = ma \text{ for all } a \in A\}.
$$

In the special case $M = A$, it follows that $\operatorname{HH}^0(A, A) \cong Z(A)$, the center of the algebra A.

Similarly, Hochschild homology in degree 0 is

$$
(1.2.1) \qquad \operatorname{HH}_0(A, M) \cong M / \operatorname{Span}_k\{am - ma \mid a \in A,\ m \in M\}.
$$

Degree 1. By Definition 1.1.13, $\operatorname{HH}^1(A, M) = \operatorname{Ker}(d_2^*)/\operatorname{Im}(d_1^*)$. Let $f \in \operatorname{Ker}(d_2^*)$, that is, $f \in \operatorname{Hom}_{A^e}(A^{\otimes 3}, M)$ and fd_2 is the zero map on $A^{\otimes 4}$. This condition is equivalent to

$$
\begin{aligned}
0 &= d_2^*(f)(1 \otimes a \otimes b \otimes 1) \\
&= f(d_2(1 \otimes a \otimes b \otimes 1)) \\
&= f(a \otimes b \otimes 1 - 1 \otimes ab \otimes 1 + 1 \otimes a \otimes b) \\
&= af(1 \otimes b \otimes 1) - f(1 \otimes ab \otimes 1) + f(1 \otimes a \otimes 1)b
\end{aligned}
$$

for all $a, b \in A$. By abuse of notation, we identify f with a function in $\mathrm{Hom}_k(A, M)$ under the isomorphism $\mathrm{Hom}_{A^e}(A^{\otimes 3}, M) \cong \mathrm{Hom}_k(A, M)$ of (1.1.11), and the above equation becomes $0 = af(b) - f(ab) + f(a)b$, or

$$
f(ab) = af(b) + f(a)b
$$

for all $a, b \in A$. This is precisely the definition of a *k-derivation* from A to M (also called more simply a *derivation* from A to M when k is understood). The k-module of all k-derivations from A to M is denoted

$$
\mathrm{Der}(A, M).
$$

We also write $\mathrm{Der}_k(A, M)$ when we want to emphasize dependence on the ground ring k. Suppose in addition that $f \in \mathrm{Im}(d_1^*)$, that is, $f = d_1^*(g)$ for some g in $\mathrm{Hom}_{A^e}(A^{\otimes 2}, M)$. The function g is determined by its value on $1 \otimes 1$, say m. Then

$$
\begin{aligned}
d_1^*(g)(1 \otimes a \otimes 1) &= g(d_1(1 \otimes a \otimes 1)) \\
&= g(a \otimes 1 - 1 \otimes a) \\
&= ag(1 \otimes 1) - g(1 \otimes 1)a \;=\; am - ma.
\end{aligned}
$$

That is, $d_1^*(g)$ corresponds to the *inner k-derivation* from A to M defined by the element m. Conversely, any inner k-derivation will determine an element of $\mathrm{Im}(d_1^*)$. The k-module of all inner k-derivations from A to M is denoted

$$
\mathrm{InnDer}(A, M).
$$

We have shown that

$$
\mathrm{HH}^1(A, M) \cong \mathrm{Der}(A, M)/\mathrm{InnDer}(A, M).
$$

In particular, if $M = A$, then $\mathrm{HH}^1(A)$ is isomorphic to the k-module of derivations of A modulo inner derivations, also called *outer derivations*. If A is commutative, the zero function is the only inner derivation, so in this case, the k-module of outer derivations, $\mathrm{HH}^1(A)$, is simply the k-module of derivations of A. Some results on when $\mathrm{HH}^1(A)$ vanishes and further discussion are in a paper of Buchweitz and Liu [**40**].

It can be shown that Hochschild homology in degree 1, $\mathrm{HH}_1(A, M)$, is isomorphic to the kernel of the canonical map $I \otimes_{A^e} M \to IM$, where I is the kernel of multiplication $\pi : A \otimes A \to A$. See Exercise 1.2.8 in case k is a field. For more details in the general case, see [**223**, §9.2], where a connection to Kähler differentials of commutative algebras is also given. In Section 4.2, we will identify I with the space $\Omega^1_{nc}A$ of noncommutative Kähler differentials.

Degree 2. By Definition 1.1.13, $\mathrm{HH}^2(A,M) = \mathrm{Ker}(d_3^*)/\mathrm{Im}(d_2^*)$. Let $f \in \mathrm{Hom}_{A^e}(A^{\otimes 4}, M)$. Then f is in $\mathrm{Ker}(d_3^*)$ if and only if for all $a, b, c \in A$,

$$
\begin{aligned}
0 &= d_3^*(f)(1 \otimes a \otimes b \otimes c \otimes 1) \\
&= f(d_3(1 \otimes a \otimes b \otimes c \otimes 1)) \\
&= f(a \otimes b \otimes c \otimes 1 - 1 \otimes ab \otimes c \otimes 1 + 1 \otimes a \otimes bc \otimes 1 - 1 \otimes a \otimes b \otimes c) \\
&= af(1 \otimes b \otimes c \otimes 1) - f(1 \otimes ab \otimes c \otimes 1) + f(1 \otimes a \otimes bc \otimes 1) \\
&\quad - f(1 \otimes a \otimes b \otimes 1)c.
\end{aligned}
$$

Identifying f with a function in $\mathrm{Hom}_k(A^{\otimes 2}, M)$ under the isomorphism $\mathrm{Hom}_{A^e}(A^{\otimes 4}, M) \cong \mathrm{Hom}_k(A^{\otimes 2}, M)$ of (1.1.11), we find that $f \in \mathrm{Ker}(d_3^*)$ if and only if

$$(1.2.2) \qquad af(b \otimes c) + f(a \otimes bc) = f(ab \otimes c) + f(a \otimes b)c$$

for all $a, b, c \in A$. A calculation shows that the image of d_2^* may be identified with the k-module of all functions f in $\mathrm{Hom}_k(A \otimes A, M)$ given by

$$(1.2.3) \qquad f(a \otimes b) = ag(b) - g(ab) + g(a)b$$

for some $g \in \mathrm{Hom}_k(A, M)$.

We will see in Section 4.2 that Hochschild 2-cocycles, that is, functions satisfying equation (1.2.2), define algebra structures on the A-bimodule $A \oplus M$. These algebras are called square-zero extensions, and they arise in connection with smoothness of algebras. In the case that $M = A$, equation (1.2.2) gives rise to infinitesimal deformations of A as discussed in Chapter 5. There we will develop algebraic deformation theory and we will see that obstructions to lifting a Hochschild 2-cocycle to a formal deformation lie in $\mathrm{HH}^3(A)$. We will also see that functions f satisfying (1.2.3) for some g, that is, Hochschild 2-coboundaries, correspond to deformations isomorphic to the original algebra. In this way, each formal deformation, up to isomorphism, will have associated to it an element of $\mathrm{HH}^2(A)$.

Actions of the center of A. In the interpretation of $\mathrm{HH}^0(A)$ as the center $Z(A)$ of A, some actions of $Z(A)$ take on a broader meaning. We present these actions of $Z(A)$ here, and extend them to actions of $\mathrm{HH}^*(A)$ in Sections 1.3 and 2.5. First note that $Z(A)$ acts on $\mathrm{Hom}_{A^e}(U, V)$ for any two A^e-modules U, V by

$$(1.2.4) \qquad (a \cdot f)(u) = af(u)$$

for all $a \in Z(A)$, $u \in U$, and $f \in \mathrm{Hom}_{A^e}(U, V)$. Taking $U = A^{\otimes(n+2)}$ and $V = A$, this action commutes with the differentials on the bar complex, inducing an action of $Z(A)$ on $\mathrm{HH}^n(A)$ under which $\mathrm{HH}^n(A)$ becomes a $Z(A)$-module. Identifying $Z(A)$ with $\mathrm{HH}^0(A)$ as described above, this is an action of $\mathrm{HH}^0(A)$ on $\mathrm{HH}^n(A)$ for each n, thus on $\mathrm{HH}^*(A)$. In the next section, this action will be extended to a graded product on $\mathrm{HH}^*(A)$. In Section 2.5, the action of $Z(A)$ on $\mathrm{Hom}_{A^e}(U, V)$ will be extended to an action of $\mathrm{HH}^*(A)$ on $\mathrm{Ext}^*_{A^e}(U, V)$.

The action of $Z(A)$ on $\mathrm{HH}^*(A)$ described above has some useful consequences. For example, let $1 = e_1 + \cdots + e_i$ be an expansion of the multiplicative identity of A as the sum of a set of orthogonal central idempotents $e_1, \ldots, e_i$. (That is, each e_j is central in A and $e_j e_l = \delta_{j,l} e_j$ for all j, l, where $\delta_{j,l}$ is the Kronecker delta.) Then $A = \bigoplus_{j=1}^{i} A e_j$, a direct sum of ideals $A e_j = e_j A = e_j A e_j$ of A. This leads to a similar decomposition $\mathrm{Hom}_{A^e}(U, A) = \bigoplus_{j=1}^{i} \mathrm{Hom}_{A^e}(U, e_j A)$ for any A^e-module U, and further,

$$(1.2.5) \qquad \mathrm{HH}^*(A) \cong \bigoplus_{j=1}^{i} \mathrm{HH}^*(e_j A) \cong \bigoplus_{j=1}^{i} e_j \, \mathrm{HH}^*(A).$$

Here we view the ideal $e_j A$ of A itself as an algebra with multiplicative identity e_j, and we may identify $e_j \, \mathrm{HH}^*(A)$ with $\mathrm{HH}^*(e_j A)$ in this expansion, under the action of $Z(A)$ on $\mathrm{HH}^*(A)$.

Exercise 1.2.6. Verify the isomorphism (1.2.1).

Exercise 1.2.7. Describe all k-derivations of $k[x]$ with the help of Example 1.1.18 and the connection to Hochschild cohomology explained in this section.

Exercise 1.2.8. Let k be a field, and let I be the kernel of the multiplication map $\pi : A \otimes A \to A$. Show that $\mathrm{HH}_1(A, M)$ is isomorphic to the kernel of the map $I \otimes_{A^e} M \to IM$ given by $\sum_i (a_i \otimes b_i) \otimes_{A^e} m \mapsto \sum_i a_i m b_i$. (*Hint*: Consider the short exact sequence $0 \to I \to A^e \xrightarrow{\pi} A \to 0$ and use Exercise 1.1.26, noting that $\mathrm{HH}_1(A, A^e) = 0$ since A^e is flat as an A^e-module.)

Exercise 1.2.9. Find the action of $Z(A)$ on $\mathrm{HH}^*(A)$ in each case:

    (a) $A = k[x]$ as in Example 1.1.18.

    (b) $A = k[x]/(x^n)$ as in Example 1.1.21.

1.3. Cup product

Hochschild cohomology $\mathrm{HH}^*(A)$ is a graded k-module by its definition. (That is, it is graded by $\mathbb{N}$, which we understand to include 0.) We will see next that it has an associative product making it into a graded commutative algebra, that is, homogeneous elements commute up to a sign determined by homological degrees. (See Theorem 1.4.6 below.) We define this product at the chain level for functions on the bar complex (1.1.4) in Definition 1.3.1 below. In fact the cup product is the unique associative product on $HH^*(A)$ satisfying some basic conditions (see Sanada [188]). There are many equivalent definitions of this associative product on $HH^*(A)$, particularly in case k is a field, making it very versatile. We give some of these other definitions in Chapter 2.

We again use the isomorphism (1.1.11) to identify $\operatorname{Hom}_{A^e}(A^{\otimes(n+2)}, M)$ with $\operatorname{Hom}_k(A^{\otimes n}, M)$ as a k-module. As in (1.1.12),

$$C^*(A, M) \cong \bigoplus_{n \geq 0} \operatorname{Hom}_k(A^{\otimes n}, M),$$

the k-module of Hochschild cochains on A with coefficients in M. In what follows we will take $M = A$.

Definition 1.3.1. Let $f \in \operatorname{Hom}_k(A^{\otimes m}, A)$ and $g \in \operatorname{Hom}_k(A^{\otimes n}, A)$. The *cup product* $f \smile g$ is the element of $\operatorname{Hom}_k(A^{\otimes(m+n)}, A)$ defined by
(1.3.2)
$$(f \smile g)(a_1 \otimes \cdots \otimes a_{m+n}) = (-1)^{mn} f(a_1 \otimes \cdots \otimes a_m) g(a_{m+1} \otimes \cdots \otimes a_{m+n})$$

for all $a_1, \ldots, a_{m+n} \in A$. If $m = 0$, we interpret this formula to be

$$(f \smile g)(a_1 \otimes \cdots \otimes a_n) = f(1) g(a_1 \otimes \cdots \otimes a_n),$$

and similarly if $n = 0$.

Remark 1.3.3. The historical definition of cup product, used by many authors, does not include the factor $(-1)^{mn}$. Our choice here will be more convenient with respect to our sign conventions, and also agrees with much of the literature. The difference is not so essential, since at the level of cohomology, this product is graded commutative, as we will see.

By its definition, the cup product is associative. A sign modification makes the k-module $C^*(A, A)$ of Hochschild cochains a *differential graded algebra*, that is, a graded algebra (i.e., $C^i(A, A) \smile C^j(A, A) \subseteq C^{i+j}(A, A)$) with a graded derivation of degree 1 and square 0. Precisely, for an m-cochain f, let $\partial(f) = (-1)^m d^*_{m+1}(f)$ and similarly for g, $f \smile g$. Then

$$(1.3.4) \qquad \partial(f \smile g) = (\partial(f)) \smile g + (-1)^m f \smile (\partial(g)).$$

(If we omit the factor $(-1)^{mn}$ in the definition (1.3.2) of cup product, then $C^*(A, A)$ is a differential graded algebra without modifying the sign of the differential. Note that the cohomology is the same in either case.) A consequence of equation (1.3.4) is that this cup product $\smile$ induces a well-defined graded associative product on Hochschild cohomology, which we denote by the same symbol:

$$\smile \; : \; \mathrm{HH}^m(A) \times \mathrm{HH}^n(A) \to \mathrm{HH}^{m+n}(A).$$

Remark 1.3.5. More generally, if B is an A-bimodule that is also an algebra for which $a(bb') = (ab)b'$, $(bb')a = b(b'a)$, and $(ba)b' = b(ab')$ for all $a \in A$ and $b, b' \in B$, a calculation shows that a formula analogous to (1.3.2) induces a product on $\mathrm{HH}^*(A, B)$. This condition is satisfied, for example, if A is a subalgebra of B and the A-bimodule structure of B is given by left and right multiplication.

We have seen that in degree 0, the Hochschild cohomology k-module $\mathrm{HH}^0(A)$ is isomorphic to $Z(A)$, the center of the algebra A. As a consequence of the definition, the cup product of two elements in degree 0 is precisely the product of the corresponding elements in $Z(A)$, and the cup product of an element in arbitrary degree n with a degree 0 element corresponds to multiplying the values of a representative function by an element in $Z(A)$. This agrees with the $\mathrm{HH}^0(A)$-module structure on Hochschild cohomology $\mathrm{HH}^*(A)$ given by equation (1.2.4). Sometimes this is enough information to determine the full structure of $\mathrm{HH}^*(A)$ as a ring under cup product, such as in the following example.

Example 1.3.6. We return to Example 1.1.18, letting k be a field and $A = k[x]$. We describe the cup product using only general properties and the graded vector space structure. We found that $\mathrm{HH}^0(k[x]) \cong k[x]$, $\mathrm{HH}^1(k[x]) \cong k[x]$, and $\mathrm{HH}^n(k[x]) = 0$ for $n > 1$, and so as a graded vector space,

$$(1.3.7) \qquad \mathrm{HH}^*(k[x]) \cong k[x] \oplus k[x],$$

with the first copy of $k[x]$ in degree 0 and the second in degree 1. We will describe cup products on $\mathrm{HH}^*(k[x])$ in view of this expression. In degree 0, the cup product is simply multiplication on $k[x]$. Likewise, the product of an element in degree 0 with an element in degree 1 corresponds to multiplication on $k[x]$, with the result having degree 1 (see Exercise 1.2.9(a)). Since $\mathrm{HH}^2(k[x]) = 0$, the product of two elements in degree 1 is 0. Thus we know all cup products on $\mathrm{HH}^*(k[x])$. To express this algebraic structure more compactly, denote by y the copy of the multiplicative identity of $k[x]$ that is in the degree 1 component in expression (1.3.7). We may then rewrite (1.3.7) as

$$\mathrm{HH}^*(k[x]) \cong k[x] \oplus k[x]y$$

with the first summand $k[x]$ the degree 0 component and the second summand $k[x]y$ the degree 1 component (treating y as a place holder). By our above description of products, we now see that $y^2 = 0$ and

$$\mathrm{HH}^*(k[x]) \cong k[x,y]/(y^2)$$

as a k-algebra, where $|x| = 0$ and $|y| = 1$. (The notation $|x|, |y|$ here refers to their homological degrees.)

We have noted that the cup product is associative as a direct consequence of formula (1.3.2). Associativity can also be deduced readily from each of the equivalent definitions of cup product that will be given in Chapter 2.

We will next turn to our claim that the cup product is graded commutative. This may be shown in many different ways. It may be proven by induction, as in [188]. Another proof uses two of the equivalent definitions of the product, namely the Yoneda product of Section 2.2 and the tensor product of Section 2.3, and the observation that the latter is an algebra homomorphism over the former (see [211]). Yet another proof uses tensor products of generalized extensions and an argument similar to the proof of Theorem 2.5.5 below (see [201, Theorem 1.1] and some discussion in Section 2.4). Our proof in the next section takes the more concrete historical approach of Gerstenhaber [82]. This proof is connected to the first appearance of the graded Lie bracket on Hochschild cohomology, also defined in the next section.

Exercise 1.3.8. Verify formula (1.3.4). For comparison, derive a similar formula for $d^*_{m+n+1}(f \smile g)$.

Exercise 1.3.9. Let B be an algebra and A a subalgebra of B. Consider B to be an A-bimodule under left and right multiplication. Let $f \in \mathrm{Hom}_k(A^{\otimes m}, A)$, $g \in \mathrm{Hom}_k(A^{\otimes n}, B)$. Define $f \smile g \in \mathrm{Hom}_k(A^{\otimes(m+n)}, B)$ by a formula analogous to (1.3.2). Show that this induces a well-defined product on $\mathrm{HH}^*(A, B)$. (More generally, see Remark 1.3.5.)

Exercise 1.3.10. Let M be an A-bimodule. Let $f \in \mathrm{Hom}_k(A^{\otimes m}, A)$ and $g \in \mathrm{Hom}_k(A^{\otimes n}, M)$. Define $f \cdot g \in \mathrm{Hom}_k(A^{\otimes(m+n)}, M)$ by

$$(f \cdot g)(a_1 \otimes \cdots \otimes a_{m+n}) = (-1)^{mn} f(a_1 \otimes \cdots \otimes a_m) g(a_{m+1} \otimes \cdots \otimes a_{m+n})$$

for all $a_1, \ldots, a_{m+n} \in A$. Show that this induces a well-defined action of $\mathrm{HH}^*(A)$ on $\mathrm{HH}^*(A, M)$ (cf. Section 2.5 below).

1.4. Gerstenhaber bracket

Together with addition and cup product, Hochschild cohomology $\mathrm{HH}^*(A)$ has another binary operation. We define this operation at the chain level on the bar complex (1.1.4) in this section, allowing k to be an arbitrary commutative ring. In Chapter 6, under the assumption that k is a field, we will examine equivalent definitions by way of other projective resolutions and exact sequences.

Definition 1.4.1. Let $f \in \mathrm{Hom}_k(A^{\otimes m}, A)$ and $g \in \mathrm{Hom}_k(A^{\otimes n}, A)$. Their *Gerstenhaber bracket* $[f, g]$ is defined at the chain level to be the element of $\mathrm{Hom}_k(A^{\otimes(m+n-1)}, A)$ given by

$$[f, g] = f \circ g - (-1)^{(m-1)(n-1)} g \circ f,$$

where the *circle product* $f \circ g$ generalizes composition of functions and is defined by

$$(f \circ g)(a_1 \otimes \cdots \otimes a_{m+n-1})$$
$$= \sum_{i=1}^{m} (-1)^u f(a_1 \otimes \cdots \otimes a_{i-1} \otimes g(a_i \otimes \cdots \otimes a_{i+n-1}) \otimes a_{i+n} \otimes \cdots \otimes a_{m+n-1}),$$

in which $u = (n-1)(i-1)$, and similarly for $g \circ f$. If $m = 0$, then $f \circ g = 0$ (as indicated by the empty sum), while if $n = 0$, then the formula should be interpreted as taking the value $g(1)$ in place of $g(a_i \otimes \cdots \otimes a_{i+n-1})$ (as indicated by the empty tensor product).

Related to the Gerstenhaber bracket is the divided square operation.

Definition 1.4.2. Let m be a nonnegative integer and $f \in \mathrm{Hom}_k(A^{\otimes m}, A)$. Assume that the characteristic of k is 2 or m is even. The *divided square* $\mathrm{Sq}(f)$ of f is defined by

$$\mathrm{Sq}(f) = f \circ f,$$

where the circle product is as in Definition 1.4.1.

The following lemma may be proven by direct computation on the bar complex (1.1.4) as in [**82**]. For ease of notation, as is common, we leave out subscripts on the differentials.

Lemma 1.4.3. *Let* $f \in \mathrm{Hom}_k(A^{\otimes m}, A)$, $g \in \mathrm{Hom}_k(A^{\otimes n}, A)$, *and* $h \in \mathrm{Hom}_k(A^{\otimes p}, A)$. *Then*

    (i) $[f, g] = -(-1)^{(m-1)(n-1)}[g, f]$,

    (ii) $(-1)^{(m-1)(p-1)}[f, [g, h]] + (-1)^{(n-1)(m-1)}[g, [h, f]]$
        $+(-1)^{(p-1)(n-1)}[h, [f, g]] = 0$,

    (iii) $d^*([f, g]) = (-1)^{n-1}[d^*(f), g] + [f, d^*(g)]$.

Property (i) is graded anticommutativity of the bracket, where we shift the homological degrees of f, g by -1. Property (ii) is the *graded Jacobi identity*. From these properties and Definition 1.4.1, we see that the k-module of Hochschild cochains $C^*(A, A) = \bigoplus_{n \geq 0} \mathrm{Hom}_k(A^{\otimes n}, A)$ is a *graded Lie algebra*. Property (iii) and a sign modification further make it a *differential graded Lie algebra*, that is, a graded Lie algebra with a graded derivation δ of degree 1 and square 0: letting $\delta(f) = (-1)^{m-1}d^*(f)$ for all $f \in \mathrm{Hom}_k(A^{\otimes m}, A)$, by Lemma 1.4.3(iii) we have

$$\delta([f,g]) = [\delta(f), g] + (-1)^{m-1}[f, \delta(g)].$$

It follows from Lemma 1.4.3 that Hochschild cohomology $\mathrm{HH}^*(A)$ is a graded Lie algebra. Moreover, $\mathrm{HH}^1(A)$ is a Lie algebra over which $\mathrm{HH}^*(A)$ is a module.

Remark 1.4.4. We emphasize that the degree of a cochain here is shifted by 1 from its homological degree, so that the cochain f in $\mathrm{Hom}_k(A^{\otimes m}, A)$ has degree $m - 1$ when considering the Lie structure. Some authors choose notation to clarify this distinction, introducing a shift operator that shifts degree when needed. We will instead point out whenever we are using this shifted degree for functions f and corresponding elements of Hochschild cohomology, and will always denote this shifted degree by $|f| - 1$, reserving the notation $|f|$ exclusively for the homological degree m of the corresponding element of $\mathrm{HH}^m(A)$.

Gerstenhaber [**82**] more generally developed the notion of a pre-Lie algebra for handling the circle product and bracket operations and proving results about the Lie structure. The details are very informative and may be found in his paper.

Recall that $\pi : A \otimes A \to A$ denotes the multiplication map. The following lemma may be proven by tedious direct computation. See also [**82**], taking care to convert the cup product there to our cup product as in Definition 1.3.1 (see Remark 1.3.3).

Lemma 1.4.5. *Let* $f \in \mathrm{Hom}_k(A^{\otimes m}, A)$ *and* $g \in \mathrm{Hom}_k(A^{\otimes n}, A)$. *Then*

(i) $(-1)^{mn}f \smile g - g \smile f$
$\quad = (d^*(g)) \circ f + (-1)^m d^*(g \circ f) + (-1)^{m-1} g \circ (d^*(f)),$

(ii) $[f, \pi] = -d^*(f).$

The following theorem is a consequence of Lemma 1.4.5(i).

Theorem 1.4.6. *Let A be an associative algebra over the commutative ring k. The cup product on $\mathrm{HH}^*(A)$ is graded commutative, that is, for all $\alpha \in \mathrm{HH}^m(A)$ and $\beta \in \mathrm{HH}^n(A)$,*

$$\alpha \smile \beta = (-1)^{mn} \beta \smile \alpha.$$

Proof. Let $f \in \mathrm{Hom}_k(A^{\otimes m}, A)$ and $g \in \mathrm{Hom}_k(A^{\otimes n}, A)$ be cocycles, that is, $d^*(f) = 0$ and $d^*(g) = 0$. By Lemma 1.4.5(i),

$$(-1)^{mn}f \smile g = g \smile f + (-1)^m d^*(g \circ f).$$

The images α and β in $\mathrm{HH}^*(A)$ of f and g thus satisfy

$$\alpha \smile \beta = (-1)^{mn}\beta \smile \alpha.$$

$\square$

As we have seen, a consequence of Lemma 1.4.3(iii) is that the bracket $[\,,\,]$, as defined at the cochain level, induces a well-defined operation on $\mathrm{HH}^*(A)$. Now we note that a consequence of Lemma 1.4.5(i) is that the divided square operation Sq, as defined at the cochain level, induces a well-defined operation on $\mathrm{HH}^*(A)$ in case $\mathrm{char}(k) = 2$ and on its subalgebra generated by homogeneous even degree elements in case $\mathrm{char}(k) \neq 2$. Next we state a further property satisfied by the bracket on Hochschild cohomology. For a proof, see [82, Corollary 1 of Theorem 5], where at the cochain level, the difference of the left and right sides of the stated equation in the lemma is shown to be a specific coboundary.

Lemma 1.4.7. *Let $\alpha \in \mathrm{HH}^m(A)$, $\beta \in \mathrm{HH}^n(A)$, and $\gamma \in \mathrm{HH}^p(A)$. Then*

$$[\gamma, \alpha \smile \beta] = [\gamma, \alpha] \smile \beta + (-1)^{m(p-1)}\alpha \smile [\gamma, \beta].$$

As a consequence of the lemma, for each γ, the operation $[\gamma, -]$ is a graded derivation with respect to cup product. Moreover, Hochschild cohomology is a Gerstenhaber algebra (sometimes also called a G-algebra), as we define next.

Definition 1.4.8. A *Gerstenhaber algebra* $(H, \smile, [\,,\,])$ is a free $\mathbb{Z}$-graded k-module H for which $(H, \smile)$ is a graded commutative associative algebra, $(H, [\,,\,])$ is a graded Lie algebra with bracket $[\,,\,]$ of degree -1 and corresponding degree shift by -1 on elements, and

$$[\gamma, \alpha \smile \beta] = [\gamma, \alpha] \smile \beta + (-1)^{|\alpha|(|\gamma|-1)}\alpha \smile [\gamma, \beta]$$

for all homogeneous α, β, γ in H.

Theorem 1.4.9. *Hochschild cohomology $\mathrm{HH}^*(A)$ is a Gerstenhaber algebra.*

Proof. The main properties to prove are dealt with in Theorem 1.4.6 and Lemmas 1.4.3 and 1.4.7. $\square$

In Chapter 6 we will examine the Gerstenhaber bracket in more detail, including ways to define it for functions on an arbitrary resolution independent of the bar resolution, and on generalized extensions.

Exercise 1.4.10. Show that if $\mathrm{char}(k) \neq 2$ and f is a Hochschild cocycle of homogeneous even degree, then $\mathrm{Sq}(f) = \frac{1}{2}[f, f]$.

Exercise 1.4.11. Verify Lemma 1.4.3(i) and (ii) by direct computation.

Exercise 1.4.12. Verify that the bracket $[\,,\,]$, as defined at the cochain level, induces a well-defined operation on $\mathrm{HH}^*(A)$ by Lemma 1.4.3(iii). Similarly verify that the divided square operation Sq is well-defined on cohomology.

Exercise 1.4.13. Verify all other properties of the Gerstenhaber bracket stated in this section either by direct computation or by reading the verifications in [**82**], taking care to convert the cup product there to our cup product as in Definition 1.3.1 (see Remark 1.3.3).

1.5. Cap product and shuffle product

The cap product is an action of $\mathrm{HH}^*(A)$ on $\mathrm{HH}_*(A)$. The shuffle product is an associative and graded commutative product on $\mathrm{HH}_*(A)$ for commutative algebras A. We define these products in this section.

The cap product is a specific pairing between Hochschild homology and cohomology modules, that is, a function

$$\mathrm{HH}_n(A) \otimes \mathrm{HH}^m(A) \xrightarrow{\frown} \mathrm{HH}_{n-m}(A),$$

defined at the chain level as follows. (We set $\mathrm{HH}_i(A)$ equal to 0 for all $i < 0$.) Identify $A \otimes_{A^e} A^{\otimes(n+2)}$ with $A \otimes A^{\otimes n}$ via the isomorphism (1.1.8) with $M = A$, symbolically representing an element of $\mathrm{HH}_n(A)$ at the chain level by a sum of elements of the form $a_0 \otimes \cdots \otimes a_n$ in $A \otimes A^{\otimes n}$.

Definition 1.5.1. Let $f \in \mathrm{Hom}_k(A^{\otimes m}, A)$ be a function representing an element of $\mathrm{HH}^m(A)$. The *cap product* of f with $a_0 \otimes \cdots \otimes a_n$ is defined by

$$(a_0 \otimes a_1 \otimes \cdots \otimes a_n) \frown f = (-1)^{m(n-m)} a_0 f(a_1 \otimes \cdots \otimes a_m) \otimes a_{m+1} \otimes \cdots \otimes a_n.$$

This formula induces a well-defined function from $\mathrm{HH}_n(A) \otimes \mathrm{HH}^m(A)$ to $\mathrm{HH}_{n-m}(A)$ as claimed: assuming that $\sum_i a_0^i \otimes \cdots \otimes a_n^i$ is a cycle for some $a_j^i \in A$, and that $f \in \mathrm{Hom}_k(A^{\otimes m}, A)$ is a cocycle, one sees that their cap product is a cycle by rewriting the image of the differential on $\sum_i a_0^i \otimes \cdots \otimes a_n^i$ in such a way as to take advantage of the relation $d^*_{m+1}(f) = 0$. Thus the cap product of a cocycle with a cycle is a cycle. Similarly one sees that the cap product of a coboundary with a cycle, or of a cocycle with a boundary, is a boundary. By its definition, the cap product gives $\mathrm{HH}_*(A)$ the structure of a right $\mathrm{HH}^*(A)$-module.

Example 1.5.2. Let k be a field and $A = k[x]$, as in Example 1.1.18. By our work in that example, we see that $\mathrm{HH}_*(A)$ is a free A-module. In degree 0, $\mathrm{HH}^0(A) \cong A$ acts on this free A-module in the canonical way. Let V be the one-dimensional vector space with basis x. We identify $\mathrm{HH}^1(A)$ with $A \otimes V^*$, where $V^* = \mathrm{Hom}_k(V, k)$ is the dual vector space, and $\mathrm{HH}_1(A)$ with $A \otimes V$ (see Exercise 1.1.25). The resolution (1.1.19) may be embedded in the bar resolution (1.1.4) by identifying the two resolutions in degree 0 and by mapping $a \otimes b$ in $k[x] \otimes k[x]$ in degree 1 to $a \otimes x \otimes b$ in the bar resolution. A calculation shows that this is a chain map. (See Section 3.4 for a more general embedding for Koszul algebras.) A further calculation shows that the action of $\mathrm{HH}^1(A) \cong A \otimes V^*$ on $\mathrm{HH}_1(A) \cong A \otimes V$ under cap product is multiplication in the first tensor factor and evaluation in the second.

For the rest of this section, assume that A is a commutative k-algebra. The shuffle product, defined next, is a product on Hochschild homology $\mathrm{HH}_*(A)$ in this commutative case. Let S_n denote the symmetric group on n symbols, and let sgn denote its sign character, that is, sgn takes even permutations to 1 and odd permutations to -1. We will need the subset $S_{p,q}$ of (p,q)-shuffles of the symmetric group S_{p+q}, defined next.

Definition 1.5.3. For nonnegative integers p and q, a (p,q)-*shuffle* is an element σ of the symmetric group S_{p+q} for which $\sigma(i) < \sigma(j)$ whenever $1 \leq i < j \leq p$ or $p+1 \leq i < j \leq p+q$. Let $S_{p,q}$ denote the subset of S_{p+q} consisting of all (p,q)-shuffles.

Identify $A \otimes_{A^e} A^{\otimes(n+2)}$ with $A^{\otimes(n+1)}$ via the isomorphism (1.1.8), taking $M = A$.

Definition 1.5.4. Let A be a commutative k-algebra. The *shuffle product* on $\mathrm{HH}_*(A)$ is defined at the chain level by

$$(a_0 \otimes a_1 \otimes \cdots \otimes a_p) \cdot (a_0' \otimes a_{p+1} \otimes \cdots \otimes a_{p+q})$$
$$= \sum_{\sigma \in S_{p,q}} (\mathrm{sgn}\,\sigma) a_0 a_0' \otimes a_{\sigma^{-1}(1)} \otimes \cdots \otimes a_{\sigma^{-1}(p+q)}$$

for all $a_0, \ldots, a_{p+q} \in A$.

The shuffle product is indeed well-defined on $\mathrm{HH}_*(A)$: the given multiplication formula can be shown to define a chain map from the tensor product complex $C_*(A,A) \otimes C_*(A,A)$ (see Section A.5) to $C_*(A,A)$. It follows that the product of a cycle with a cycle is again a cycle, and the product of a cycle with a boundary is a boundary. Also see [**223**, §9.4.1] for a more general context.

Theorem 1.5.5. *Let A be a commutative k-algebra. Then $\mathrm{HH}_*(A)$ is a graded commutative k-algebra under the shuffle product.*

Proof. Compose each σ in $S_{p,q}$ with the (p,q)-shuffle τ given by

$$\tau(1) = p+1, \ \ldots, \ \tau(q) = p+q, \ \tau(q+1) = 1, \ \ldots, \ \tau(p+q) = p.$$

This interchanges a shuffle product and its opposite, up to the sign $(-1)^{pq} = \mathrm{sgn}\,\tau$. $\qquad\square$

Exercise 1.5.6. Verify that the cap product is well-defined and gives $\mathrm{HH}_*(A)$ the structure of a right $\mathrm{HH}^*(A)$-module.

Exercise 1.5.7. Verify the details in Example 1.5.2 by explicitly calculating the action of $\mathrm{HH}^1(A)$ on $\mathrm{HH}_1(A)$.

Exercise 1.5.8. Verify that the shuffle product is well-defined.

Exercise 1.5.9. Let k be a field and let $A = k[x]$, as in Example 1.1.18. Show that $\mathrm{HH}_*(A) \cong k[x,y]/(y^2)$ as a graded algebra under shuffle product, where $|x| = 0$, $|y| = 1$.

Exercise 1.5.10. Let $A = k[x]/(x^2)$. Find the structure of $\mathrm{HH}_*(A)$ under shuffle product.

1.6. Harrison cohomology and Hodge decomposition

In this section, let k be a field of characteristic 0, and let A be a commutative k-algebra. We use the bar complex to define Harrison cohomology, a variant of Hochschild cohomology defined specifically in this setting. In turn, Harrison cohomology is one summand in the Hodge decomposition of Hochschild cohomology, also defined in this section.

Recall Definition 1.5.3 of (p, q)-shuffles $S_{p,q}$.

Definition 1.6.1. Let k be a field of characteristic 0, let A be a commutative k-algebra, and let $f \in \mathrm{Hom}_k(A^{\otimes n}, A)$. We call f a *Harrison cochain* if

$$\sum_{\sigma \in S_{p,q}} (\mathrm{sgn}\,\sigma) f(a_{\sigma^{-1}(1)} \otimes \cdots \otimes a_{\sigma^{-1}(n)}) = 0$$

for each pair p, q for which $p + q = n$ and all $a_1, \ldots, a_n \in A$. It can be shown that the Harrison cochains form a subcomplex of the complex of Hochschild cochains. The cohomology of this subcomplex is the *Harrison cohomology* of A.

Details on Harrison cohomology may be found in [85, §4] or [107]. Barr [19] first discovered that Harrison cohomology is a direct summand of Hochschild cohomology. The corresponding decomposition of Hochschild cohomology can be refined to obtain the Hodge decomposition (1.6.2) below, as we see next.

For each r, consider the following element of the group algebra kS_n:

$$s_{r,n-r} = \sum_{\sigma \in S_{r,n-r}} (\mathrm{sgn}\,\sigma)\sigma,$$

where the sum is taken over all $(r, n - r)$-shuffles as in Definition 1.5.3. Let

$$s_n = \sum_{r=1}^{n-1} s_{r,n-r}.$$

By [**84**, Theorem 1.2], each s_n may be written as a sum,

$$s_n = \lambda_1 e_n(1) + \cdots + \lambda_n e_n(n),$$

where $\lambda_i = 2^i - 2$ and for each j,

$$e_n(j) = \prod_{i \neq j} (\lambda_j - \lambda_i)^{-1} \prod_{i \neq j} (s_n - \lambda_i)$$

in the group algebra kS_n. The elements $e_n(j)$ are the *Eulerian idempotents*; for each n, the set $\{e_n(1), \ldots, e_n(n)\}$ is a set of orthogonal idempotents whose sum is 1 in kS_n. These idempotents may be defined equivalently by a generating function due to Garsia [**80**]:

$$\sum_{j=0}^{n} e_n(j) x^j = \frac{1}{n!} \sum_{\sigma \in S_n} (\operatorname{sgn} \sigma)(x - d_\sigma)(x - d_\sigma + 1) \cdots (x - d_\sigma + n - 1)\sigma,$$

where d_σ denotes the number of descents in σ, that is, the number of elements i with $\sigma(i) > \sigma(i+1)$. See [**146**, §4.5] for Eulerian idempotents in a general setting and resulting decompositions of Hochschild homology.

Consider the action of the symmetric group S_n on $A^{\otimes(n+2)}$ by permutation of n tensor factors (excluding the outermost two), that is,

$$\sigma(a_0 \otimes a_1 \otimes \cdots \otimes a_n \otimes a_{n+1}) = a_0 \otimes a_{\sigma^{-1}(1)} \otimes \cdots \otimes a_{\sigma^{-1}(n)} \otimes a_{n+1}$$

for all $\sigma \in S_n$ and $a_0, \ldots, a_{n+1} \in A$. The resulting actions of the Eulerian idempotents on the bar complex commute with the differentials in the sense that $d_n e_n(j) = e_{n-1}(j) d_n$ for all n, j as functions on $A^{\otimes(n+2)}$. Consider the corresponding right action of S_n on $\operatorname{Hom}_{A^e}(A^{\otimes(n+2)}, A)$. The complex $C^*(A, A)$ of (1.1.10) has a direct sum decomposition induced by this action and the Eulerian idempotents, and so does Hochschild cohomology $\mathrm{HH}^*(A)$: for each n,

$$(1.6.2) \qquad \mathrm{HH}^n(A) = \bigoplus_{j=1}^{n} \mathrm{HH}^n(A) e_n(j),$$

called the *Hodge decomposition* (or the *λ-decomposition*) [**146**, Theorem 4.5.10]. The component $\mathrm{HH}^n(A)e_n(1)$ is precisely the Harrison cohomology in degree n [**146**, Proposition 4.5.13]. The component $\mathrm{HH}^n(A)e_n(n)$ consists of the skew multiderivations $\bigwedge_A(\operatorname{Der}(A))$ [**146**, Theorem 4.5.12].

The Hodge decomposition (1.6.2) is due independently to Gerstenhaber and Schack [**84**] and Loday [**145**]. Ronco [**186**] explained this decomposition in the larger setting of André-Quillen homology [**179**]. Vigué-Poirrier [**218**] generalized it to differential graded commutative algebras. Other group actions on the cochains $C^*(A, A)$ lead to other decompositions of Hochschild cohomology, for example, Bergeron and Bergeron [**27**] considered an action of signed permutations and thereby obtained a decomposition analogous to (1.6.2) for algebras with involution.

Exercise 1.6.3. For small values of n, find the Eulerian idempotents $e_n(j)$ using either of the two equivalent definitions given in this section. Verify that $\mathrm{HH}^n(A)e_n(1)$ is indeed isomorphic to Harrison cohomology in degree n.

Chapter 2

Cup Product and Actions

We now examine in much greater detail the cup product on Hochschild cohomology, given in Definition 1.3.1 on the bar resolution. In this book, we will often work with resolutions other than bar resolutions, or with generalized extensions (that is, the n-extensions of Section A.3), so definitions of the cup product involving arbitrary resolutions or extensions are essential. We give several such equivalent definitions in this chapter. We also define actions of Hochschild cohomology rings on Ext spaces of modules.

We make the simplifying assumption from now on that k is a field, and we use the resulting identification (1.1.15) of Hochschild homology and cohomology spaces with Tor and Ext spaces.

2.1. From cocycles to chain maps

The first equivalent definition of cup product, in the next section, uses a construction of chain maps from cocycles that we recall in this section. This construction is general in that the same technique can be used to obtain a chain map corresponding to a homogeneous element of any Ext space (see Exercise 2.1.2). We present here the more specific case that we will use in this chapter and frequently throughout the book.

Let $P_{\bullet}$ be any projective resolution of A as an A^e-module with differential $d_{\bullet}$ (see Section A.2). Let $g \in \mathrm{Hom}_{A^e}(P_n, A)$ be a cocycle, that is, $gd_{n+1} = 0$, viewed in a commutative diagram as:

$$\cdots \longrightarrow P_{n+1} \xrightarrow{d_{n+1}} P_n \xrightarrow{d_n} P_{n-1} \xrightarrow{d_{n-1}} \cdots \xrightarrow{d_1} P_0 \xrightarrow{\varepsilon} A \longrightarrow 0$$

with the zero map 0 from P_{n+1} and the map $g : P_n \to A$.

Extend g to a chain map $g_. : P_. \to P_.$ as follows. (More precisely, it will be a chain map from $P_.$ to the shifted complex $P_.[-n]$ where we take the zero map on P_i if $i < n$; see Section A.1.) Let $K_n = \operatorname{Ker}(d_{n-1}) = \operatorname{Im}(d_n)$, and note that $K_n \cong P_n / \operatorname{Im}(d_{n+1})$. As in diagram (A.2.4), let $\varepsilon_n : P_n \to K_n$ denote the quotient map and $i_n : K_n \to P_{n-1}$ the inclusion map. Since g is a cocycle, it factors through K_n. Denote by $\overline{g}$ the map from K_n to A for which $\overline{g}\varepsilon_n = g$. We illustrate these maps by modifying the earlier commutative diagram:

(2.1.1)

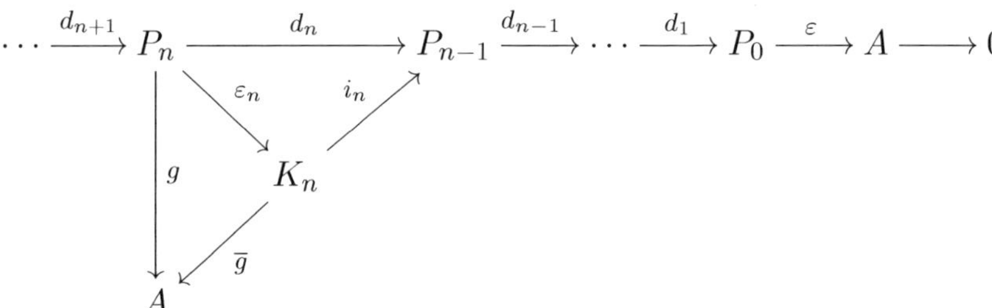

From there we extract the sequence

$$\cdots \xrightarrow{d_{n+3}} P_{n+2} \xrightarrow{d_{n+2}} P_{n+1} \xrightarrow{d_{n+1}} P_n \xrightarrow{\varepsilon_n} K_n \longrightarrow 0,$$

which is a projective resolution of K_n as an A^e-module. We rewrite diagram (2.1.1) to focus on this part, and apply the Comparison Theorem (Theorem A.2.7) which guarantees existence of maps $g_i : P_{n+i} \to P_i$ $(i \geq 0)$ that commute with the differentials, or more precisely, $g_i : P_{n+i} \to P[-n]_{n+i}$, where $P[-n]_.$ is the shifted complex with differentials $(-1)^n d_i$ (see Section A.1). We obtain the following commutative diagram:

$$
\begin{array}{ccccccccc}
\cdots \longrightarrow & P_{n+2} & \xrightarrow{d_{n+2}} & P_{n+1} & \xrightarrow{d_{n+1}} & P_n & \xrightarrow{\varepsilon_n} & K_n & \longrightarrow 0 \\
 & \downarrow{g_2} & & \downarrow{g_1} & & \downarrow{g_0} & & \downarrow{\overline{g}} & \\
\cdots \longrightarrow & P_2 & \xrightarrow{(-1)^n d_2} & P_1 & \xrightarrow{(-1)^n d_1} & P_0 & \xrightarrow{\varepsilon} & A & \longrightarrow 0
\end{array}
$$

We give some details of the maps g_i that are guaranteed by the Comparison Theorem, as we will use them often. Take g_i to be the zero map if $i < 0$. Since ε maps P_0 surjectively onto A, and P_n is projective, there is a map $g_0 : P_n \to P_0$ such that $\varepsilon g_0 = g$. We claim that the image of $g_0 d_{n+1}$ is contained in K_1. To see this, note that $K_1 = \operatorname{Ker} \varepsilon$ and that

$0 = gd_{n+1} = \varepsilon g_0 d_{n+1}$, so the image of $g_0 d_{n+1}$ is contained in $\operatorname{Ker} \varepsilon = K_1$. Now, since $g_0 d_{n+1}$ maps to K_1, P_1 surjects onto K_1 via d_1, and P_{n+1} is projective, there is a map $g_1 : P_{n+1} \to P_1$ such that $(-1)^n d_1 g_1 = g_0 d_{n+1}$. Once again, we see that the image of $g_1 d_{n+2}$ is contained in K_2, and this ensures existence of $g_2 : P_{n+2} \to P_2$, and so on. By the Comparison Theorem, the chain map $g_{\bullet}$ is unique up to chain homotopy.

Exercise 2.1.2. Generalize the construction given in this section. Let B be a ring. Let $P_{\bullet}$ and $Q_{\bullet}$ be projective resolutions of B-modules U and V, and let $g \in \operatorname{Hom}_B(P_n, V)$ be a cocycle, so that g represents an element of $\operatorname{Ext}_B^n(U, V)$. Let $\varepsilon : Q_0 \to V$ be the augmentation map of the resolution $Q_{\bullet}$. Show that there exists a chain map $g_{\bullet} : P_{\bullet} \to Q_{\bullet}[-n]$ for which $g = \varepsilon g_0$.

Exercise 2.1.3. Let $A = k[x]$, and let $P_{\bullet}$ be the resolution in Example 1.1.18. Letting g be any nonzero 1-cocycle, find g_0. Explain that g_i is the zero map for all $i \geq 1$.

2.2. Yoneda product

In this section, we give a second definition of product on Hochschild cohomology $HH^*(A)$ that is sometimes called the *Yoneda product*. We will define this product at the chain level on any resolution as a composition of chain maps. Then we will show that for the bar resolution, this definition is indeed equivalent to the cup product as defined by equation (1.3.2). We will use the same notation $\smile$ for this product in anticipation of our proof that it is equivalent to that in Definition 1.3.1.

Let $P_{\bullet}$ be any projective resolution of A as an A^e-module. Let $f \in \operatorname{Hom}_{A^e}(P_m, A)$ and $g \in \operatorname{Hom}_{A^e}(P_n, A)$ be cocycles. Extend g to a chain map $g_{\bullet} : P_{\bullet} \to P[-n]_{\bullet}$ as described in Section 2.1. The map $f \smile g \in \operatorname{Hom}_{A^e}(P_{m+n}, A)$ is defined to be the composition fg_m:

$$(2.2.1) \qquad\qquad f \smile g = fg_m.$$

This composition may be viewed in a diagram:

$$
\begin{array}{ccccccc}
\cdots \longrightarrow P_{m+n} & \xrightarrow{d_{m+n}} \cdots \xrightarrow{d_{n+2}} & P_{n+1} & \xrightarrow{d_{n+1}} & P_n & & \\
\downarrow{\scriptstyle g_m} & & \downarrow{\scriptstyle g_1} & & \downarrow{\scriptstyle g_0} & \searrow^{g} & \\
\cdots \longrightarrow P_m & \xrightarrow[(-1)^n d_m]{} \cdots \xrightarrow[(-1)^n d_2]{} & P_1 & \xrightarrow[(-1)^n d_1]{} & P_0 & \xrightarrow{\varepsilon} A & \longrightarrow 0 \\
\downarrow{\scriptstyle f} & & & & & & \\
A & & & & & &
\end{array}
$$

Note that $f \smile g$ is a cocycle because f is a cocycle and $g_{\bullet}$ is a chain map. Since $g_{\bullet}$ is unique up to chain homotopy, a different such chain map $g'_{\bullet}$ results in a cohomologous cocycle fg'_m. If f is a coboundary, then $f \smile g$

is a coboundary since $g_\bullet$ is a chain map. If g is a coboundary, we claim that we may set $g_m = 0$ for all $m \geq 1$. To see this, write $g = hd_n$ for a cochain $h \in \mathrm{Hom}_{A^e}(P_{n-1}, A)$. By projectivity of P_{n-1}, there is a map $h' : P_{n-1} \to P_0$ such that $\varepsilon h' = h$. Setting $g_0 = h'd_n$ and $g_1 = 0$, we find that $-d_1 g_1 = g_0 d_{n+1} = h'd_n d_{n+1} = 0$. Now we may set $g_m = 0$ for all $m > 1$ as well. Hence $f \smile g = 0$ at the chain level in this case when $m \geq 1$. We thus see that this cup product, given at the chain level, induces a well-defined product on cohomology.

Equivalently, we may identify both cocycles f, g with choices of chain maps $f_\bullet, g_\bullet$ that they induce as described above, in which case $f \smile g$ may be identified at the chain level with a composition of chain maps, as indicated by a commuting diagram:

$$
\begin{array}{ccccccccc}
\cdots \longrightarrow & P_{m+n+1} & \xrightarrow{d_{m+n+1}} & P_{m+n} & \xrightarrow{d_{m+n}} & \cdots \xrightarrow{d_{n+2}} & P_{n+1} & \xrightarrow{d_{n+1}} & P_n \\
& \downarrow{\scriptstyle g_{m+1}} & & \downarrow{\scriptstyle g_m} & & & \downarrow{\scriptstyle g_1} & & \downarrow{\scriptstyle g_0} \\
\cdots \longrightarrow & P_{m+1} & \xrightarrow{(-1)^n d_{m+1}} & P_m & \xrightarrow{(-1)^n d_m} & \cdots \xrightarrow{(-1)^n d_2} & P_1 & \xrightarrow{(-1)^n d_1} & P_0 \\
& \downarrow{\scriptstyle f_1} & & \downarrow{\scriptstyle f_0} & & & & & \\
\cdots \longrightarrow & P_1 & \xrightarrow{(-1)^{m+n} d_1} & P_0 & & & & &
\end{array}
$$

Again, at the level of cohomology, this does not depend on choices. Filling in missing components and arrows, we obtain a chain map $(f \smile g)_\bullet$, an element of $\mathrm{Hom}_{A^e}(P_\bullet, P_\bullet[-m-n])$ defined componentwise by

$$(f \smile g)_i = f_i g_{m+i}$$

as a map from P_{m+n+i} to P_i for all $i \geq 0$. We map P_j to 0 for $j < m + n$.

Returning to the definition (2.2.1) of product, we claim that it does not depend on choice of resolution $P_\bullet$. To see this, apply the Comparison Theorem (Theorem A.2.7) to $P_\bullet$ and any other projective A^e-resolution $Q_\bullet$ of A. Composition with a comparison map takes the chain map $g_\bullet$ to a corresponding chain map on $Q_\bullet$ and the composition fg_m of (2.2.1) to a corresponding composition as a function on Q_{m+n}.

Taking $P_\bullet$ to be the bar resolution $B(A)$ given by (1.1.4), we will show that the product defined by equation (2.2.1) is the same as that defined by equation (1.3.2). Then by the Comparison Theorem (Theorem A.2.7), comparing any other projective resolution to $B(A)$, definition (2.2.1) will then indeed be an equivalent definition of cup product on Hochschild cohomology $\mathrm{HH}^*(A)$. Let us apply the definition (2.2.1) when $P_n = A^{\otimes(n+2)}$ for all n. Given $g : A^{\otimes(n+2)} \to A$, a cocycle, we must choose $g_0 : A^{\otimes(n+2)} \to A \otimes A$ for which $\pi g_0 = g$, that is,

$$\pi g_0(1 \otimes a_1 \otimes \cdots \otimes a_n \otimes 1) = g(1 \otimes a_1 \otimes \cdots \otimes a_n \otimes 1)$$

for all $a_1, \ldots, a_n \in A$. One such choice is given by

$$g_0(1 \otimes a_1 \otimes \cdots \otimes a_n \otimes 1) = 1 \otimes g(1 \otimes a_1 \otimes \cdots \otimes a_n \otimes 1),$$

where we extend to an A^e-module map by defining

$$g_0(a_0 \otimes a_1 \otimes \cdots \otimes a_{n+1}) = a_0 \otimes g(1 \otimes a_1 \otimes \cdots \otimes a_n \otimes 1)a_{n+1}$$

for all $a_0, \ldots, a_{n+1} \in A$. Then a choice of $g_1 : A^{\otimes(n+3)} \to A^{\otimes 3}$ can be given by

$$g_1(1 \otimes a_1 \otimes \cdots \otimes a_{n+1} \otimes 1) = (-1)^n \otimes a_1 \otimes g(1 \otimes a_2 \otimes \cdots \otimes a_{n+1} \otimes 1),$$

extended to an A^e-module map; a calculation shows that $g_0 d_{n+1} = (-1)^n d_1 g_1$ using the definition of g_0 and the assumption that g is a cocycle. In general we may choose

$$g_i(1 \otimes a_1 \otimes \cdots \otimes a_{n+i} \otimes 1) = (-1)^{in} \otimes a_1 \otimes \cdots \otimes a_i \otimes g(1 \otimes a_{i+1} \otimes \cdots \otimes a_{n+i} \otimes 1).$$

We see that if $f \in \mathrm{Hom}_{A^e}(A^{\otimes(m+2)}, A)$, then under definition (2.2.1),

$$
\begin{aligned}
(f \smile g)&(1 \otimes a_1 \otimes \cdots \otimes a_{m+n} \otimes 1) \\
&= f g_m(1 \otimes a_1 \otimes \cdots \otimes a_{m+n} \otimes 1) \\
&= (-1)^{mn} f(1 \otimes a_1 \otimes \cdots \otimes a_m \otimes g(1 \otimes a_{m+1} \otimes \cdots \otimes a_{m+n} \otimes 1)) \\
&= (-1)^{mn} f(1 \otimes a_1 \otimes \cdots \otimes a_m \otimes 1)g(1 \otimes a_{m+1} \otimes \cdots \otimes a_{m+n} \otimes 1)
\end{aligned}
$$

for all $a_1, \ldots, a_{m+n} \in A$. This is equivalent to formula (1.3.2) under the identification of $\mathrm{Hom}_{A^e}(A^{\otimes(m+n+2)}, A)$ and $\mathrm{Hom}_k(A^{\otimes(m+n)}, A)$ via isomorphism (1.1.11). There is another choice of lifting $g_\bullet$ of g to which a comparison leads to another proof that the product on Hochschild cohomology is graded commutative; see, e.g., Solberg [**203**, Theorem 2.1] where the definition of cup product is the historical one (see Remark 1.3.3 and cf. Theorem 1.4.6).

We may use equation (2.2.1) to compute cup products for more examples, starting with the truncated polynomial rings $k[x]/(x^n)$ as we see next.

Example 2.2.2. We return to Example 1.1.21, in which $A = k[x]/(x^n)$, to compute cup products. Let $p = \mathrm{char}(k)$, and assume that p does not divide n. (The case where p divides n is different, and is the next example.) In degree 1, let $g \in \mathrm{Hom}_{A^e}(A^e, A)$ be the function given by $g(1 \otimes 1) = x$. By the description of the cohomology in Example 1.1.21 and of the action of $\mathrm{HH}^0(A) \cong Z(A)$ described at the end of Section 1.2, g represents an element in $\mathrm{HH}^1(A)$ and generates $\mathrm{HH}^1(A)$ as an $\mathrm{HH}^0(A)$-module. In degree 2, let $f \in \mathrm{Hom}_{A^e}(A^e, A)$ be the function given by $f(1 \otimes 1) = 1$. Then by similar reasoning, f represents an element in $\mathrm{HH}^2(A)$ and generates $\mathrm{HH}^2(A)$ as an $\mathrm{HH}^0(A)$-module. We will find cup products of these functions as compositions of chain maps, and show that other cup products are determined by these.

In relation to the resolution (1.1.22), let $K_1 = A^e / \operatorname{Im}(v\cdot)$. Since g is a cocycle, it factors through K_1 as in the following diagram. The map from A^e to K_1 in the diagram can be taken to be the quotient map, which may be identified with the map $u\cdot$ to $\operatorname{Im}(u\cdot) \cong K_1$ as a submodule of A^e in degree 0.

$$
\begin{array}{ccccccccc}
\cdots & \longrightarrow & A^e & \xrightarrow{\,u\cdot\,} & A^e & \xrightarrow{\,v\cdot\,} & A^e & \longrightarrow & K_1 & \longrightarrow & \cdots \\
& & \downarrow{\scriptstyle g_2} & & \downarrow{\scriptstyle g_1} & & \downarrow{\scriptstyle g_0} & \searrow{\scriptstyle g} & \downarrow{\scriptstyle \overline{g}} & & \\
\cdots & \longrightarrow & A^e & \xrightarrow{\,-v\cdot\,} & A^e & \xrightarrow{\,-u\cdot\,} & A^e & \xrightarrow{\,\pi\,} & A & \longrightarrow & 0
\end{array}
$$

We may find $g_0, g_1, g_2, \ldots$ via the technique described in Section 2.1, starting by finding g_0, then g_1, and so on. Set $g_{2m}(1 \otimes 1) = x \otimes 1$ and

$$
g_{2m+1}(1 \otimes 1) = - \sum_{i=1}^{n-1} i x^{n-1-i} \otimes x^i
$$

for all $m \geq 0$. Extend to A^e-module homomorphisms. A calculation then shows that the above diagram commutes.

We use these maps g_j to find some cup products. First note $gg_1 = 0$ since $x^n = 0$ in A, and so $g \smile g = 0$. We similarly find that

$$
(f \smile g)(1 \otimes 1) = fg_2(1 \otimes 1) = f(x \otimes 1) = x.
$$

Comparing to the results of Example 1.1.21, we find that $f \smile g$ represents a generator for $\mathrm{HH}^3(A)$ as an $\mathrm{HH}^0(A)$-module.

Now let $K_2 = A^e / \operatorname{Im}(u\cdot)$. Since f is a cocycle, it factors through K_2 as shown in the following diagram:

$$
\begin{array}{ccccccccc}
\cdots & \longrightarrow & A^e & \xrightarrow{\,v\cdot\,} & A^e & \xrightarrow{\,u\cdot\,} & A^e & \longrightarrow & K_2 & \longrightarrow & \cdots \\
& & \downarrow{\scriptstyle f_2} & & \downarrow{\scriptstyle f_1} & & \downarrow{\scriptstyle f_0} & \searrow{\scriptstyle f} & \downarrow{\scriptstyle \overline{f}} & & \\
\cdots & \longrightarrow & A^e & \xrightarrow{\,v\cdot\,} & A^e & \xrightarrow{\,u\cdot\,} & A^e & \xrightarrow{\,\pi\,} & A & \longrightarrow & 0
\end{array}
$$

Set $f_j(1 \otimes 1) = 1 \otimes 1$ for all $j \geq 0$. A calculation then shows that the above diagram commutes. We find that $(f \smile f)(1 \otimes 1) = ff_2(1 \otimes 1) = 1$, and similarly for all powers i of f, by induction:

$$
f^i(1 \otimes 1) = f^{i-1}f_2(1 \otimes 1) = 1,
$$

where $f^i = f \smile \cdots \smile f$ (i factors of f). Again comparing to the results of Example 1.1.21, we see that in each even degree, $\mathrm{HH}^{2i}(A)$ is generated by the image $\overline{f^i}$ of f^i in $\mathrm{HH}^{2i}(A)$ as an $\mathrm{HH}^0(A)$-module. Similarly, the image of $f^i \smile g = f^i g_2$ generates $\mathrm{HH}^{2i+1}(A)$ as an $\mathrm{HH}^0(A)$-module. Due to the structure of the resolution (1.1.22), $x^{n-1} \smile g$ and $x^{n-1} \smile f$ are coboundaries. Taking advantage of graded commutativity (Theorem 1.4.6), we claim that we have now computed all products needed to conclude the ring structure of $\mathrm{HH}^*(A)$. To see this, map the polynomial ring $k[x, y, z]$

onto $\mathrm{HH}^*(A)$ by sending x to a copy of itself in $\mathrm{HH}^0(A)$, and y and z to the elements of cohomology represented by f and g, respectively. Our observations so far imply that this map factors through a quotient as follows:

$$k[x, y, z]/(x^n,\ x^{n-1}y,\ y^2,\ x^{n-1}z) \longrightarrow \mathrm{HH}^*(A).$$

This map is surjective since x, f, g generate $\mathrm{HH}^*(A)$ as we have seen. Comparing vector space dimensions in each degree (setting $|x| = 0$, $|y| = 1$, and $|z| = 2$), we see that this map must in fact be an isomorphism:

$$\mathrm{HH}^*(A) \cong k[x, y, z]/(x^n,\ x^{n-1}y,\ y^2,\ x^{n-1}z).$$

Example 2.2.3. Let $A = k[x]/(x^n)$ as in the previous example but now under the assumption that the characteristic p of k divides n. We again use the description of Hochschild cohomology in Example 1.1.21. At the chain level in degree 1, let $g \in \mathrm{Hom}_{A^e}(A^e, A)$ be the function defined by $g(1 \otimes 1) = 1$. In degree 2, let $f \in \mathrm{Hom}_{A^e}(A^e, A)$ be the function defined by $f(1 \otimes 1) = 1$. Then f, g are cocycles. Let $K_1 = A^e/\mathrm{Im}(v\cdot)$. Set $g_{2m}(1 \otimes 1) = 1 \otimes 1$ and

$$g_{2m+1}(1 \otimes 1) = -\sum_{i=1}^{n-1} i x^{n-1-i} \otimes x^{i-1}$$

for all $m \geq 0$. (Note that if $n = 2$, this formula yields $g_{2m+1}(1 \otimes 1) = -1 \otimes 1$.) A calculation shows, since $n \equiv 0$ in k, that $g_\bullet$ is a chain map. We use equation (2.2.1) to compute in case p is odd:

$$(2.2.4) \qquad \begin{aligned} (g \smile g)(1 \otimes 1) &= gg_1(1 \otimes 1) = -\frac{n(n-1)}{2}x^{n-2} = 0, \\ (f \smile g)(1 \otimes 1) &= fg_2(1 \otimes 1) = 1. \end{aligned}$$

Further calculations show as in Example 2.2.2 that the images of f and g generate $\mathrm{HH}^*(A)$ as an algebra over its degree 0 component $\mathrm{HH}^0(A) \cong A$ in this case. If $p = 2$ and $n = 2s$ for some even integer s, similar computations show that $g \smile g = 0$ and $(f \smile g)(1 \otimes 1) = 1$. Thus we find, similarly to Example 2.2.2, that in odd characteristic p dividing n, or in case $p = 2$ and $n = 2s$ for some even integer s, there is an isomorphism of algebras,

$$\mathrm{HH}^*(A) \cong k[x, y, z]/(x^n,\ y^2),$$

where $|x| = 0$, $|y| = 1$, $|z| = 2$. If $p = 2$ and $n = 2s$ for some odd integer s, then $(g \smile g)(1 \otimes 1) = x^{n-2}$. If $n = 2$, this implies that the image of $g \smile g$ generates $\mathrm{HH}^2(A)$ as an $\mathrm{HH}^0(A)$-module, and similarly the image of g generates $\mathrm{HH}^*(A)$ as an algebra over $A \cong \mathrm{HH}^0(A)$. If $n \neq 2$, it does not, and there is a relation among the generators. Thus we find that if $p = 2$ and $n = 2$, there is an isomorphism of algebras,

$$\mathrm{HH}^*(A) \cong k[x, y]/(x^n),$$

where $|x| = 0$, $|y| = 1$. If $p = 2$ and $n = 2s$ for some odd integer $s > 1$, then

$$\mathrm{HH}^*(A) \cong k[x, y, z]/(x^n, \, x^2 y^2),$$

where $|x| = 0$, $|y| = 1$, $|z| = 2$.

More generally, Hochschild cohomology rings of algebras of the form $k[x]/(f)$, where (f) is the ideal generated by a polynomial f, were computed by Holm [**118**].

Exercise 2.2.5. Verify the claims made right after equation (2.2.1). Specifically, verify that $f \smile g$ as defined in (2.2.1) is indeed a cocycle. If $g'_{\bullet} : P_{\bullet} \to P[-n]_{\bullet}$ is another chain map extending g, show that $f g'_m$ is cohomologous to $f g_m$.

Exercise 2.2.6. Verify that the definition (2.2.1) of cup product on $\mathrm{HH}^*(A)$ does not depend on choice of resolution. That is, given two projective resolutions $P_{\bullet}$ and $Q_{\bullet}$ of A as an A^e-module, and a comparison map $\phi : Q_{\bullet} \to P_{\bullet}$ lifting the identity map from A to A (whose existence is guaranteed by the Comparison Theorem, Theorem A.2.7), the cup product $f \smile g$ defined on $P_{\bullet}$ and the cup product $(f \phi_m) \smile (g \phi_n)$ defined on $Q_{\bullet}$ represent the same element of $\mathrm{HH}^{m+n}(A)$.

Exercise 2.2.7. Verify that $f_{\bullet}$, $g_{\bullet}$ of Example 2.2.2 and of Example 2.2.3 are indeed chain maps. Verify the claimed structure of Hochschild cohomology in each of these examples.

2.3. Tensor product of complexes

Another definition of associative product on Hochschild cohomology $\mathrm{HH}^*(A)$ that is also equivalent to the cup product of Definition 1.3.1 is a convolution product arising from a tensor product of complexes and a diagonal map. We present this definition in this section.

We will use a standard sign convention in this chapter and throughout the rest of the book. Let V, V', W, W' be graded vector spaces (graded by $\mathbb{N}$ or $\mathbb{Z}$), and let $f : V \to V'$ and $g : W \to W'$ be k-linear graded functions, that is, there is some m for which

$$f(V_n) \subseteq V_{n+m}$$

for all n, and similarly for g. Write $|f| = m$. The function $f \otimes g$ on $V \otimes W$ is defined by

$$(2.3.1) \qquad\qquad (f \otimes g)(v \otimes w) = (-1)^{|g||v|} f(v) \otimes g(w)$$

for all homogeneous $v \in V$, $w \in W$. We use an analogous sign convention for tensor products of graded modules over other rings.

Let $P_\bullet$ be any A^e-projective resolution of A with augmentation map $\mu : P_0 \to A$. We may view $P_\bullet$ as a graded vector space with ith component P_i. Since P_i is an A^e-module, it may be viewed as an A-bimodule, and thus we may consider the tensor product $P_i \otimes_A P_j$ to be an A-bimodule (equivalently, A^e-module) for each i, j. We will work with the tensor product complex $P_\bullet \otimes_A P_\bullet$:

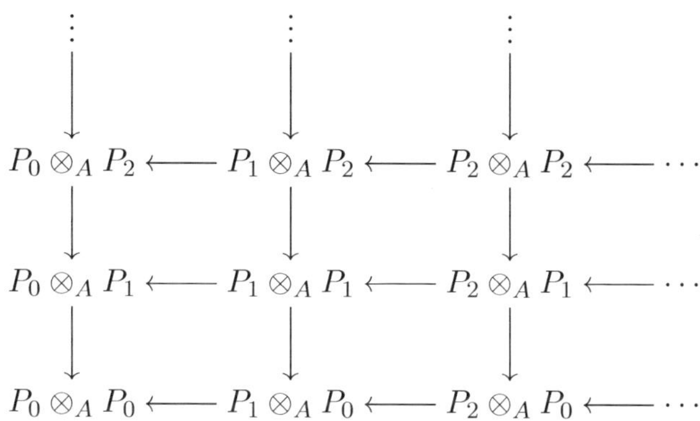

We will show that the total complex of $P_\bullet \otimes_A P_\bullet$ is also an A^e-projective resolution of A. First we must show that for each m, n, the A^e-module $P_m \otimes_A P_n$ is projective. To see this, note that since P_m, P_n are projective A^e-modules, each is a direct summand of a direct sum of copies of A^e. Thus it suffices to show that $A^e \otimes_A A^e$ is free. This follows from the isomorphism $A^e \otimes_A A^e \xrightarrow{\sim} A^e \otimes_k A$ of A^e-modules given by $(a_1 \otimes a_2) \otimes_A (a_3 \otimes a_4) \mapsto a_1 \otimes a_4 \otimes a_2 a_3$ for $a_1, a_2, a_3, a_4 \in A$ (note A^e acts only on the outermost two factors of A in $A^e \otimes_A A^e$). Since k is assumed to be a field, A is free as a k-module, and so $A^e \otimes_k A$ is free as an A^e-module.

Next, to see that the total complex $\mathrm{Tot}(P_\bullet \otimes_A P_\bullet)$ of the tensor product complex $P_\bullet \otimes_A P_\bullet$ has cohomology A concentrated in degree 0, we apply the Künneth Theorem (Theorem A.5.2): the module $A \otimes A^{\mathrm{op}}$, under right multiplication by elements of A on the right factor, is a free right A-module since A is a free k-module. It follows that $P_\bullet$ is, by restriction, a projective resolution of the free right A-module A. Considering only the right A-module structure of the resolution $P_\bullet$ now, it necessarily splits as a sequence of right A-modules. That is, the augmentation map $\mu : P_0 \to A$ splits via some map $\iota : A \to P_0$ such that $A \cong \iota(A)$ is a direct summand of P_0, so that $P_0 \cong A \oplus \mathrm{Im}(d_1) = A \oplus \mathrm{Ker}(\mu)$ as a right A-module. Then $d_1 : P_1 \to P_0$ splits so that $P_1 \cong \mathrm{Im}(d_1) \oplus \mathrm{Im}(d_2)$ as a right A-module, and so on. It follows that the boundary spaces $\mathrm{Im}(d_i)$ are also all projective right A-modules, that is, the hypotheses of the Künneth Theorem (Theorem A.5.2) hold. The Tor terms in the Künneth sequence (that is, the sequence in the Künneth Theorem statement) vanish: the only term in which both arguments are nonzero is $\mathrm{Tor}_1^A(\mathrm{H}_0(P), \mathrm{H}_0(P)) = \mathrm{Tor}_1^A(A, A) = 0$ (since A is free as an A-module). This implies that the total complex of $P_{\boldsymbol{\cdot}} \otimes_A P_{\boldsymbol{\cdot}}$ is indeed a resolution of $A \otimes_A A \cong A$ by A^e-projective modules. (We often identify $A \otimes_A A$ with A via canonical isomorphism given by multiplication π on A.)

By the Comparison Theorem (Theorem A.2.7) there is a chain map $\Delta : P_{\boldsymbol{\cdot}} \to P_{\boldsymbol{\cdot}} \otimes_A P_{\boldsymbol{\cdot}}$ lifting the canonical isomorphism $A \xrightarrow{\sim} A \otimes_A A$. Such a map Δ is unique up to chain homotopy. Sometimes Δ is called a *diagonal map*.

The cup product on Hochschild cohomology may be defined via any diagonal map $\Delta : P_{\boldsymbol{\cdot}} \to P_{\boldsymbol{\cdot}} \otimes_A P_{\boldsymbol{\cdot}}$ as a convolution product in the following way. Let $f \in \mathrm{Hom}_{A^e}(P_m, A)$, $g \in \mathrm{Hom}_{A^e}(P_n, A)$ represent elements of $\mathrm{HH}^m(A, A)$, $\mathrm{HH}^n(A, A)$, respectively. View $f \otimes g$ as a function on $P_m \otimes_A P_n$ and extend to $P_{\boldsymbol{\cdot}} \otimes_A P_{\boldsymbol{\cdot}}$ by setting $(f \otimes g)(x) = 0$ for all $x \in P_r \otimes_A P_s$ such that $(r, s) \neq (m, n)$. We define

$$(2.3.2) \qquad\qquad f \smile g = (f \otimes g)\Delta,$$

which indeed takes values in A under the identification of $A \otimes_A A$ with A via canonical isomorphism. Calculations show that this induces a well-defined operation on Hochschild cohomology: the product of cocycles is a cocycle, and the product of a cocycle with a coboundary is a coboundary. Moreover, this operation on Hochschild cohomology does not depend on choice of diagonal map Δ, since any two diagonal maps are chain homotopic. It also does not depend on choice of resolution $P_{\boldsymbol{\cdot}}$: for any two resolutions, by the Comparison Theorem (Theorem A.2.7), there are chain maps between them. These chain maps induce a diagonal map on one from that of the other, and we have seen that the product is independent of choices of diagonal map.

A word on notation: the tensor product of maps $f \otimes g$ could instead be written $f \otimes_A g$. Here and elsewhere we abuse notation to reduce clutter by omitting the subscript A on the tensor product symbol for maps and elements. It should be clear from the domain of such a function whether the tensor product is taken over k or another ring, and similarly it should be clear from the space to which an element belongs as to whether the tensor product is taken over k or another ring.

If $P_{\boldsymbol{\cdot}}$ is the bar resolution $B(A)$ given by (1.1.4), one choice of chain map Δ induces precisely the chain level cup product (1.3.2): define a diagonal map Δ on the bar resolution $B(A)$ of (1.1.4) by

$$(2.3.3) \quad \Delta(1 \otimes a_1 \otimes \cdots \otimes a_s \otimes 1) = \sum_{i=0}^{s} (1 \otimes a_1 \otimes \cdots \otimes a_i \otimes 1) \otimes (1 \otimes a_{i+1} \otimes \cdots \otimes a_s \otimes 1)$$

for all $a_1, \ldots, a_s \in A$. A calculation shows that formula (2.3.2), with these choices, is precisely formula (1.3.2); the factor $(-1)^{mn}$ in formula (1.3.2) agrees with the sign convention (2.3.1). If $P_\bullet$ is not the bar resolution, as in our earlier discussion, the product (2.3.2) will be the same at the level of cohomology. Thus on cohomology, our product in (2.3.2) is equivalent to the cup product (1.3.2), justifying our use of the same notation for this product as well.

Exercise 2.3.4. Verify details of the claims that the convolution product (2.3.2) induces a well-defined map on Hochschild cohomology and is independent of choices, that is, show that:

> (a) The product of cocycles is a cocycle, and the product of a cocycle with a coboundary is a coboundary.
>
> (b) The product induced on cohomology is independent of choice of projective resolution $P_\bullet$ and diagonal map Δ.

Exercise 2.3.5. Let $A = k[x]/(x^2)$, and let $P_\bullet$ be the resolution (1.1.22). Let $\xi_i = 1 \otimes 1$ in $P_i = A^e$. Show that the following defines a diagonal map on $P_\bullet$:

$$\Delta(\xi_i) = \sum_{j=0}^{i} \xi_j \otimes \xi_{i-j}.$$

Use this diagonal map and equation (2.3.2) to reproduce the algebra structure of $\mathrm{HH}^*(A)$ given in Examples 2.2.2 and 2.2.3 when $n = 2$.

2.4. Yoneda composition and tensor product of extensions

Two more definitions of the product on Hochschild cohomology $\mathrm{HH}^*(A)$ are described by way of generalized extensions. (See Section A.3 for a summary of n-extensions and connections with Ext.)

We begin with the *Yoneda composition* (or *Yoneda splice*). This definition uses the description of Hochschild cohomology $\mathrm{HH}^n(A) \cong \mathrm{Ext}^n_{A^e}(A, A)$ as equivalence classes of n-extensions of A by A as A^e-modules. Let **f** and **g** be m- and n-extensions,

$$\mathbf{f}: \quad 0 \longrightarrow A \xrightarrow{\alpha_m} M_{m-1} \xrightarrow{\alpha_{m-1}} \cdots \xrightarrow{\alpha_2} M_1 \xrightarrow{\alpha_1} M_0 \xrightarrow{\alpha_0} A \longrightarrow 0,$$

$$\mathbf{g}: \quad 0 \longrightarrow A \xrightarrow{\beta_n} N_{n-1} \xrightarrow{\beta_{n-1}} \cdots \xrightarrow{\beta_2} N_1 \xrightarrow{\beta_1} N_0 \xrightarrow{\beta_0} A \longrightarrow 0,$$

respectively, representing elements of $\mathrm{HH}^m(A)$ and $\mathrm{HH}^n(A)$. Consider the composition $\beta_n \alpha_0$, a map from M_0 to N_{n-1}. We may use this map to combine the two sequences (by "splicing" them) into a new sequence that we denote by $\mathbf{g} \smile \mathbf{f}$:

$$(2.4.1) \qquad \mathbf{g} \smile \mathbf{f} :$$

$$0 \longrightarrow A \xrightarrow{\alpha_m} M_{m-1} \xrightarrow{\alpha_{m-1}} \cdots \xrightarrow{\alpha_1} M_0 \xrightarrow{\beta_n \alpha_0} N_{n-1} \xrightarrow{\beta_{n-1}} \cdots$$

$$\xrightarrow{\beta_1} N_0 \xrightarrow{\beta_0} A \longrightarrow 0.$$

A calculation shows that this sequence is exact at M_0 and at N_{n-1}. Therefore it is an exact sequence, in other words, it is an $(m+n)$-extension of A by A. We have defined $\mathbf{g} \smile \mathbf{f}$ to be this $(m+n)$-extension, representing an element of $\mathrm{HH}^{m+n}(A)$. Letting $f \in \mathrm{Hom}_{A^e}(P_m, A)$ and $g \in \mathrm{Hom}_{A^e}(P_n, A)$ be maps corresponding to the sequences $\mathbf{f}$ and $\mathbf{g}$, respectively, it can be seen to follow from the definitions and the correspondence between generalized extensions and Ext (see Section A.3) that $\mathbf{g} \smile \mathbf{f}$ corresponds to $(-1)^{mn} f \smile g$, where $f \smile g$ is the Yoneda product (2.2.1). This equivalence is explained in detail for example in [**3**, Theorem 4.3], where slightly different left-right and sign conventions were chosen.

We may alternatively take the tensor product, over A, of an m-extension with an n-extension to obtain an $(m+n)$-extension from the total complex. This may be seen to be equivalent to Yoneda composition, and thus to cup product, by mapping to two edges. Specifically, the tensor product complex of the m- and n-extensions $\mathbf{f}$ and $\mathbf{g}$ as above is the following:

$$
\begin{array}{ccccccc}
A \otimes_A A & \longleftarrow & M_0 \otimes_A A & \longleftarrow \cdots \longleftarrow & M_{m-1} \otimes_A A & \longleftarrow & A \otimes_A A \\
\downarrow & & \downarrow & & \downarrow & & \downarrow \\
A \otimes_A N_{n-1} & \longleftarrow & M_0 \otimes_A N_{n-1} & \longleftarrow \cdots \longleftarrow & M_{m-1} \otimes_A N_{n-1} & \longleftarrow & A \otimes_A N_{n-1} \\
\downarrow & & \downarrow & & \downarrow & & \downarrow \\
\vdots & & \vdots & & \vdots & & \vdots \\
\downarrow & & \downarrow & & \downarrow & & \downarrow \\
A \otimes_A N_1 & \longleftarrow & M_0 \otimes_A N_1 & \longleftarrow \cdots \longleftarrow & M_{m-1} \otimes_A N_1 & \longleftarrow & A \otimes_A N_1 \\
\downarrow & & \downarrow & & \downarrow & & \downarrow \\
A \otimes_A N_0 & \longleftarrow & M_0 \otimes_A N_0 & \longleftarrow \cdots \longleftarrow & M_{m-1} \otimes_A N_0 & \longleftarrow & A \otimes_A N_0 \\
\downarrow & & \downarrow & & \downarrow & & \downarrow \\
A \otimes_A A & \longleftarrow & M_0 \otimes_A A & \longleftarrow \cdots \longleftarrow & M_{m-1} \otimes_A A & \longleftarrow & A \otimes_A A
\end{array}
$$

Consider this diagram to be an augmented double complex. The corresponding $(m+n)$-extension $\mathbf{f} \otimes_A \mathbf{g}$ of A by A is obtained by deleting the leftmost column and bottom row, taking the total complex, augmenting with A, and applying the canonical isomorphisms $A \otimes_A A \cong A$, $A \otimes_A N_i \cong N_i$, $M_i \otimes_A A \cong M_i$:

$$0 \longrightarrow A \longrightarrow M_{m-1} \oplus N_{n-1} \longrightarrow \cdots$$

$$\longrightarrow \begin{matrix}(M_1 \otimes_A N_0) \oplus \\ (M_0 \otimes_A N_1)\end{matrix} \longrightarrow M_0 \otimes_A N_0 \longrightarrow A \longrightarrow 0.$$

The Künneth Theorem (Theorem A.5.2) implies that the above sequence is exact, since all modules in the original sequences $\mathbf{f}$ and $\mathbf{g}$ are free as right A-modules.

We next show that the total complex above is equivalent, as an $(m+n)$-extension, to the Yoneda splice $\mathbf{g} \smile \mathbf{f}$ of (2.4.1). Delete the upper left and lower right corners, instead replacing them with compositions of the maps:

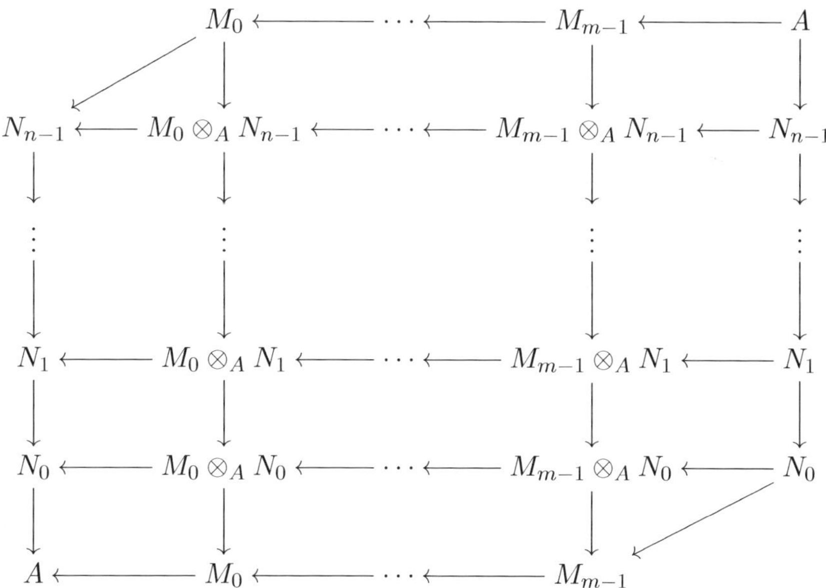

Now we see that the top and left edges form the Yoneda splice $\mathbf{g} \smile \mathbf{f}$ of (2.4.1). The total complex maps to $\mathbf{g} \smile \mathbf{f}$ simply by canonical projection to the top row and canonical projection followed by maps induced by the augmentation map $M_0 \to A$ in the left column. In this way we see that the $(m+n)$-extension arising from the total complex is equivalent to $\mathbf{g} \smile \mathbf{f}$ as defined by (2.4.1), and thus corresponds to the cup product.

This diagram also leads to another proof of graded commutativity: up to signs, the right and lower edges form the Yoneda splice $\mathbf{f} \smile \mathbf{g}$. From Section A.5, in the tensor product complex, vertical maps have signs attached so that for example the rightmost column maps take the sign $(-1)^m$. We use the convention that $-(\mathbf{f} \smile \mathbf{g})$ is the extension $\mathbf{f} \smile \mathbf{g}$ in which α_0 is replaced by $-\alpha_0$ and all other maps are the same. Thus we see that the rightmost column followed by the bottom row is equivalent to the $(m+n)$-extension $(-1)^{mn}\mathbf{f} \smile \mathbf{g}$: a map from $\mathbf{f} \otimes_A \mathbf{g}$ to $(-1)^{mn}\mathbf{f} \smile \mathbf{g}$ is given by projecting each degree i component $(m \le i \le m+n)$ onto $N_{i-m} \cong M_m \otimes_A N_{i-m}$ followed by multiplication by $(-1)^{m(m+n-i)}$, and projecting each degree i component $(0 \le i \le m-1)$ onto $M_i \otimes_A N_0$ followed by mapping to $M_i \otimes_A A \cong M_i$ via $(-1)^{mn}\beta_0$.

We will frequently exploit the agreement of the associative products on Hochschild cohomology $\mathrm{HH}^*(A)$, defined at the chain level and on generalized extensions in Sections 2.2–2.4 and in Definition 1.3.1. In practice, in each setting we will use the version that is most convenient for that setting.

Exercise 2.4.2. Verify that the sequence (2.4.1) is indeed exact.

Exercise 2.4.3. Verify directly that the product defined by (2.4.1) is equivalent to the Yoneda product of formula (2.2.1) by applying the correspondence between generalized extensions and cochains described in Section A.3.

Exercise 2.4.4. Let $A = k[x]/(x^2)$, and let V be the 2-dimensional A-module with basis v_1, v_2 for which $xv_2 = v_1$ and $xv_1 = 0$. Consider a 1-extension of A-modules,

$$\mathbf{f} : \qquad 0 \longrightarrow k \longrightarrow V \longrightarrow k \longrightarrow 0.$$

(The map $k \to V$ sends 1 to v_1 and the map $V \to k$ sends v_1 to 0 and v_2 to 1.) Find $\mathbf{f} \smile \mathbf{f}$ as a Yoneda composition. Find elements of $\mathrm{Hom}_{A^e}(A^e, V)$ corresponding to $\mathbf{f}$ and to $\mathbf{f} \smile \mathbf{f}$. (See Section A.3 for the correspondence between generalized extensions and cochains.)

Exercise 2.4.5. Verify the claim that the total complex of the tensor product of the m-extension $\mathbf{f}$ and the n-extension $\mathbf{g}$ in this section indeed results in an $(m+n)$-extension, by applying the Künneth Theorem (Theorem A.5.2).

2.5. Actions of Hochschild cohomology

For any two (left) A-modules M and N, the Hochschild cohomology ring $\mathrm{HH}^*(A)$ acts on the graded vector space $\mathrm{Ext}_A^*(M, N)$ in such a way that $\mathrm{Ext}_A^*(M, N)$ is a graded $\mathrm{HH}^*(A)$-module. (There is also an action for any two right modules.) Similarly, for any A-bimodule B, the Hochschild cohomology ring $\mathrm{HH}^*(A)$ acts on $\mathrm{HH}^*(A, B)$. We describe these actions next.

Let M, N be A-modules. Choose an A^e-projective resolution of A as an A^e-module,

$$(2.5.1) \qquad \cdots \xrightarrow{d_2} P_1 \xrightarrow{d_1} P_0 \xrightarrow{\varepsilon} A \to 0,$$

for example, we could take the bar resolution (1.1.4). Apply $- \otimes_A M$ and identify $A \otimes_A M$ with M under the canonical isomorphism $A \otimes_A M \cong M$ given by the module action:

$$(2.5.2) \qquad \cdots \xrightarrow{d_2 \otimes 1_M} P_1 \otimes_A M \xrightarrow{d_1 \otimes 1_M} P_0 \otimes_A M \xrightarrow{\varepsilon \otimes 1_M} M \to 0.$$

Each term P_i in the sequence (2.5.1) is projective as an A^e-module, thus is a direct summand of a free A^e-module. So each term $P_i \otimes_A M$ is projective as a left A-module, where the action is on the left tensor factor only. (It suffices to prove that $(A \otimes A^{\mathrm{op}}) \otimes_A M$ is projective as a left A-module, which is immediate from the isomorphism $(A \otimes A^{\mathrm{op}}) \otimes_A M \cong A \otimes M$ given by $(a \otimes b) \otimes m \mapsto a \otimes bm$ for $a, b \in A$, $m \in M$.) Taking a tensor product is right exact, so the map $\varepsilon \otimes 1_M$ is surjective. Each term in the sequence (2.5.1) is a projective right A-module, so all P_i and $\mathrm{Im}(d_i)$ are projective as right A-modules, and consequently the sequence (2.5.2) is exact. (This follows for example from Theorem A.4.6, the first long exact sequence for Tor, applied to the short exact sequences $0 \to \mathrm{Im}(d_{i+1}) \to P_i \to \mathrm{Im}(d_i) \to 0$ from which (2.5.1) may be built.) Therefore the sequence (2.5.2) is a projective resolution of M as a (left) A-module.

Let $f \in \mathrm{Hom}_{A^e}(P_i, A)$ represent an element of $\mathrm{HH}^i(A)$. Let

$$(2.5.3) \qquad \phi_M(f) = f \otimes 1_M$$

in $\mathrm{Hom}_A(P_i \otimes_A M, M)$, representing an element of $\mathrm{Ext}^i_A(M, M)$. By the Comparison Theorem (Theorem A.2.7), we may lift $\phi_M(f)$ to a chain map, that is, to maps $\phi_M(f)_j : P_{i+j} \otimes_A M \to P_j \otimes_A M$ for all $j \geq 0$:

$$
\begin{array}{ccccccc}
P_{i+j} \otimes_A M & \longrightarrow \cdots \longrightarrow & P_{i+1} \otimes_A M & \longrightarrow & P_i \otimes_A M & & \\
\Big\downarrow {\phi_M(f)_j} & & \Big\downarrow {\phi_M(f)_1} & & \Big\downarrow {\phi_M(f)_0} & \searrow^{\phi_M(f)} & \\
P_j \otimes_A M & \longrightarrow \cdots \longrightarrow & P_1 \otimes_A M & \longrightarrow & P_0 \otimes_A M & \longrightarrow & M
\end{array}
$$

Compose with any function $g \in \mathrm{Hom}_A(P_j \otimes_A M, N)$ to obtain the element $g\phi_M(f)_j$ of $\mathrm{Hom}_A(P_{i+j} \otimes_A M, N)$. This induces a well-defined map

$$(2.5.4) \qquad \mathrm{Ext}^j_A(M, N) \otimes \mathrm{HH}^i(A) \to \mathrm{Ext}^{i+j}_A(M, N)$$

that defines a right module action of $\mathrm{HH}^*(A)$ on $\mathrm{Ext}^*_A(M, N)$. Under the correspondence with generalized extensions, this is equivalent to applying $- \otimes_A M$ to the generalized extension corresponding to f, and taking the Yoneda composition with the generalized extension corresponding to g.

Similarly, there is a left action of $\mathrm{HH}^*(A)$ on $\mathrm{Ext}^*_A(M, N)$: first apply $-\otimes_A N$ to the resolution $P_\bullet$ to obtain a map ϕ_N from $\mathrm{HH}^*(A)$ to $\mathrm{Ext}^*_A(N, N)$. Then compose chain maps corresponding to elements of $\mathrm{Ext}^*_A(N, N)$ and $\mathrm{Ext}^*_A(M, N)$. Again, there is an equivalent generalized extension version of this left action. These right and left actions are the same up to a sign, as the next theorem shows. The theorem is a special case of [**201**, Theorem 1.1].

Theorem 2.5.5. *Let* $\alpha \in \mathrm{HH}^i(A)$ *and* $\beta \in \mathrm{Ext}^j_A(M, N)$. *Then*

$$\alpha \cdot \beta = (-1)^{ij} \beta \cdot \alpha.$$

That is, the left and right actions of $\mathrm{HH}^*(A)$ *on* $\mathrm{Ext}^*_A(M, N)$ *agree up to a sign.*

Proof. The element α may be represented by an i-extension of A by A:

$$(2.5.6) \qquad 0 \to A \xrightarrow{\iota} E \to P_{i-2} \to \cdots \to P_1 \to P_0 \to A \to 0$$

for an A^e-module E and projective A^e-modules P_i. (The correspondence between i-extensions and cochains described in Section A.3 indicates that we may assume $P_1, \ldots, P_{i-2}$ are projective.) We may assume that E and all P_l are projective as right A-modules since this is an extension of A by A, which is a projective right A-module.

The proof is by induction on $j = |\beta|$. Suppose $\beta \in \mathrm{Ext}^0_A(M, N) \cong \mathrm{Hom}_A(M, N)$. If $\alpha \in \mathrm{HH}^0(A) \cong Z(A)$, the action is by multiplication, and so the equation claimed in the theorem statement holds. If $\alpha \in \mathrm{HH}^1(A)$, then the 1-extension (2.5.6) is simply $0 \to A \xrightarrow{\iota} E \xrightarrow{\varepsilon} A \to 0$. We let

$$X = (N \oplus (E \otimes_A M))/\{(a\beta(m),\ -\iota(a) \otimes m) \mid a \in A,\ m \in M\}.$$

Consider the following diagram:

$$
\begin{array}{ccccccccc}
0 & \longrightarrow & A \otimes_A M & \xrightarrow{\iota \otimes 1} & E \otimes_A M & \xrightarrow{\varepsilon \otimes 1} & A \otimes_A M & \longrightarrow & 0 \\
 & & \downarrow{\scriptstyle 1 \otimes \beta} & & \downarrow{\binom{0}{1}} & & \downarrow{=} & & \\
0 & \longrightarrow & A \otimes_A N & \xrightarrow{\binom{1}{0}} & X & \xrightarrow{(0,\varepsilon \otimes 1)} & A \otimes_A M & \longrightarrow & 0 \\
 & & \downarrow{=} & & \downarrow{(\iota \otimes 1, 1 \otimes \beta)} & & \downarrow{1 \otimes \beta} & & \\
0 & \longrightarrow & A \otimes_A N & \xrightarrow{\iota \otimes 1} & E \otimes_A N & \xrightarrow{\varepsilon \otimes 1} & A \otimes_A N & \longrightarrow & 0
\end{array}
$$

Note that X is the pushout (as defined in Section A.1) of the upper left corner, and that the diagram commutes. Therefore $\alpha \cdot \beta = \beta \cdot \alpha$. A similar argument applies for any $\alpha \in \mathrm{HH}^i(A)$. Thus the formula holds when $j = 0$.

Next suppose $\beta \in \mathrm{Ext}^1_A(M, N)$ and that $0 \to N \to X \to M \to 0$ is a corresponding 1-extension. For each l, consider the short exact sequence

$$(2.5.7) \qquad 0 \to K_l \to P_{l-1} \to K_{l-1} \to 0,$$

where K_l, and K_{l-1} are the lth and $(l-1)$st syzygies in the sequence (2.5.6). (Take $K_0 = A$.) We may assume that each K_l is projective as a right A-module, since (2.5.6) consists entirely of projective right A-modules. Tensor the sequence (2.5.7) with $0 \to N \to X \to M \to 0$ over A to obtain the following commuting diagram with exact rows and columns:

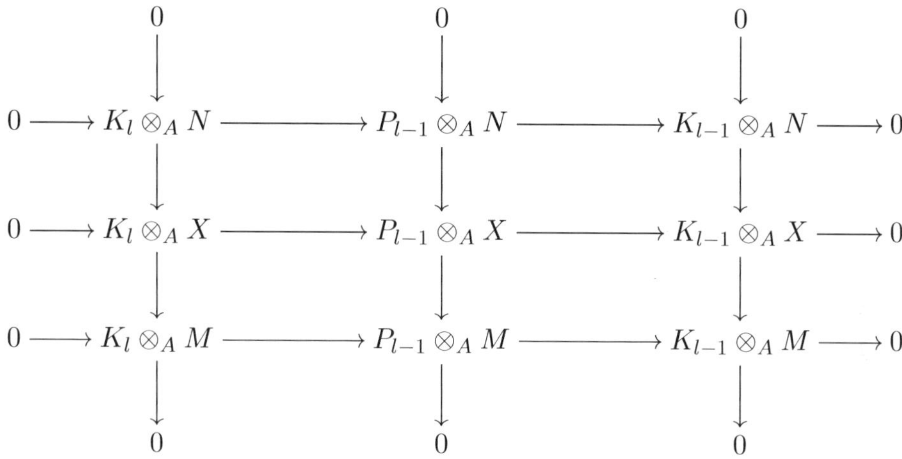

Then by [**151**, Lemma VIII.3.1], the 2-extensions obtained from the edges by composing maps across the upper right and lower left corners, namely

$$0 \to K_l \otimes_A N \longrightarrow P_{l-1} \otimes_A N \longrightarrow K_{l-1} \otimes_A X \longrightarrow K_{l-1} \otimes_A M \to 0,$$

$$0 \to K_l \otimes_A N \longrightarrow K_l \otimes_A X \longrightarrow P_{l-1} \otimes_A M \longrightarrow K_{l-1} \otimes_A M \to 0,$$

are equivalent up to multiplication by -1. (Alternatively, this can be seen directly by projecting the total complex of the tensor product complex onto each of these two sequences. In this approach, the factor -1 comes from the middle vertical maps in the tensor product complex.) Considering α to correspond to the Yoneda splice of all the short exact sequences (2.5.7), we find by induction on i that $\alpha \cdot \beta = (-1)^i \beta \cdot \alpha$. Thus the formula holds when $j = 1$.

Finally, if $\beta \in \mathrm{Ext}^j_A(M, N)$ for $j > 1$, view β as a Yoneda splice of j short exact sequences. Induction on j yields $\alpha \cdot \beta = (-1)^{ij} \beta \cdot \alpha$. $\square$

Note that the graded vector space $\mathrm{Ext}^*_A(M, M)$ is itself an associative algebra with product given by Yoneda product (that is, composition of chain maps analogous to that in Section 2.2) or equivalently Yoneda composition (that is, splice of generalized extensions analogous to that in Section 2.4). It is typically not graded commutative. However, its graded center, defined as follows, is important in relation to Hochschild cohomology.

Definition 2.5.8. Let B be a graded algebra. The *graded center* of B is the subalgebra $Z_{\mathrm{gr}}(B)$ generated by all homogeneous elements $\alpha \in B$ such that $\alpha \cdot \beta = (-1)^{|\alpha||\beta|}\beta \cdot \alpha$ for all homogeneous elements $\beta \in B$.

We may now state a corollary of Theorem 2.5.5, which follows immediately by taking $N = M$ and applying the definition of the action of $\mathrm{HH}^*(A)$ on $\mathrm{Ext}_A^*(M, M)$.

Corollary 2.5.9. *Let M be an A-module. The map*

$$\phi_M : \mathrm{HH}^*(A) \to \mathrm{Ext}_A^*(M, M),$$

defined at the chain level by $\phi_M(f) = f \otimes 1_M$ as in (2.5.3), is a ring homomorphism whose image is contained in the graded center $Z_{\mathrm{gr}}(\mathrm{Ext}_A^*(M, M))$.

For some algebras and modules, there are general results describing precisely the image of $\mathrm{HH}^*(A)$ in $\mathrm{Ext}_A^*(M, M)$. For Koszul algebras and a canonical choice of A-module M, $\mathrm{HH}^*(A)$ surjects onto $Z_{\mathrm{gr}}(\mathrm{Ext}_A^*(M, M))$. See [**38**, Theorem 4.1] for details; some discussion is in Section 3.4. For some more general algebras, the image of $\mathrm{HH}^*(A)$ in $\mathrm{Ext}_A^*(M, M)$, for a canonical choice of M, is the A_∞-center, defined in Section 7.4. Some discussion of this phenomenon is in Section 7.4, and details are in [**34**].

We give an example to illustrate Corollary 2.5.9 next.

Example 2.5.10. Let $A = k[x]/(x^n)$ as in Examples 1.1.21, 2.2.2, and 2.2.3. Let $M = k$, an A-module via the augmentation map $\varepsilon : A \to k$ given by $\varepsilon(x) = 0$. The following is a free resolution of k as an A-module:

$$\cdots \longrightarrow A \xrightarrow{x^{n-1}\cdot} A \xrightarrow{x\cdot} A \xrightarrow{x^{n-1}\cdot} A \xrightarrow{x\cdot} A \xrightarrow{\varepsilon} k \longrightarrow 0.$$

Applying $\mathrm{Hom}_A(-, k)$ after truncating by dropping the term k, and identifying each space $\mathrm{Hom}_A(A, k)$ with k, we have:

$$\cdots \xleftarrow{0} k \xleftarrow{0} k \xleftarrow{0} k \xleftarrow{0} k \xleftarrow{0} k \longleftarrow 0.$$

The maps are indeed all zero maps since $\varepsilon(x) = 0$ and $\varepsilon(x^{n-1}) = 0$. Therefore $\mathrm{Ext}_A^i(k, k) \cong k$ for all i. Under a Yoneda/cup product defined in an analogous way to that for Hochschild cohomology in Section 2.2, similar calculations to those in Examples 2.2.2 and 2.2.3 show that

$$\mathrm{Ext}_A^*(k, k) \cong \begin{cases} k[y], & \text{if } n = 2, \\ k[y, z]/(y^2), & \text{otherwise,} \end{cases}$$

where $|y| = 1$ and $|z| = 2$, independent of the characteristic of k. We will determine the image of $\mathrm{HH}^*(A)$ in $\mathrm{Ext}_A^*(k, k)$ under the map ϕ_k defined by (2.5.3).

First assume that n is not divisible by $p = \mathrm{char}(k)$ and note that by the definition of g in Example 2.2.2, $\phi_k(g) = 0$. If f is the function defined in Example 2.2.2 and $n = 2$, then $\phi_k(f) = y^2$, and if $n > 2$, then $\phi_k(f) = z$. Thus the image of $\mathrm{HH}^*(A)$ under ϕ_k is the subalgebra of even degree elements. Next assume that p divides n as in Example 2.2.3. If $n = 2$, then $\phi_k(g) = y$ and $\phi_k(f) = y^2$. If $n > 2$, then $\phi_k(g) = y$ and $\phi_k(f) = z$. Thus ϕ_k maps $\mathrm{HH}^*(A)$ surjectively onto $\mathrm{Ext}^*_A(k, k)$ in this case. Note that in case $n = 2$, independently of the characteristic p of k, the image of $\mathrm{HH}^*(A)$ under ϕ_k is the full graded center of $\mathrm{Ext}^*_A(k, k)$ as in Definition 2.5.8 (in case $p \neq 2$, this is the subspace of even degree elements due to signs involved in multiplying by odd degree elements). In case $n > 2$ and n is not divisible by $\mathrm{char}(k)$, the image of $\mathrm{HH}^*(A)$ under ϕ_k is not the full graded center. We will return to this example in Section 7.4. (See Theorem 7.4.4 and Example 7.4.5.)

If B is an A-bimodule, then $\mathrm{HH}^*(A)$ acts on $\mathrm{HH}^*(A, B)$ similarly to how it acts on $\mathrm{Ext}^*_A(M, N)$ as described earlier in this section: take a Yoneda product, or first tensor with B and then take a Yoneda product. Again these two actions agree up to a sign by a proof similar to that of Theorem 2.5.5. See [**201**, Theorem 1.1] for a more general setting that includes as special cases both this statement and Theorem 2.5.5 (by taking one of the two rings to be the field in [**201**]).

Exercise 2.5.11. Verify that formula (2.5.3) indeed leads to a well-defined map (2.5.4).

Exercise 2.5.12. Verify that (2.5.4) indeed defines a right module action of $\mathrm{HH}^*(A)$ on $\mathrm{Ext}^*_A(M, N)$.

Exercise 2.5.13. Let $A = k[x]$. Find $\mathrm{Ext}^*_A(k, k)$ and the image of $\mathrm{HH}^*(A)$ under the map ϕ_k defined by (2.5.3).

Exercise 2.5.14. Let M be an A-module. Verify the claims made about a product on $\mathrm{Ext}^*_A(M, M)$:

  (a) Define a Yoneda product on cochains, analogous to the definition in Section 2.2, and show that it induces a well-defined product on $\mathrm{Ext}^*_A(M, M)$.

  (b) Define a Yoneda composition on generalized extensions, analogous to the definition in Section 2.4, and show that it induces a well-defined product on $\mathrm{Ext}^*_A(M, M)$.

  (c) Show that the products defined in (a) and (b) agree.

Chapter 3

Examples

In this chapter we look at resolutions, designed for particular types of algebras, that aid both theoretical understanding and computation of Hochschild homology and cohomology spaces and rings. These algebras and resolutions in turn will provide a rich assortment of examples on which to draw in later chapters.

We begin by working with tensor products and twisted tensor products of algebras. These are constructions of larger algebras out of component parts, and we describe their resulting homological behavior. Also featured are skew group algebras, another class of algebras built from component parts, important in noncommutative algebra and geometry. We introduce Koszul complexes and use them to prove the classical Hochschild-Kostant-Rosenberg Theorem that gives the Hochschild homology and cohomology rings of smooth commutative algebras. We define Koszul algebras and monomial algebras and standard bimodule resolutions tailored to them.

The algebras in this chapter are some important classes both of commutative and of noncommutative algebras. For some, we completely describe their Hochschild cohomology rings here. For others, we simply construct resolutions specific to the algebras that can be used to find homological information, and provide references where more details may be found.

We take k to be a field.

3.1. Tensor product of algebras

Let A and B be algebras over k. The *tensor product algebra* of A and B is $A \otimes B$ as a vector space, with multiplication given by

$$(a \otimes b) \cdot (a' \otimes b') = aa' \otimes bb'$$

for all $a, a' \in A$ and $b, b' \in B$, extended linearly to all elements in $A \otimes B$. If A and B are graded algebras (graded by $\mathbb{N}$ or by $\mathbb{Z}$), their *graded tensor product algebra* is $A \otimes B$ as a vector space, with multiplication given by

$$(a \otimes b) \cdot (a' \otimes b') = (-1)^{|a'||b|} aa' \otimes bb'$$

for all homogeneous $a, a' \in A$ and $b, b' \in B$, where $|a'|$ denotes the degree of a' in $\mathbb{Z}$, and similarly for $|b|$. (We have used the same notation for homological degree, but this should cause no confusion, as it will be clear in each context which degree is meant. In Theorem 3.1.2 below, they are the same, considering Hochschild cohomology to be graded by homological degree.)

Under some finiteness conditions, the Hochschild cohomology ring of the tensor product of two algebras is the graded tensor product of their Hochschild cohomology rings, as we will see in Theorem 3.1.2 below. This theorem will allow us to understand the Hochschild cohomology rings of algebras that are tensor products of others. First we look at properties of $A \otimes B$-bimodules and construct a needed resolution.

For any A-bimodule M and B-bimodule N, the vector space $M \otimes N$ is an $A \otimes B$-bimodule (equivalently, $(A \otimes B)^e$-module):

$$(a \otimes b) \cdot (m \otimes n) \cdot (a' \otimes b') = ama' \otimes bnb'$$

for all $a, a' \in A$, $b, b' \in B$, $m \in M$, $n \in N$. We claim that if the A^e-module M and the B^e-module N are both projective, then $M \otimes N$ is a projective $(A \otimes B)^e$-module. To see this, note that since M is a projective A^e-module, there is an A^e-module M' such that $M \oplus M' \cong (A^e)^{\oplus I}$ for some indexing set I. Similarly, there is a B^e-module N' such that $N \oplus N' \cong (B^e)^{\oplus J}$ for some indexing set J. Since tensor product distributes over direct sum, as a vector space, $M \otimes N$ is a direct summand of

$$(A^e)^{\oplus I} \otimes (B^e)^{\oplus J} \cong (A^e \otimes B^e)^{\oplus(I \times J)} \cong ((A \otimes B)^e)^{\oplus(I \times J)}.$$

The module actions of A and B commute, and so $M \otimes N$ is a direct summand of $((A \otimes B)^e)^{\oplus(I \times J)}$ as an $(A \otimes B)^e$-module. Thus $M \otimes N$ is indeed a projective $(A \otimes B)^e$-module.

Now let $P_\bullet$ be a projective A^e-resolution of A, and $Q_\bullet$ a projective B^e-resolution of B. Taking their tensor product over k, consider the bicomplex

$$(3.1.1) \qquad\qquad\qquad P_\bullet \otimes Q_\bullet$$

and its corresponding total complex $\mathrm{Tot}(P_\bullet \otimes Q_\bullet)$ which we sometimes denote more simply by $P_\bullet \otimes Q_\bullet$ when no confusion will arise. Each vector space $P_i \otimes Q_j$ is a projective $(A \otimes B)^e$-module as explained above. By the Künneth Theorem (Theorem A.5.2), since the tensor product is over the field k, the total complex of $P_\bullet \otimes Q_\bullet$ has homology concentrated in degree 0, where it is $\mathrm{H}_0(P_\bullet \otimes Q_\bullet) \cong \mathrm{H}_0(P_\bullet) \otimes \mathrm{H}_0(Q_\bullet) \cong A \otimes B$. Finally we note that the differentials on the tensor product complex $P_\bullet \otimes Q_\bullet$ are in fact $(A \otimes B)^e$-module homomorphisms by their definitions. Thus $P_\bullet \otimes Q_\bullet$ is a projective resolution of $A \otimes B$ as an $(A \otimes B)^e$-module.

In the following theorem, the algebra structures on the Hochschild cohomology spaces are those given by cup product (see Sections 1.3 and 2.2–2.4).

Theorem 3.1.2. *Let A and B be k-algebras for which there exists a free A^e-resolution $P_\bullet$ of A and a free B^e-resolution $Q_\bullet$ of B such that for each i, P_i is a finitely generated A^e-module and Q_i is a finitely generated B^e-module. Then*

$$\mathrm{HH}^*(A \otimes B) \cong \mathrm{HH}^*(A) \otimes \mathrm{HH}^*(B)$$

as algebras, where the right side is a graded tensor product algebra.

If A and B are finite-dimensional algebras (that is, finite-dimensional as vector spaces), then their bar resolutions consist of finitely generated bimodules and so the hypothesis in the theorem holds. There are many other types of algebras for which the hypothesis holds as well. The isomorphism in the theorem is in fact an isomorphism of Gerstenhaber algebras—see [**140**] for the definition of Gerstenhaber bracket on a graded tensor product of Gerstenhaber algebras, and a proof of this more general statement.

Proof of Theorem 3.1.2. Let $P_\bullet \otimes Q_\bullet$ be the bicomplex (3.1.1) discussed above. The Hochschild cohomology space $\mathrm{HH}^*(A \otimes B)$ is the cohomology of the total complex of the bicomplex $\mathrm{Hom}_{(A \otimes B)^e}(P_\bullet \otimes Q_\bullet, A \otimes B)$. Since $P_\bullet$ and $Q_\bullet$ are free resolutions consisting of finitely generated modules, $P_i \otimes Q_j \cong (A^e)^{\oplus I} \otimes (B^e)^{\oplus J} \cong ((A \otimes B)^e)^{\oplus(I \times J)}$ for some finite indexing sets I, J, and there are finite-dimensional vector spaces P_i', Q_j' such that $P_i \cong A \otimes P_i' \otimes A$, $Q_j \cong B \otimes Q_j' \otimes B$. It follows that

$$\mathrm{Hom}_{(A \otimes B)^e}(P_i \otimes Q_j, A \otimes B) \cong \mathrm{Hom}_k(P_i' \otimes Q_j', A \otimes B).$$

Consider embedding $\mathrm{Hom}_k(P_i', A) \otimes \mathrm{Hom}_k(Q_j', B) \hookrightarrow \mathrm{Hom}_k(P_i' \otimes Q_j', A \otimes B)$ via tensor product of functions. Since all P_i' and Q_j' are finite-dimensional vector spaces, this is an isomorphism, and moreover the differentials correspond under this isomorphism. By the Künneth Theorem (Theorem A.5.2), since the tensor product is taken over the field k, the cohomology is $\mathrm{HH}^*(A) \otimes \mathrm{HH}^*(B)$ as a vector space.

Next we determine the algebra structure, using definition (2.3.2) of product. Define a diagonal map $\Delta : P_\bullet \otimes Q_\bullet \to (P_\bullet \otimes Q_\bullet) \otimes_{A \otimes B} (P_\bullet \otimes Q_\bullet)$ by

$$P_\bullet \otimes Q_\bullet \xrightarrow{\Delta_A \otimes \Delta_B} (P_\bullet \otimes_A P_\bullet) \otimes (Q_\bullet \otimes_B Q_\bullet) \xrightarrow{\sim} (P_\bullet \otimes Q_\bullet) \otimes_{A \otimes B} (P_\bullet \otimes Q_\bullet),$$

where Δ_A, Δ_B are diagonal maps for $P_\bullet$, $Q_\bullet$, respectively. The second map is the isomorphism given by interchanging factors, with a sign, that is,

$$(x \otimes x') \otimes (y \otimes y') \mapsto (-1)^{|x'||y|}(x \otimes y) \otimes (x' \otimes y')$$

for all homogeneous $x, x' \in P_{\bullet}$ and $y, y' \in Q_{\bullet}$. (Standard arguments show that the second map is well-defined and is indeed an isomorphism.) Then Δ is a chain map by the definition of the differential of a tensor product of complexes. Now let f, g be homogeneous elements of degrees m, n in $\operatorname{Hom}_{A^e}(P_{\bullet}, A)$, and let f', g' be homogeneous elements of degrees m', n' in $\operatorname{Hom}_{B^e}(Q_{\bullet}, B)$. Consider the element of $\operatorname{HH}^*(A) \otimes \operatorname{HH}^*(B)$ represented by $f \otimes f'$, which is an element in

$$
\begin{aligned}
\operatorname{Hom}_{A^e}(P_m, A) \otimes \operatorname{Hom}_{B^e}(Q_{m'}, B) \ &\cong\ \operatorname{Hom}_k(P'_m, A) \otimes \operatorname{Hom}_k(Q'_{m'}, B) \\
&\cong\ \operatorname{Hom}_k(P'_m \otimes Q'_{m'}, A \otimes B) \\
&\cong\ \operatorname{Hom}_{(A \otimes B)^e}(P_m \otimes Q_{m'}, A \otimes B).
\end{aligned}
$$

Map $\operatorname{HH}^*(A) \otimes \operatorname{HH}^*(B)$ to $\operatorname{HH}^*(A \otimes B)$ by sending $f \otimes f'$ to the function $f \otimes f'$ in $\operatorname{Hom}_{(A \otimes B)^e}(P_m \otimes Q_{m'}, A \otimes B)$, and similarly map $g \otimes g'$ to $g \otimes g'$ in $\operatorname{Hom}_{(A \otimes B)^e}(P_n \otimes Q_{n'}, A \otimes B)$. We determine their cup product by using the diagonal map Δ and formula (2.3.2): let $x \in P_r$, $y \in Q_s$ for some r, s with $r + s = m + n + m' + n'$, and suppose $\Delta_A(x)$ has component in $P_m \otimes P_n$ given by $\sum_s x'_s \otimes x''_s$ and $\Delta_B(y)$ has component in $Q_{m'} \otimes Q_{n'}$ given by $\sum_t y'_t \otimes y''_t$. (This is the only component on which the cup product potentially takes a nonzero value.) Then

$$
\begin{aligned}
((f &\otimes f') \smile (g \otimes g'))(x \otimes y) \\
&= (f \otimes f' \otimes g \otimes g')\Delta(x \otimes y) \\
&= (-1)^{m(m'+n)+n'(m+m'+n)} \sum_{s,t} f(x'_s) \otimes f'(y'_t) \otimes g(x''_s) \otimes g'(y''_t) \\
&= (-1)^{m(m'+n)+n'(m+m'+n)} \sum_{s} f(x'_s)g(x''_s) \otimes \sum_{t} f'(y'_t)g'(y''_t) \\
&= (-1)^{mm'+(m+n)n'} \big((f \otimes g)\Delta_A(x)\big) \otimes \big((f' \otimes g')\Delta_B(y)\big) \\
&= (-1)^{mm'+(m+n)n'} \big((f \smile g)(x)\big) \otimes \big((f' \smile g')(y)\big) \\
&= (-1)^{m'n} \big((f \smile g) \otimes (f' \smile g')\big)(x \otimes y).
\end{aligned}
$$

Therefore $(f \otimes f') \smile (g \otimes g')$ is equal to

$$
(-1)^{m'n}(f \smile g) \otimes (f' \smile g').
$$

We have thus shown that the images of $f \otimes f'$ and $g \otimes g'$ in $\operatorname{HH}^*(A \otimes B)$ have cup product that is the image of $(-1)^{m'n}(f \smile g) \otimes (f' \smile g')$, as desired. $\square$

The theorem allows us to understand the Hochschild cohomology rings of many more algebras, as the following examples show. For these examples, we need the notions of tensor algebra, symmetric algebra, and exterior algebra generated by a vector space. We recall these notions and fix our notation here.

Let V be a vector space. The *tensor algebra* of V over k is denoted $T(V)$ (or $T_k(V)$), and as a graded vector space is

$$T(V) = k \oplus V \oplus (V \otimes V) \oplus (V \otimes V \otimes V) \oplus \cdots .$$

We write $T^n(V)$ or $T(V)_n$ for the nth tensor power of V, that is, $V \otimes \cdots \otimes V$ (n tensor factors). Multiplication is given by $(v_1 \otimes \cdots \otimes v_r) \cdot (v_{r+1} \otimes \cdots \otimes v_s) = v_1 \otimes \cdots \otimes v_s$ for all $v_1, \ldots, v_s \in V$. Sometimes we write $v_1 \cdots v_s$ in place of $v_1 \otimes \cdots \otimes v_s$ for simplicity of notation. If V is finite dimensional with basis $x_1, \ldots, x_m$, we sometimes write

$$k \langle x_1, \ldots, x_m \rangle = T(V),$$

the free algebra generated by $x_1, \ldots, x_m$. The *symmetric algebra* $S(V)$ (or $S_k(V)$) and *exterior algebra* $\bigwedge(V)$ (or $\bigwedge_k(V)$) are defined by

$$S(V) = T(V)/(v \otimes w - w \otimes v \mid v, w \in V)$$

$$\text{and} \quad \bigwedge(V) = T(V)/(v \otimes w + w \otimes v \mid v, w \in V)$$

provided $\operatorname{char}(k) \neq 2$. In case $\operatorname{char}(k) = 2$, define $\bigwedge(V)$ instead as the quotient of $T(V)$ by the ideal $(v \otimes v \mid v \in V)$. We write $S^n(V)$ (respectively, $\bigwedge^n(V)$) for the image of $T^n(V)$ in $S(V)$ (respectively, in $\bigwedge(V)$). These definitions work equally well in the more general setting where k is taken to be any commutative ring and V any k-module.

Example 3.1.3. Let $A_1 = k[x_1]$, $A_2 = k[x_2]$, and $A = A_1 \otimes A_2$. Then $A \cong k[x_1, x_2]$, a polynomial ring in two indeterminates. In Example 1.3.6, we found that $\mathrm{HH}^*(A_i) \cong k[x_i, y_i]/(y_i^2)$ as a k-algebra, where $|x_i| = 0$, $|y_i| = 1$ for $i = 1, 2$. By Theorem 3.1.2, $\mathrm{HH}^*(k[x_1, x_2])$ is isomorphic to the graded tensor product $\mathrm{HH}^*(A_1) \otimes \mathrm{HH}^*(A_2)$. Since it is a graded tensor product and $|y_i| = 1$, we thus see that

$$\mathrm{HH}^*(k[x_1, x_2]) \cong k[x_1, x_2] \otimes \bigwedge(y_1, y_2)$$

as a graded algebra, where by $\bigwedge(y_1, y_2)$ we mean the exterior algebra on the vector space with basis y_1, y_2. We could instead write $k[x_1, x_2]$ as $S(x_1, x_2)$, the symmetric algebra on the vector space with basis x_1, x_2 to which it is isomorphic.

More generally, by induction on the number m of indeterminates,

$$\mathrm{HH}^*(k[x_1, \ldots, x_m]) \cong k[x_1, \ldots, x_m] \otimes \bigwedge(y_1, \ldots, y_m)$$

as a graded algebra, with $|x_i| = 0$, $|y_i| = 1$. We explain another way to see this structure on Hochschild cohomology. Let V be the vector space with basis $x_1, \ldots, x_m$, and let W be the vector space with basis $y_1, \ldots, y_m$. Identify W with the dual space V^* of V. We may view the Hochschild cohomology of $A = k[x_1, \ldots, x_m]$ as $A \otimes \bigwedge(V^*)$. Equivalently, this view arises naturally from the tensor product complex approach to cup products of Section 2.3, as we will explain next.

The A^e-projective resolution of $A = k[x_1, \ldots, x_m]$ obtained by a tensor product of m copies of resolution (1.1.19), one for each x_i, may be described as follows. As a graded vector space, let

$$(3.1.4) \qquad P_{\scriptscriptstyle\bullet} = A \otimes \textstyle\bigwedge^{\scriptscriptstyle\bullet}(V) \otimes A,$$

that is, $P_n = A \otimes \bigwedge^n(V) \otimes A$ for each n. As a free A^e-module, we may canonically identify each P_n with the degree n component of the tensor product of m copies of resolution (1.1.19), one for each of $x_1, \ldots, x_m$. We identify $1 \otimes x_{i_1} \wedge \cdots \wedge x_{i_n} \otimes 1$ in P_n with $(1 \otimes 1)^{\otimes m}$ in the tensor product complex, where the tensor factor $1 \otimes 1$ is taken to be in degree 1 when it is in any of the positions $i_1, \ldots, i_n$, and is in degree 0 otherwise. Under this identification, the differential on $P_{\scriptscriptstyle\bullet}$ is given by

$$d_n(1 \otimes x_{i_1} \wedge \cdots \wedge x_{i_n} \otimes 1) = \sum_{j=1}^{n} (-1)^{j-1}\big(x_{i_j} \otimes x_{i_1} \wedge \cdots \wedge \hat{x}_{i_j} \wedge \cdots \wedge x_{i_n} \otimes 1$$
$$- 1 \otimes x_{i_y1} \wedge \cdots \wedge \hat{x}_{i_j} \wedge \cdots \wedge x_{i_n} \otimes x_{i_j}\big),$$

where the notation $\hat{x}_{i_j}$ indicates a deleted exterior factor. This extends linearly to the formula:

$$d_n(1 \otimes v_1 \wedge \cdots \wedge v_n \otimes 1) = \sum_{j=1}^{n} (-1)^{j-1}\big(v_j \otimes v_1 \wedge \cdots \wedge \hat{v}_j \wedge \cdots \wedge v_n \otimes 1$$
$$- 1 \otimes v_1 \wedge \cdots \wedge \hat{v}_j \wedge \cdots \wedge v_n \otimes v_j\big)$$

for all $v_1, \ldots, v_n \in V$. Applying $\operatorname{Hom}_{A^e}(-, A)$, we see that the induced differentials are all 0 since they involve differences of left and right multiplication by v_j and A is commutative. Therefore the Hochschild cohomology space in each degree n is

$$\operatorname{Hom}_{A^e}(A \otimes \textstyle\bigwedge^n(V) \otimes A, A) \;\cong\; \operatorname{Hom}_k(\textstyle\bigwedge^n(V), A)$$
$$\cong\; A \otimes (\textstyle\bigwedge^n(V))^* \;\cong\; A \otimes \textstyle\bigwedge^n(V^*),$$

since V is finite dimensional, which agrees with our earlier observation. Now, there is a diagonal map $\Delta : P_{\scriptscriptstyle\bullet} \to P_{\scriptscriptstyle\bullet} \otimes_A P_{\scriptscriptstyle\bullet}$ arising from an embedding of $P_{\scriptscriptstyle\bullet}$ into the bar resolution $B(A)$ that will be given by (3.4.12). This diagonal map on $P_{\scriptscriptstyle\bullet}$ will be the composition of this embedding with the diagonal map on $B(A)$ given by formula (2.3.3). This and definition (2.3.2) of cup product yield another proof that the product on Hochschild cohomology is indeed that of the tensor product $A \otimes \bigwedge(V^*)$ of the two algebras A and $\bigwedge(V^*)$.

Our next examples are truncated polynomial rings.

Example 3.1.5. Let $n_1, n_2 \geq 2$. Let $A_1 = k[x_1]/(x_1^{n_1})$, $A_2 = k[x_2]/(x_2^{n_2})$, and $A = A_1 \otimes A_2$. Then $A \cong k[x_1, x_2]/(x_1^{n_1}, x_2^{n_2})$. For each i, we found in

Example 2.2.2 that if n_i is not divisible by char(k), then

$$\mathrm{HH}^*(A_i) \cong k[x_i, y_i, z_i]/(x_i^{n_i},\ x_i^{n_i-1}y_i,\ y_i^2,\ x_i^{n_i-1}z_i),$$

where $|x_i| = 0$, $|y_i| = 1$, $|z_i| = 2$. If neither n_1 nor n_2 is divisible by char(k), then by Theorem 3.1.2,

$$\mathrm{HH}^*(k[x_1, x_2]/(x_1^{n_1}, x_2^{n_2}))$$
$$\cong k[x_1, x_2, z_1, z_2] \otimes \bigwedge(y_1, y_2)/(x_i^{n_i} \otimes 1,\ x_i^{n_i-1} \otimes y_i,\ x_i^{n_i-1}z_i \otimes 1).$$

The cases where one or both of n_1, n_2 is divisible by char(k) may be treated similarly by applying results of Example 2.2.3. More generally, we see by induction on the number m of generators that if none of $n_1, \ldots, n_m$ is divisible by char(k), then

$$\mathrm{HH}^*(k[x_1, \ldots, x_m]/(x_1^{n_1}, \ldots, x_m^{n_m}))$$
$$\cong k[x_1, \ldots, x_m, z_1, \ldots, z_m] \otimes \bigwedge(y_1, \ldots, y_m)/J,$$

where $J = (x_i^{n_i} \otimes 1,\ x_i^{n_i-1} \otimes y_i,\ x_i^{n_i-1}z_i \otimes 1)$ and $|x_i| = 0$, $|y_i| = 1$, $|z_i| = 2$. If one or more of $n_1, \ldots, n_m$ is divisible by char(k), then we obtain the Hochschild cohomology ring by applying Theorem 3.1.2 repeatedly, using the results of Example 2.2.3, for a similar expression.

It is useful to describe the resolution of $A = k[x_1, \ldots, x_m]/(x_1^{n_1}, \ldots, x_m^{n_m})$ obtained as a tensor product of complexes for the factors $A_i = k[x_i]/(x_i^{n_i})$: let V be the vector space with basis $\xi_1, \ldots, \xi_m$, and let $S(V)$ be the symmetric algebra on the vector space with basis $\xi_1, \ldots, \xi_m$ so $S(V) \cong k[\xi_1, \ldots, \xi_m]$. Let

$$P_{\bullet} = A \otimes S(V) \otimes A,$$

where $|\xi_i| = 1$ for each i. So for each j, P_j is the free A^e-module on standard monomials (that is, monomials with scalar coefficient 1) in $S(V)$ of degree j, since these form a basis for the degree j subspace of $S(V)$ as a vector space over k. Then $P_{\bullet}$ may be identified with the total complex of the tensor product of m copies of resolution (1.1.22), one for each x_i: the monomial $1 \otimes \xi_1^{i_1} \cdots \xi_m^{i_m} \otimes 1$ is identified with the tensor product of m copies of $1 \otimes 1$, one in each $A_i \otimes A_i$, where the exponents $i_1, \ldots, i_m$ indicate the homological degree of the corresponding factor. We see then that the differential on $P_{\bullet}$ corresponding to the differential on the tensor product of complexes is given by $d = \sum_{j=1}^m d_j$, where

(3.1.6)
$$d_j(1 \otimes \xi_1^{i_1} \cdots \xi_m^{i_m} \otimes 1)$$
$$= (-1)^l \begin{cases} 0, & \text{if } i_j = 0, \\ u_j \cdot (1 \otimes \xi_1^{i_1} \cdots \xi_j^{i_j-1} \cdots \xi_m^{i_m} \otimes 1), & \text{if } i_j \text{ is odd}, \\ v_j \cdot (1 \otimes \xi_1^{i_1} \cdots \xi_j^{i_j-1} \cdots \xi_m^{i_m} \otimes 1), & \text{if } i_j \text{ is even and } i_j > 0, \end{cases}$$

with $u_j = x_j \otimes 1 - 1 \otimes x_j$, $v_j = x_j^{n_j-1} \otimes 1 + x_j^{n_j-2} \otimes x_j + \cdots + 1 \otimes x_j^{n_j-1}$, and $l = i_1 + \cdots + i_{j-1}$. Applying $\mathrm{Hom}_{A^e}(-, A)$, we find that since A is commutative, the induced differentials in odd degrees are all 0 while in even degrees they are multiplication by $n_j x_j^{n_j-1}$ (which is 0 if and only if n_j is divisible by $\mathrm{char}(k)$). This is consistent with the graded vector space structure of $\mathrm{HH}^*(A)$ described above.

Exercise 3.1.7. Let V be a vector space of dimension m, and let $n \leq m$. What is the vector space dimension of $\bigwedge^n(V)$?

Exercise 3.1.8. Verify that the formula for the differential in Example 3.1.3 is indeed that of the tensor product of m copies of resolution (1.1.19), and that after applying $\mathrm{Hom}_{A^e}(-, A)$, the induced differentials are indeed all 0.

Exercise 3.1.9. Verify that the formula for the differential in Example 3.1.5 is indeed that of the tensor product of m copies of resolution (1.1.22), and that after applying $\mathrm{Hom}_{A^e}(-, A)$, the induced differentials are as claimed.

Exercise 3.1.10. Let M be an A-bimodule and N a B-bimodule. Consider $M \otimes N$ to be an $A \otimes B$-bimodule with action defined by $(a \otimes b)(m \otimes n)(a' \otimes b') = ama' \otimes bnb'$ for all $a, a' \in A$, $b, b' \in B$, $m \in M$, $n \in N$.

(a) Use the techniques of the proof of Theorem 3.1.2 to prove that under suitable finiteness hypotheses,
$$\mathrm{HH}^*(A \otimes B, M \otimes N) \cong \mathrm{HH}^*(A, M) \otimes \mathrm{HH}^*(B, N)$$
as graded vector spaces.

(b) Prove a homology version of Theorem 3.1.2, using the same proof techniques:
$$\mathrm{HH}_*(A \otimes B, M \otimes N) \cong \mathrm{HH}_*(A, M) \otimes \mathrm{HH}_*(B, N)$$
as graded vector spaces. Does this isomorphism require any finiteness hypotheses?

Exercise 3.1.11. Use Exercise 3.1.10(b) to obtain an expression for the Hochschild homology space $\mathrm{HH}_*(k[x_1, \ldots, x_m])$ similar to that for Hochschild cohomology found in Example 3.1.3.

3.2. Twisted tensor product of algebras

Twisted tensor products generalize the tensor products and graded tensor products discussed in Section 3.1. Here we present the work of Bergh and Oppermann [**30**] on a twisted tensor product of algebras where the multiplication is twisted by a bicharacter on grading groups.

Let A_1, A_2 be k-algebras that are graded by abelian groups Γ_1, Γ_2, respectively. That is, $A_1 = \bigoplus_{\gamma \in \Gamma_1}(A_1)_\gamma$, a direct sum of vector spaces, such that $(A_1)_\gamma (A_1)_{\gamma'} \subseteq (A_1)_{\gamma+\gamma'}$ for all $\gamma, \gamma' \in \Gamma_1$, and similarly for A_2, Γ_2. Let $t : \Gamma_1 \times \Gamma_2 \to k^\times$ be a *bicharacter*, that is,

$$t(0, \gamma_2) = t(\gamma_1, 0) = 1,$$
$$t(\gamma_1 + \gamma_1', \gamma_2) = t(\gamma_1, \gamma_2) t(\gamma_1', \gamma_2),$$
$$t(\gamma_1, \gamma_2 + \gamma_2') = t(\gamma_1, \gamma_2) t(\gamma_1, \gamma_2')$$

for all $\gamma_i, \gamma_i' \in \Gamma_i$. The *twisted tensor product algebra*

$$A = A_1 \otimes^t A_2$$

is $A_1 \otimes A_2$ as a vector space, and multiplication is determined by

$$(a_1 \otimes a_2) \cdot (a_1' \otimes a_2') = t(|a_1'|, |a_2|) a_1 a_1' \otimes a_2 a_2'$$

for all homogeneous elements $a_1, a_1' \in A_1$ and $a_2, a_2' \in A_2$, where $|a_1'|$ denotes the degree of a_1' in Γ_1 and similarly $|a_2|$ in Γ_2. By its definition, $A_1 \otimes^t A_2$ is graded by the group $\Gamma_1 \times \Gamma_2$: $(A_1 \otimes^t A_2)_{(\gamma_1, \gamma_2)} = (A_1)_{\gamma_1} \otimes (A_2)_{\gamma_2}$ for all $\gamma_1 \in \Gamma_1, \gamma_2 \in \Gamma_2$.

Our first example is the quantum plane.

Example 3.2.1. Let $A_1 = k[x_1]$ and $A_2 = k[x_2]$, each graded by $\mathbb{Z}$ in the standard way: let $|x_1| = |x_2| = 1$. Let q be any nonzero scalar. Define $t : \mathbb{Z} \times \mathbb{Z} \to k^\times$ by $t(m, n) = q^{-mn}$ for all $m, n \in \mathbb{Z}$. Recall from Section 3.1 that $k\langle x_1, x_2 \rangle$ denotes the free k-algebra on x_1, x_2, so that $k\langle x_1, x_2 \rangle$ is identified with the tensor algebra $T(V)$, where V is the vector space with basis x_1, x_2. The twisted tensor product $A_1 \otimes^t A_2$ of A_1 and A_2 is given by

$$A = A_1 \otimes^t A_2 \cong k\langle x_1, x_2 \rangle / (x_1 x_2 - q x_2 x_1),$$

called a *skew polynomial ring* or *quantum plane*, and is denoted $k_q[x_1, x_2]$. The latter terminology recalls the affine plane, which may be identified with the set of maximal ideals of the commutative polynomial ring $k[x_1, x_2]$ (the case $q = 1$) if k is algebraically closed. More generally we iterate this construction to obtain a *skew polynomial ring* or *quantum affine space*

$$k_{\mathbf{q}}[x_1, \ldots, x_m] = k\langle x_1, \ldots, x_m \rangle / (x_i x_j - q_{ij} x_j x_i \mid 1 \le i, j \le m)$$

determined by a set $\mathbf{q} = \{q_{ij}\}$ of nonzero scalars for which $q_{ii} = 1$ and $q_{ji} = q_{ij}^{-1}$ for all i, j $(1 \le i, j \le m)$. Sometimes $k_{\mathbf{q}}[x_1, \ldots, x_m]$ is also called a *quantum symmetric algebra*, however, this term is also used both more generally and slightly differently in other contexts, so we will not use it here.

Our next example is a noncommutative version of a truncated polynomial ring.

Example 3.2.2. Let $n_1, n_2 \geq 2$, $A_1 = k[x_1]/(x_1^{n_1})$, and $A_2 = k[x_2]/(x_2^{n_2})$. Each of A_1, A_2 is $\mathbb{Z}$-graded just as in Example 3.2.1. Define a bicharacter t as in that example. The twisted tensor product algebra $A_1 \otimes^t A_2$ is isomorphic to $k\langle x_1, x_2 \rangle/(x_1 x_2 - q x_2 x_1, \; x_1^{n_1}, \; x_2^{n_2})$. More generally we iterate this construction to obtain a *truncated skew polynomial ring,*

$$k\langle x_1, \ldots, x_m \rangle/(x_i x_j - q_{ij} x_j x_i, \; x_i^{n_i} \mid 1 \leq i, j \leq m)$$

determined by a set $\mathbf{q} = \{q_{ij}\}$ of scalars satisfying the conditions given in Example 3.2.1, and positive integers $n_1, \ldots, n_m$. Such an algebra is also called a *quantum complete intersection*, recalling the special case in which all $q_{ij} = 1$ (a commutative algebra that is a complete intersection).

We return to the general case of a twisted tensor product $A_1 \otimes^t A_2$ of algebras A_1, A_2 determined by abelian grading groups Γ_1, Γ_2 and a bicharacter $t : \Gamma_1 \times \Gamma_2 \to k^\times$. We construct a projective resolution of the $(A_1 \otimes^t A_2)^e$-module $A_1 \otimes^t A_2$ from those of A_1 and A_2. Let $P_{\bullet}$ be a graded A_1^e-projective resolution of A_1, that is, each A_1^e-module P_i is graded by Γ_1 in such a way that $(A_1)_\gamma (P_i)_{\gamma'} (A_1)_{\gamma''} \subseteq (P_i)_{\gamma + \gamma' + \gamma''}$ for all $\gamma, \gamma', \gamma'' \in \Gamma_1$, the differentials preserve the grading, and P_i embeds as a direct summand of a free module via a graded map (that is, a map that preserves Γ_1-degree). For example, the bar resolution (1.1.4) is graded (by grading a tensor product in the usual way, $(A \otimes A)_\gamma = \bigoplus_{\gamma' + \gamma'' = \gamma} (A_{\gamma'} \otimes A_{\gamma''})$). Similarly, let $Q_{\bullet}$ be a graded projective resolution of the A_2^e-module A_2.

Consider the tensor product complex $P_{\bullet} \otimes Q_{\bullet}$ as a complex of vector spaces. By the Künneth Theorem (Theorem A.5.2), since the tensor product is over the field k, the total complex has homology 0 in all positive degrees, and in degree 0 its homology is $A_1 \otimes A_2$. We will put the structure of an $(A_1 \otimes^t A_2)^e$-module on each $P_i \otimes Q_j$ in such a way that it is projective and the differentials are $(A_1 \otimes^t A_2)^e$-module homomorphisms. It will follow that the total complex of $P_{\bullet} \otimes Q_{\bullet}$ is an $(A_1 \otimes^t A_2)^e$-projective resolution of $A_1 \otimes^t A_2$. For all homogeneous $x \in P_i$, $y \in Q_j$, $a_1, a_1' \in A_1$, and $a_2, a_2' \in A_2$, set

$$(a_1 \otimes a_2)(x \otimes y)(a_1' \otimes a_2') = t(|x|, |a_2|) t(|a_1'|, |a_2|) t(|a_1'|, |y|) a_1 x a_1' \otimes a_2 y a_2'.$$

A calculation shows that this gives $P_i \otimes Q_j$ the structure of an $(A_1 \otimes^t A_2)^e$-module. Moreover, it can be shown to be projective by the same argument used on page 46. Finally, a calculation shows that the differentials on the tensor product complex $P_{\bullet} \otimes Q_{\bullet}$ are $(A_1 \otimes^t A_2)^e$-module homomorphisms since the differentials on $P_{\bullet}$ and $Q_{\bullet}$ preserve gradings.

We have proven the following theorem of Bergh and Oppermann [**30**].

Theorem 3.2.3. *Let A_1 and A_2 be k-algebras graded by abelian groups Γ_1 and Γ_2, respectively. Let $P_{\bullet}$ (respectively, $Q_{\bullet}$) be a graded projective resolution of A_1 as an A_1^e-module (respectively, of A_2 as an A_2^e-module).*

Then $\mathrm{Tot}(P_\bullet \otimes Q_\bullet)$ *is a projective resolution of the twisted tensor product algebra* $A_1 \otimes^t A_2$ *as an* $(A_1 \otimes^t A_2)^e$-*module.*

Remark 3.2.4. Suppose that A_1, A_2 are *augmented algebras*, that is, for each i, there is an algebra homomorphism $\varepsilon_i : A_i \to k$ (called an *augmentation map*). If these augmentation maps ε_1, ε_2 are graded maps, then the twisted tensor product algebra $A_1 \otimes^t A_2$ is augmented by $\varepsilon = \varepsilon_1 \otimes \varepsilon_2$. Consider k itself to be a module for each of A_1, A_2, $A_1 \otimes^t A_2$ via the augmentation maps $\varepsilon_1, \varepsilon_2, \varepsilon$, respectively. A construction similar to that above leads to a projective resolution of k as an $A_1 \otimes^t A_2$-module from projective resolutions of k as an A_1-module and as an A_2-module. See [**30**] for details.

The resolution given by the total complex of $P_\bullet \otimes Q_\bullet$ constructed above may be used to understand Hochschild cohomology of the twisted tensor product algebra $A_1 \otimes^t A_2$. There is however not a result as straightforward as Theorem 3.1.2 that describes the Hochschild cohomology ring of $A_1 \otimes^t A_2$ in terms of that of A_1 and A_2. (For one thing, Hochschild cohomology is graded commutative, while the twisted tensor product generally is not, and so even in degree 0, the Hochschild cohomology space of the twisted tensor product algebra may not be the twisted tensor product of the degree 0 Hochschild cohomology spaces of the two algebras.) The grading on $P_\bullet$, $Q_\bullet$ by Γ_1, Γ_2 does impart some structure to the Hochschild cohomology ring of $A_1 \otimes^t A_2$, and one consequence is that there is a result analogous to Theorem 3.1.2 for the part of Hochschild cohomology graded by a subgroup of $\Gamma_1 \times \Gamma_2$ contained in the kernel of the bicharacter t. This is proven by retracing the steps of the proof of Theorem 3.1.2, noting that the sequence of isomorphisms of Hom spaces given there works for subspaces graded by such a subgroup of $\Gamma_1 \times \Gamma_2$. See [**30**, Theorem 4.7].

We give details of the resolution of the twisted tensor product $A_1 \otimes^t A_2$ constructed above for our Examples 3.2.1 and 3.2.2.

Example 3.2.5. Let $A = k_q[x_1, x_2] \cong A_1 \otimes^t A_2$ as in Example 3.2.1. Let $P_\bullet$ be the resolution (1.1.19) of Example 1.1.18 for A_1. In order for $P_\bullet$ to be a *graded* resolution, the differentials must preserve grading, and accordingly we shift the $\mathbb{Z}$-grading of A_1^e in the homological degree 1 component so that $1 \otimes 1$ there has graded degree 1. Then the differential is indeed a graded map. Let $Q_\bullet$ be a similar resolution for A_2. By Theorem 3.2.3, $\mathrm{Tot}(P_\bullet \otimes Q_\bullet)$ is an A^e-projective resolution of A. Apply $\mathrm{Hom}_{A^e}(-, A)$ to obtain

$$0 \longleftarrow \mathrm{Hom}_{A^e}(P_1 \otimes Q_1, A) \longleftarrow \mathrm{Hom}_{A^e}\big((P_0 \otimes Q_1) \oplus (P_1 \otimes Q_0), A\big)$$

$$\longleftarrow \mathrm{Hom}_{A^e}(A^e, A) \longleftarrow 0,$$

which is equivalent, under standard isomorphisms, to

$$0 \longleftarrow A \xleftarrow{\ d_2^*\ } A \oplus A \xleftarrow{\ d_1^*\ } A \longleftarrow 0.$$

Under our identifications and degree shifts, a calculation shows that the differentials d_1^*, d_2^* are given by

$$d_1^*(a \otimes b) = ((q^{-|a|} - 1)a \otimes bx_2,\ (1 - q^{-|b|})ax_1 \otimes b),$$
$$d_2^*(a \otimes b,\ a' \otimes b') = (1 - q^{-|b|-1})ax_1 \otimes b + (1 - q^{-|a'|-1})a' \otimes b'x_2$$

for all homogeneous $a, a' \in A_1$ and $b, b' \in A_2$. (Note that in case $q = 1$, these differentials are indeed 0, in accordance with Example 3.1.3.) We may iterate this construction to obtain a free resolution of A as an A^e-module for $A = k_{\mathbf{q}}[x_1, \ldots, x_m]$.

Example 3.2.6. Another important family of algebras that may be constructed as twisted tensor products are the quantum complete intersections of Example 3.2.2. For simplicity, we focus here on the case $m = 2$ and $n_1 = n_2 = 2$: $A = k\langle x_1, x_2 \rangle / (x_1 x_2 - q x_2 x_1,\ x_1^2,\ x_2^2)$. Similar techniques yield information about the more general case. Buchweitz, Green, Madsen, and Solberg [38] used precisely these examples to answer a question of Happel [105], showing that some of these are finite-dimensional algebras that have finite-dimensional Hochschild cohomology yet infinite global dimension. We give some details next.

Let $P.$ (respectively, $Q.$) be resolution (1.1.22) for $A_1 = k[x_1]/(x_1^{n_1})$ (respectively, $A_2 = k[x_2]/(x_2^{n_2})$). The total complex of $P. \otimes Q.$ has the structure of an A^e-projective resolution of A, where $A = A_1 \otimes^t A_2$. This is equivalent to the projective resolution constructed in [38].

Now consider k to be an A-module on which each of x_1, x_2 acts as 0, that is, A is augmented with augmentation map $\varepsilon : A \to k$ given by the algebra homomorphism sending each of x_1, x_2 to 0. Applying $- \otimes_A k$ to $\mathrm{Tot}(P. \otimes Q.)$, we obtain a projective resolution of k as an A-module. This resolution may be used to show that $\mathrm{Ext}_A^n(k, k) \neq 0$ for all n, independent of the characteristic of k and of the value of q. It follows that A has infinite global dimension, since existence of a projective resolution of finite length would imply $\mathrm{Ext}_A^n(k, k) = 0$ for all n larger than the length of the resolution.

If q *is not a root of unity* and $\mathrm{char}(k) \neq 2$, we claim the Hochschild cohomology ring $\mathrm{HH}^*(A)$ is 5-dimensional as a vector space: in this case, computations using our resolution $P. \otimes Q.$ show that $\mathrm{HH}^0(A)$ is a vector space spanned by 1 and $x_1 x_2$ (the center of A), $\mathrm{HH}^1(A)$ is a vector space spanned by elements y_1 and y_2 that arise from functions at the chain level taking $(1 \otimes 1) \otimes (1 \otimes 1)$ in degree $(1, 0)$ to $x_1 \otimes 1$ and $(1 \otimes 1) \otimes (1 \otimes 1)$ in degree $(0, 1)$ to $1 \otimes x_2$, respectively, $\mathrm{HH}^2(A)$ is a vector space spanned by $y_1 \smile y_2$, and $\mathrm{HH}^i(A) = 0$ for all $i \geq 3$. Therefore $\mathrm{HH}^*(A)$ is a 5-dimensional vector space as claimed. In this way we see that A has infinite global dimension and yet finite-dimensional Hochschild cohomology. Generalizations to larger classes of examples of algebras with these properties are given by Bergh and Erdmann [**28**] and by Parker and Snashall [**170**].

We next determine the algebra structure of the Hochschild cohomology ring in this case that q is not a root of unity and $\mathrm{char}(k) \neq 2$. The cup product of the degree 0 element $x_1 x_2$ with y_i is 0 for $i = 1, 2$. Considering the vector space dimension in each degree, it now follows that there is an isomorphism of algebras

$$\mathrm{HH}^*(A) \cong k[x_1 x_2]/((x_1 x_2)^2) \times_k \bigwedge(y_1, y_2),$$

where the latter is a fiber product. (The *fiber product* $R_1 \times_k R_2$ of two augmented k-algebras R_1, R_2 is the subring of $R_1 \oplus R_2$ consisting of pairs (r_1, r_2) such that the images of r_1 and r_2 under the respective augmentation maps are equal, i.e., the fiber product is the pullback of the two augmentation maps.) See [**38**] for details, as well as the cases where $\mathrm{char}(k) = 2$ or where $q = 0$.

If q *is a root of unity*, the Hochschild cohomology ring has a completely different structure. In this case, it is infinite dimensional, yet there are gaps: it is 0 in infinitely many degrees. See [**38**] for details.

Related examples are the more general quantum complete intersections of Example 3.2.2, appearing in several papers, and more general algebras for which the relation $x_1 x_2 - q x_2 x_1$ is replaced by $(x_1 x_2)^l - q(x_2 x_1)^l$ for some positive integer l. See, for example, [**30, 63, 64, 167**].

For comparison, we briefly present a more general definition of twisted tensor product algebra given by Čap, Schichl, and Vanžura [**43**]: again let A_1, A_2 be k-algebras. Let $\tau : A_2 \otimes A_1 \to A_1 \otimes A_2$ be a bijective k-linear map for which $\tau(1_{A_2} \otimes a_1) = a_1 \otimes 1_{A_2}$ and $\tau(a_2 \otimes 1_{A_1}) = 1_{A_1} \otimes a_2$ for all $a_1 \in A_1$, $a_2 \in A_2$, and $\tau(\pi_{A_2} \otimes \pi_{A_1}) = (\pi_{A_1} \otimes \pi_{A_2})(1 \otimes \tau \otimes 1)(\tau \otimes \tau)(1 \otimes \tau \otimes 1)$ as maps from $A_2 \otimes A_2 \otimes A_1 \otimes A_1$ to $A_1 \otimes A_2$. Call τ a *twisting map*. The *twisted tensor product algebra* $A_1 \otimes_\tau A_2$ is defined to be $A_1 \otimes A_2$ as a vector space with multiplication given by $(\pi_{A_1} \otimes \pi_{A_2})(1 \otimes \tau \otimes 1)$ as a map from $(A_1 \otimes A_2) \otimes (A_1 \otimes A_2)$ to $A_1 \otimes A_2$. This is a more general notion than our earlier twisted tensor product of algebras arising from a bicharacter. In that case we may set $\tau(a_2 \otimes a_1) = t(|a_1|, |a_2|)a_1 \otimes a_2$ for all homogeneous $a_1 \in A_1$, $a_2 \in A_2$, and then $A_1 \otimes^t A_2 \cong A_1 \otimes_\tau A_2$. In fact, any algebra that is isomorphic as a vector space to $A_1 \otimes A_2$ for some subalgebras A_1 and A_2 is isomorphic to a twisted tensor product algebra under this more general definition. For more details, see [**43**], and for resolutions for these more general twisted tensor product algebras, see [**103, 148, 196**].

Exercise 3.2.7. Let $A = A_1 \otimes^t A_2$ be the twisted tensor product given in Example 3.2.1. For each positive integer n, let

$$[n]_q = 1 + q + q^2 + \cdots + q^{n-1}, \quad [n]_q! = [1]_q[2]_q \cdots [n]_q, \quad \begin{bmatrix} n \\ r \end{bmatrix}_q = \frac{[n]_q!}{[r]_q![n-r]_q!}$$

for all r $(0 \leq r \leq n)$, where $[0]_q! = 1$ and the last expression is defined only when $[r]_q![n-r]_q! \neq 0$.

   (a) Show that when the coefficients in the expression below are defined, for all $n \geq 1$,

$$(x_1 + x_2)^n = \sum_{r=0}^{n} \begin{bmatrix} n \\ r \end{bmatrix}_{q^{-1}} x_1^r x_2^{n-r}.$$

   (b) Suppose that q is a primitive nth root of unity. Show that in A,

$$(x_1 + x_2)^n = x_1^n + x_2^n.$$

Exercise 3.2.8. Verify that the formulas for d_1^* and d_2^* given in Example 3.2.5 are indeed those arising from the differential on the tensor product of the resolutions $P_\bullet$ and $Q_\bullet$.

Exercise 3.2.9. Suppose q is not a root of unity, and let $A = k_q[x_1, x_2]$, defined as in Example 3.2.5. Use the resolution given there to find the structure of $\mathrm{HH}^*(A)$ as a graded vector space.

Exercise 3.2.10. Let $A = k_{\mathbf{q}}[x_1, \ldots, x_m]$. Use the methods of Example 3.2.5 to find a formula for the differential of a resolution of A as an A^e-module, constructed as described there.

Exercise 3.2.11. Verify the claimed structure of $\mathrm{HH}^*(A)$ in Example 3.2.6 in case q is not a root of unity and $\mathrm{char}(k) \neq 2$.

Exercise 3.2.12. Let $A = k\langle x_1, x_2 \rangle / (x_1 x_2 - q x_2 x_1, \; x_1^{n_1}, \; x_2^{n_2})$ for some nonzero scalar q and positive integers $n_1, n_2 \geq 2$. Use the resolution in Example 3.2.6 to find (a) $\mathrm{Ext}_A^*(k, k)$ and (b) the Hochschild homology $\mathrm{HH}_*(A)$.

3.3. Koszul complexes and the HKR Theorem

In this section we define Koszul complexes associated to regular sequences of central elements in an algebra. We use them to prove the Hochschild-Kostant-Rosenberg (HKR) Theorem [**116**] that describes Hochschild homology and cohomology rings of smooth commutative algebras (Theorem 3.3.6).

Definition 3.3.1. Let x be a central element of an algebra A. The *Koszul complex* associated to x is

$$K(x): \qquad\qquad 0 \longrightarrow A \overset{x\cdot}{\longrightarrow} A \longrightarrow 0,$$

concentrated in degrees 1 and 0. More generally, let $\mathbf{x} = (x_1, \ldots, x_n)$ be a sequence of central elements $x_1, \ldots, x_n$ in A. The *Koszul complex* associated to $\mathbf{x}$ is the total complex

$$(3.3.2) \qquad K(\mathbf{x}): \qquad \mathrm{Tot}(K(x_1) \otimes_A \cdots \otimes_A K(x_n)).$$

Note that if x is not a zero divisor of A, then by the above definition of the Koszul complex $K(x)$ associated to the element x, $\mathrm{H}_1(K(x)) = 0$ and $\mathrm{H}_0(K(x)) \cong A/(x)$. Consequently $K(x)$ is a (free) resolution of the A-module $A/(x)$ with action of A given by projection onto $A/(x)$ followed by multiplication. This observation generalizes to a statement about the Koszul complex $K(\mathbf{x})$ associated to the sequence $\mathbf{x}$ provided it is a regular sequence, as we define next. Given an A-module M, we say that a nonzero element $x \in A$ is a *zero divisor* of M if $xm = 0$ for some $m \neq 0$.

Definition 3.3.3. A sequence $(x_1, \ldots, x_n)$ of central elements in an algebra A is a *regular sequence* if x_1 is not a zero divisor of A and for each i, x_i is not a zero divisor of the A-module $A/(x_1, \ldots, x_{i-1})$, where this time the notation $(x_1, \ldots, x_{i-1})$ refers to the ideal generated by these elements.

Theorem 3.3.4. *If* $\mathbf{x} = (x_1, \ldots, x_n)$ *is a regular sequence in* A, *then the Koszul complex* $K(\mathbf{x})$ *is a free resolution of the A-module* $A/(x_1, \ldots, x_n)$.

Proof. We will use induction on n. Suppose $n = 1$. Since x_1 is not a zero divisor of A, augmenting the sequence $K(x_1)$ by the term $A/(x_1)$ in degree -1 yields an exact sequence. Clearly the terms in nonnegative degrees are free as (left) A modules. Thus the statement holds when $n = 1$.

Since $\mathbf{x}$ is a regular sequence by hypothesis, $(x_1, \ldots, x_{n-1})$ is a regular sequence. Assume that $K(x_1, \ldots, x_{n-1})$ is a free resolution of the A-module $A/(x_1, \ldots, x_{n-1})$. Consider A to be a complex concentrated in degree 0, and accordingly, $A[-1]$ to be the complex consisting of A in degree 1. We form a short exact sequence of complexes $0 \to A \to K(x_n) \to A[-1] \to 0$:

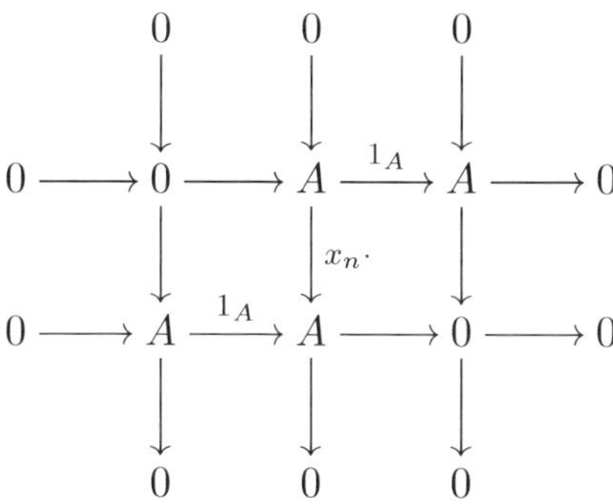

Take the tensor product, over A, with $K(x_1, \ldots, x_{n-1})$. There is a canonical isomorphism $K(x_1, \ldots, x_{n-1}) \otimes_A A \cong K(x_1, \ldots, x_{n-1})$, and so this results in a short exact sequence of complexes

$$0 \longrightarrow K(x_1, \ldots, x_{n-1}) \longrightarrow K(\mathbf{x}) \longrightarrow K(x_1, \ldots, x_{n-1})[-1] \longrightarrow 0.$$

By the induction hypothesis, the corresponding homology long exact sequence (see Theorem A.4.2) ends in

$$0 \longrightarrow \mathrm{H}_1(K(\mathbf{x})) \longrightarrow \mathrm{H}_1(K(x_1, \ldots, x_{n-1})[-1])$$

$$\xrightarrow{\partial_1} \mathrm{H}_0(K(x_1, \ldots, x_{n-1})) \longrightarrow \mathrm{H}_0(K(\mathbf{x})) \longrightarrow 0.$$

The two middle terms are simply $A/(x_1, \ldots, x_{n-1})$, again as a consequence of the induction hypothesis. A diagram chase shows that the connecting homomorphism ∂_1 is multiplication by x_n. Since x_n is not a zero divisor of $A/(x_1, \ldots, x_{n-1})$ by hypothesis, the sequence

$$0 \to A/(x_1, \ldots, x_{n-1}) \xrightarrow{x_n \cdot} A/(x_1, \ldots, x_{n-1}) \to A/(x_1, \ldots, x_n) \to 0$$

is exact. Comparing the two sequences, we conclude that $\mathrm{H}_1(K(\mathbf{x})) = 0$ and $\mathrm{H}_0(K(\mathbf{x})) \cong A/(x_1, \ldots, x_n)$. A look at the rest of the long exact sequence shows that by the induction hypothesis, $\mathrm{H}_i(K(\mathbf{x})) = 0$ for all $i \geq 1$. Finally, we note that the terms in $K(\mathbf{x})$ are all direct sums of (left) A-modules of the form $A \otimes_A A \otimes_A \cdots \otimes_A A \cong A$, and so are free. $\qquad\square$

See [**223**, Section 4.5] for a more general version of the above theorem.

Example 3.3.5. Let $A = k[x_1, \ldots, x_n]$. Then

$$(x_1 \otimes 1 - 1 \otimes x_1, \ldots, x_n \otimes 1 - 1 \otimes x_n)$$

is a regular sequence in A^e. The associated Koszul complex

$$K(x_1 \otimes 1 - 1 \otimes x_1, \ldots, x_n \otimes 1 - 1 \otimes x_n)$$

can be identified with resolution (3.1.4) by its definition, as a consequence of an isomorphism $A^e/(x_1 \otimes 1 - 1 \otimes x_1, \ldots, x_n \otimes 1 - 1 \otimes x_n) \cong A$. Similarly, $(x_1, \ldots, x_n)$ is a regular sequence of A and $K(x_1, \ldots, x_n)$ is a resolution of $A/(x_1, \ldots, x_n) \cong k$ as an A-module.

A classical theorem of Hochschild, Kostant, and Rosenberg [**116**] gives the structure of Hochschild (co)homology of a smooth finitely generated commutative algebra. The theorem has more general versions than what we present here; see, for example, [**223**, Theorem 9.4.7]. Here we can take as a definition of *smooth algebra* A one for which there is a finite projective resolution by finitely generated projective A^e-modules (see the general Definition 4.1.4 of smoothness for not necessarily commutative algebras). For the commutative rings of interest here, this is equivalent to many other definitions of smoothness; see, e.g., [**146**, Proposition 3.4.2] and [**216**]. See also [**156**]. In Chapter 4, we will explore smooth commutative and noncommutative algebras in some detail.

Recall from Section 1.2 that for a commutative algebra A, the Hochschild cohomology in degree 1 may be identified with the space of k-derivations: $\mathrm{HH}^1(A) \cong \mathrm{Der}(A)$. We will use here the standard notation $\Omega^1_{com}(A)$ for the Hochschild homology space in degree 1:

$$\Omega^1_{com}(A) = \mathrm{HH}_1(A).$$

In Section 4.2 we will use this notation $\Omega^1_{com}(A)$ for the A-module of Kähler differentials defined there and show that it is indeed isomorphic to $\mathrm{HH}_1(A)$. Thus one consequence of the following theorem is that for a smooth finitely generated commutative algebra, Hochschild homology $\mathrm{HH}_*(A)$ and cohomology $\mathrm{HH}^*(A)$ are generated as algebras by their degree 1 components $\mathrm{HH}_1(A)$ and $\mathrm{HH}^1(A)$, respectively. The algebra structure on Hochschild homology is that given by the shuffle product (Section 1.5), and that on Hochschild cohomology by cup product (Section 1.3 and Chapter 2).

Theorem 3.3.6 (Hochschild-Kostant-Rosenberg Theorem). *Let k be a field, and let A be a smooth finitely generated commutative algebra over k. There are isomorphisms of graded algebras,*

$$\mathrm{HH}^*(A) \cong \textstyle\bigwedge_A(\mathrm{Der}(A)) \quad and \quad \mathrm{HH}_*(A) \cong \textstyle\bigwedge_A(\Omega^1_{com}(A)).$$

Proof. We will prove the cohomology isomorphism. The homology isomorphism is similar; see, e.g., [**223**, Theorem 9.4.7]. There is a map

$$\psi : \textstyle\bigwedge_A(\mathrm{Der}(A)) \to \mathrm{HH}^*(A)$$

given by identifying $\bigwedge^0_A(\mathrm{Der}(A))$ with $A \cong \mathrm{HH}^0(A)$ and $\bigwedge^1_A(\mathrm{Der}(A))$ with $\mathrm{Der}(A) \cong \mathrm{HH}^1(A)$, and then extending to an algebra homomorphism. Note that ψ is well-defined due to the graded commutativity of $\mathrm{HH}^*(A)$. Further, ψ is an isomorphism if and only if $\psi \otimes_A A_{\mathfrak{m}}$ is an isomorphism for every maximal ideal $\mathfrak{m}$ of A. (See Matsumura [**156**] for needed properties of localization such as [**156**, Theorem 4.6].) We will prove this latter statement.

Since Ext commutes with localization [**223**, Proposition 3.3.10], there is an isomorphism $\mathrm{HH}^*(A_{\mathfrak{m}}) \cong \mathrm{HH}^*(A) \otimes_A A_{\mathfrak{m}}$ for any maximal ideal $\mathfrak{m}$ of A. We will show that $\mathrm{HH}^*(A_{\mathfrak{m}}) \cong \bigwedge_{A_{\mathfrak{m}}}(\mathrm{Der}(A_{\mathfrak{m}}))$ for every maximal ideal $\mathfrak{m}$ of A. The claimed isomorphism will follow. This will be the case once we have shown that ψ is an isomorphism whenever A is a local ring.

Now assume A is local with unique maximal ideal $\mathfrak{m}$. Let M be the inverse image of $\mathfrak{m}$ in A^e under the multiplication map π, so that M is a maximal ideal of A^e. Since A is smooth, A^e is a regular ring [**223**, Proposition 9.4.6]. (See [**156**] or [**223**] for properties of regular rings.) Since A is local, $A \cong A_{\mathfrak{m}}$. Now $A \cong A_{\mathfrak{m}} \cong (A^e)_M/(\mathrm{Ker}\,\pi)_M$ [**156**, Theorem 4.2] and since $(A^e)_M$ is a regular local ring, $(\mathrm{Ker}\,\pi)_M$ is generated by a regular sequence $\mathbf{x}$ [**223**, Exercise 4.4.2]. By Theorem 3.3.4, $K(\mathbf{x})$ is a free resolution of the $(A^e)_M$-module $A \cong A_{\mathfrak{m}}$. Restrict to A^e and apply $\mathrm{Hom}_{A^e}(-, A)$. The differentials are 0 since $\mathbf{x}$ is a regular sequence of elements in $(\operatorname{Ker}\pi)_M$, and upon taking homomorphisms to A, we multiply. Thus the cohomology is as claimed. The multiplicative structure is as stated by applying a diagonal map and formula (2.3.2) for cup product, analogous to the argument at the end of Example 3.1.3. $\qquad\qquad\square$

Remark 3.3.7. It follows from Theorem 3.3.6 that for a smooth finitely generated commutative algebra A, $\mathrm{HH}^n(A) = \mathrm{HH}^n(A)e_n(n)$, a component of the Hodge decomposition (1.6.2) of $\mathrm{HH}^n(A)$; see [**146**, Theorem 4.5.12]. The Harrison cohomology of A in degrees 2 and higher is thus 0 since it is in a different component; see the Hodge decomposition (1.6.2) and comments following it. The vanishing of Harrison cohomology in degree 2 has implications for the algebraic deformation theory of A (see Section 5.2).

There are converse statements to Theorem 3.3.6. Let A be a finitely generated commutative algebra. If the Hochschild homology of A vanishes in one positive even degree and in one positive odd degree, then A is smooth [**14, 15**]. If its Hochschild cohomology vanishes for sufficiently many positive degrees, then A is smooth [**13**]. If its Hochschild homology is finitely generated, then A is smooth [**12**].

Exercise 3.3.8. Verify the claims made in Example 3.3.5, that is, the Koszul complex associated to the regular sequence

$$(x_1 \otimes 1 - 1 \otimes x_1, \ldots, x_n \otimes 1 - 1 \otimes x_n)$$

is equivalent to resolution (3.1.4).

Exercise 3.3.9. Let $A = k[x_1, \ldots, x_m]$. Explain how Theorem 3.3.6 is consistent with our earlier calculation of $\mathrm{HH}^*(A)$ in Example 3.1.3. What is $\mathrm{HH}_*(A)$ (cf. Exercise 3.1.11)?

3.4. Koszul algebras

Some of the algebras we have seen in this chapter turn out to be Koszul algebras, as introduced by Priddy [**177**]. In this section, we will give several equivalent definitions of Koszul algebras, one of which is existence of a bimodule resolution of a particular type. This resolution can be used to compute Hochschild cohomology, and we revisit earlier examples in this light. We assume here that the algebra A is graded (by $\mathbb{N}$) and *connected*, that is, $A_0 = k$. More general Koszul algebras are considered in the literature. (See, e.g., [**154**] for an introduction to Koszul algebras defined by quivers and relations, and further references therein.)

Let V be a finite-dimensional vector space, and let $T(V)$ be the tensor algebra of V as defined in Section 3.1. So $T(V) = \bigoplus_{n \geq 0} T^n(V)$, where

$T^0(V) = k$, $T^1(V) = V$, and $T^n(V) = V \otimes \cdots \otimes V$ (n tensor factors). Then $T(V)$ is a graded algebra with $|v| = 1$ for all $v \in V$. Let R be a subspace of $T^2(V)$, that is,

$$R \subseteq V \otimes V,$$

and let

$$A = T(V)/(R),$$

where (R) denotes the ideal generated by R in $T(V)$. We call R the space of *relations* for A. By definition, A is a *quadratic algebra*, that is, A is a graded algebra generated by elements in degree 1 and with relations in degree 2.

Let $V^* = \mathrm{Hom}_k(V, k)$ be the dual vector space to V. Since V is finite dimensional, we may identify $(V \otimes V)^*$ with $V^* \otimes V^*$. Let

$$R^\perp = \{u \in V^* \otimes V^* \mid u(r) = 0 \ \text{ for all } \ r \in R\}.$$

The *quadratic dual* (or *Koszul dual*) of A is the quadratic algebra

$$A^! = T(V^*)/(R^\perp).$$

Example 3.4.1. Let V be a vector space with basis $x_1, \ldots, x_m$. Let $\mathbf{q} = \{q_{ij}\}$ be a set of nonzero scalars with $q_{ii} = 1$ and $q_{ji} = q_{ij}^{-1}$ for all i, j as in Example 3.2.1. Let

$$R = \mathrm{Span}_k\{x_i \otimes x_j - q_{ij}x_j \otimes x_i \mid 1 \leq i < j \leq m\}.$$

Then $A = T(V)/(R) \cong k_{\mathbf{q}}[x_1, \ldots, x_m]$, the skew polynomial ring of Example 3.2.1. Let $x_1^*, \ldots, x_m^*$ denote the dual basis to the basis $x_1, \ldots, x_m$ of V. Calculations show that the subspace $R^\perp$ of $V^* \otimes V^*$ is

$$R^\perp = \mathrm{Span}_k\{x_i^* \otimes x_j^* + q_{ij}^{-1}x_j^* \otimes x_i^*, \ x_i^* \otimes x_i^* \mid 1 \leq i < j \leq m\}.$$

Setting $y_i = x_i^*$ for each i, the Koszul dual of A is thus

$$A^! \cong k\langle y_1, \ldots, y_m\rangle/(y_iy_j + q_{ij}^{-1}y_jy_i, \ y_i^2 \mid 1 \leq i, j \leq m),$$

sometimes denoted $\bigwedge_{\mathbf{q}^{-1}}(V^*)$ and called a *quantum exterior algebra*. (Here $k\langle y_1, \ldots, y_m\rangle$ denotes the tensor algebra on the vector space V^*.) As a special case, if $q_{ij} = 1$ for all i, j, then $A \cong k[x_1, \ldots, x_m]$ and $A^! \cong \bigwedge(V^*)$.

Example 3.4.2. Let V be a finite-dimensional vector space, and let $R = 0$, the zero subspace of $V \otimes V$. Then $A = T(V)/(R) \cong T(V)$ and $A^! \cong k \oplus V^*$ since $R^\perp = V^* \otimes V^*$.

Next we will define Koszul algebras to be connected quadratic algebras having particular types of resolutions. To this end, identify the field k with the quotient space A/A_+, where

$$A_+ = \bigoplus_{n>0} A_n.$$

Let $\varepsilon : A \to k$ be the corresponding quotient map. Then A is an augmented algebra, as defined in Remark 3.2.4, via this augmentation map ε, and k is an A-module via ε. Consider a graded free resolution $P_\bullet$ of the A-module k, that is, each A-module P_n is free and thus graded via the grading on A, shifted in such a way that the differentials are graded maps. We further require each P_n to be *locally finite-dimensional* (i.e., having finite-dimensional homogeneous components). Note that such a $P_\bullet$ exists since V is finite dimensional, as for example we could take $P_\bullet$ to be $B(A) \otimes_A k$, where $B(A)$ is the bar resolution (1.1.4). For each P_n, choose a free basis $\{p_l^n \mid l \in L_n\}$ for L_n some indexing set in order to identify it with the free module $A^{\oplus L_n} \cong \bigoplus_{l \in L_n} A p_l^n$. The differentials may then be viewed as matrices with entries in A. Since we have chosen each P_n to be finite dimensional in each degree, the differential on each homogeneous subspace is given by a finite matrix. We say that $P_\bullet$ is *minimal* if the matrix entries are all in A_+, and *linear* if the matrix entries are all in A_1. In this context, this definition of a minimal resolution agrees with the more general definition of Section A.2 (since, for example, applying $\mathrm{Hom}_A(-, k)$ to $P_\bullet$ results in a complex with differentials all 0).

Definition 3.4.3. A graded connected quadratic algebra $A = T(V)/(R)$ is a *Koszul algebra* if the A-module k has a linear minimal graded free resolution.

There are many equivalent definitions of Koszul algebras, as we will see in the next theorem below. In fact, for any algebra that is graded, connected, and locally finite dimensional (i.e., having finite-dimensional homogeneous components), existence of a linear minimal graded free resolution implies it is quadratic (see [**177**]). Thus the assumption that the algebra is quadratic need not be part of the definition of Koszul algebra, and we include it only for convenience, as is common to do. Some of the equivalent definitions of Koszul algebras involve specific complexes, which we will construct next. For more general constructions of minimal resolutions in particular, see [**42**] for algebras defined by generators and relations and [**105**] for finite-dimensional algebras.

Consider the following sequence:

$$\cdots \xrightarrow{d_4} K_3(A) \xrightarrow{d_3} A \otimes R \otimes A \xrightarrow{d_2} A \otimes V \otimes A \xrightarrow{d_1}$$

(3.4.4)

$$A \otimes A \xrightarrow{\ \ \pi\ \ } A \longrightarrow 0,$$

where for each $n \geq 2$, $K_n(A) = A \otimes K_n'(A) \otimes A$ with

$$K_n'(A) = \bigcap_{i+j=n-2} (V^{\otimes i} \otimes R \otimes V^{\otimes j}).$$

Write $K_0(A) = A \otimes A$ and $K_1(A) = A \otimes V \otimes A$. The differentials d_n are those of the bar resolution $B(A)$ defined in (1.1.4) under the canonical embedding of $K(A)$ into $B(A)$; a calculation shows that the sequence (3.4.4) is indeed closed under this differential. Then (3.4.4) is a chain complex. It may or may not be a resolution of A, and it turns out that Koszul algebras are precisely those for which it is, as Theorem 3.4.6 below states.

We need some more notation to state the theorem. For each $n \geq 0$, let $\widetilde{K}_n(A) = A \otimes K'_n(A)$, a left A-module under multiplication by the leftmost factor. Note that $\widetilde{K}_n(A) \cong K_n(A) \otimes_A k$ as a (left) A-module. We will be interested in the resulting sequence:

$$(3.4.5) \qquad \cdots \longrightarrow \widetilde{K}_2(A) \longrightarrow \widetilde{K}_1(A) \longrightarrow \widetilde{K}_0(A) \longrightarrow k \longrightarrow 0.$$

Theorem 3.4.6. *Let $A = T(V)/(R)$ be a finitely generated graded connected quadratic algebra. The following are equivalent:*

   (i) *A is a Koszul algebra.*

   (ii) *$\mathrm{Ext}^*_A(k, k) \cong A^!$ as graded algebras.*

   (iii) *$\mathrm{Ext}^*_A(k, k)$ is generated by $\mathrm{Ext}^1_A(k, k)$ as an algebra.*

   (iv) *$\widetilde{K}(A)$ is a resolution of k as an A-module.*

   (v) *$K(A)$ is a resolution of A as an A^e-module.*

When the equivalent conditions of the theorem hold, we call $\widetilde{K}(A)$ a *Koszul resolution*, and $K(A)$ a *Koszul bimodule resolution* (also called a Koszul resolution when it is clear from context that bimodules are intended).

To prove Theorem 3.4.6, we will largely follow notes of Krähmer [137]. More details may be found there and in earlier literature such as [177]. We will first introduce a second grading on $\mathrm{Ext}^*_A(k, k)$ arising from the grading on the algebra A, and prove a lemma about the relationship between the quadratic dual $A^!$ and this grading. We assume for the rest of this section that $A = T(V)/(R)$ is a graded connected quadratic algebra.

To define the second grading on $\mathrm{Ext}^*_A(k, k)$, we consider the reduced bar resolution $\widetilde{B}_\bullet$ of k as an A-module, where $\widetilde{B}_n = A \otimes (A_+)^{\otimes n}$ and the differentials are given by

$$d_n(a_0 \otimes \cdots \otimes a_n) = \sum_{i=0}^{n-1} (-1)^i a_0 \otimes \cdots \otimes a_i a_{i+1} \otimes \cdots \otimes a_n$$

for $n \geq 1$. (Note that $\widetilde{B}_n = B_n(A) \otimes_A k$ and these differentials are induced by those on the reduced bar resolution $\overline{B}_\bullet(A)$ of A as an A^e-module introduced in Definition 1.1.17.) Then $\widetilde{B}_\bullet$ is graded by the grading on A:

$$(\widetilde{B}_n)_m = \bigoplus_{j_0 + \cdots + j_n = m} (A_{j_0} \otimes (A_+)_{j_1} \otimes \cdots \otimes (A_+)_{j_n}).$$

This leads to a grading on $\mathrm{Hom}_A(\widetilde{B}_n, k)$, where

$$\mathrm{Hom}_A(\widetilde{B}_n, k)_{-m} = \{ f \in \mathrm{Hom}_A(\widetilde{B}_n, k) \mid f((\widetilde{B}_n)_i) = 0 \text{ if } i \neq m \}.$$

The bigrading on $\mathrm{Ext}_A^*(k, k)$ is now given as

$$\mathrm{Ext}_A^n(k, k) = \bigoplus_{m \geq 0} \mathrm{Ext}_A^{n, -m}(k, k),$$

where the vector space $\mathrm{Ext}_A^{n, -m}(k, k)$ is the subquotient $\mathrm{Ker}(d_{n+1}^*)/\mathrm{Im}(d_n^*)$ of $\mathrm{Hom}_A(\widetilde{B}_n, k)_{-m}$; we have identified the maps d_{n+1}^*, d_n^* with their restrictions to the indicated subspaces.

There is a cup product on $\mathrm{Ext}_A^*(k, k)$, defined via the reduced bar resolution $\widetilde{B}_\bullet$ just as in (1.3.2) for Hochschild cohomology, or equivalently as a Yoneda product analogous to that on Hochschild cohomology in Section 2.2. Calculations show that on bigraded components, this cup product is a map

$$\mathrm{Ext}_A^{m, -n}(k, k) \times \mathrm{Ext}_A^{r, -s}(k, k) \xrightarrow{\smile} \mathrm{Ext}_A^{m+r, -n-s}(k, k).$$

We will next make some further observations about this bigrading.

In degree 0, $\mathrm{Ext}_A^0(k, k) = \mathrm{Ext}_A^{0,0}(k, k) \cong \mathrm{Hom}_A(k, k) \cong k$ since $\mathrm{Ker}(d_1^*)$ consists of the functions that are 0 on all positive degree elements. Next note that $\mathrm{Ext}_A^{1, -1}(k, k)$ consists of those elements of $\mathrm{Ext}_A^1(k, k)$ whose representative functions in $\mathrm{Hom}_A(A \otimes A_+, k)$ have degree -1 as functions. The only elements of $A \otimes A_+$ not sent to 0 by such a function are the elements of the subspace $k \otimes V \cong V$ of $A \otimes A_+$. Thus $\mathrm{Ext}_A^{1, -1}(k, k)$ has a basis $\{f_i\}$ dual to a basis $\{v_i\}$ of V. It follows that

$$(3.4.7) \qquad\qquad \mathrm{Ext}_A^1(k, k) = \mathrm{Ext}_A^{1, -1}(k, k),$$

since a calculation shows that elements of $\mathrm{Ker}(d_2^*)$ must be 0 on elements of $A \otimes A_+$ of degree at least 2. The next lemma identifies the subalgebra of $\mathrm{Ext}_A^*(k, k)$ generated by these degree -1 elements, and will be a key part of the proof of Theorem 3.4.6.

Lemma 3.4.8. *Let A be a graded connected quadratic algebra. Then*

$$\mathrm{Ext}_A^{n, -n}(k, k) = \mathrm{Ext}_A^{1, -1}(k, k) \smile \cdots \smile \mathrm{Ext}_A^{1, -1}(k, k)$$

(n factors) for all $n \geq 2$, and

$$A^! \cong \bigoplus_{n \geq 0} \mathrm{Ext}_A^{n, -n}(k, k).$$

Proof. First let $n = 2$. Note that by definition, representative cochains of elements of $\mathrm{Ext}_A^{2,-2}(k,k)$ are in $\mathrm{Hom}_A^2(A \otimes A_+ \otimes A_+, k)$ and so are determined by their values on $1 \otimes v \otimes w$ for all $v, w \in V$. Let $\{f_{i,j}\}$ be the dual basis to the basis $\{1 \otimes v_i \otimes v_j\}$ of $k \otimes V \otimes V$. As before, let $\{f_i\}$ be a dual basis to the basis $\{v_i\}$ of V. By definition of cup product, $f_{i,j} = -f_i \smile f_j$, and so $\mathrm{Ext}_A^{2,-2}(k,k) = \mathrm{Ext}_A^{1,-1}(k,k) \smile \mathrm{Ext}_A^{1,-1}(k,k)$. Now by similar arguments and induction,

$$(3.4.9) \qquad \mathrm{Ext}_A^{n,-n}(k,k) = \mathrm{Ext}_A^{1,-1}(k,k) \smile \cdots \smile \mathrm{Ext}_A^{1,-1}(k,k).$$

Next note that by equation (3.4.9), the elements of $\mathrm{Ext}_A^{n,-n}(k,k)$ can be expressed as iterated cup products on elements of the space $\mathrm{Ext}_A^{1,-1}(k,k) \cong V^*$, that is, $\bigoplus_{n \geq 0} \mathrm{Ext}_A^{n,-n}(k,k)$ is isomorphic to a quotient of $T(V^*)$. We determine the ideal given by $\mathrm{Im}(d_n^*)$: first suppose $n = 2$. Then for all $v, w \in V$,

$$d_2^*(f)(1 \otimes v \otimes w) = f(v \otimes w - 1 \otimes vw) = f(1 \otimes vw)$$

since f is an A-module homomorphism and elements of A_+ act as 0 on k. We claim that $\mathrm{Im}(d_2^*) = R^\perp$. First note that $d_2^*(f)(1 \otimes r) = f(0) = 0$ for all $r \in R$ since the multiplication map takes r to 0. Thus $\mathrm{Im}(d_2^*) \subseteq R^\perp$. Next, given $g \in R^\perp$, we see that g may be identified with a function on the vector space $(V \otimes V)/R$. There is a well-defined A-module homomorphism f from $A \otimes A_+$ to k corresponding to g: take $f(1 \otimes vw) = g(1 \otimes v \otimes w)$ for all $v, w \in V$ and $f(1 \otimes a) = 0$ for all $a \in A_i$ if $i \neq 2$. Then $g = d_2^*(f)$. So $R^\perp \subseteq \mathrm{Im}(d_2^*)$. Therefore $\mathrm{Im}(d_2^*) = R^\perp$. By the above calculation, we may also view d_2^* as π^*, where π is multiplication, so $R^\perp = \mathrm{Im}(\pi^*)$. From this we see that $\mathrm{Ext}_A^{2,-2}(k,k) \cong T^2(V^*)/R^\perp$, and we obtain a map

$$A^! = T(V^*)/(R^\perp) \longrightarrow \bigoplus_{n \geq 0} \mathrm{Ext}_A^{n,-n}(k,k).$$

This map is surjective due to equation (3.4.9). To see that it is also injective, using a similar argument to that for $\mathrm{Im}(d_2^*)$ and the formula for d_n, we may show that $\mathrm{Im}(d_n^*)$ is generated by $\mathrm{Im}(\pi^*) = R^\perp$. It follows that

$$(3.4.10) \qquad A^! \cong \bigoplus_{n \geq 0} \mathrm{Ext}_A^{n,-n}(k,k).$$

$\square$

Proof of Theorem 3.4.6. We present an outline and some of the details; more details may be found in [137] and in [177].

If (i) holds, that is, if A is a Koszul algebra, then k has a linear minimal graded free resolution. It is always the case that $\mathrm{Ext}_A^{i,-j}(k,k) = 0$ when $j < i$ (consider the reduced bar resolution). Since the differentials are assumed to have degree 0, necessarily $\operatorname{Ext}_A^{i,-j}(k,k) = 0$ whenever $j > i$. By isomorphism (3.4.10), condition (ii) follows. In fact, this argument can be reversed and shows that A is Koszul if and only if $\operatorname{Ext}_A^*(k,k) = \bigoplus_{n\geq 0} \operatorname{Ext}_A^{n,-n}(k,k)$, and by isomorphism (3.4.10), we have shown the equivalence of conditions (i) and (ii).

By equation (3.4.9) and isomorphism (3.4.10), the quadratic dual $A^!$ of A is generated in degree 1, and so (ii) implies (iii). By (3.4.7), it is always the case that $\operatorname{Ext}_A^1(k,k) = \operatorname{Ext}_A^{1,-1}(k,k)$, and so it now also follows from isomorphism (3.4.10) that (iii) implies (ii).

Identify the Koszul complex $\widetilde{K}(A)$ with $A \otimes (A^!)^*$, that is, for each n, $\widetilde{K}_n(A) \cong A \otimes (A^!)_n^*$. By equation (3.4.9) and isomorphism (3.4.10), the sequence (3.4.5) is exact if and only if $\operatorname{Ext}_A^*(k,k)$ is generated by $\operatorname{Ext}_A^1(k,k)$. Thus we have shown equivalence of (iii) and (iv).

Now assume that (v) holds, that is, $K(A)$ is a resolution of A as an A^e-module. Since $\widetilde{K}(A) = K(A) \otimes_A k$ and sequence (3.4.4) consists of free right A-modules, it follows that $\widetilde{K}(A)$ is exact other than in degree 0 where it has homology $A \otimes_A k \cong k$. Thus $\widetilde{K}(A)$ is a resolution of k as an A-module, that is, (iv) holds.

Finally, assume that (iv) holds, that is, the sequence (3.4.5) is exact. Since (3.4.4) consists of free right A-modules, it splits as a sequence of right A-modules, and so applying $-\otimes_A k$ commutes with taking homology. Consequently, as $\widetilde{K}(A) = K(A) \otimes_A k$, for all $n \geq 1$,

$$\operatorname{H}_n(K(A)) \otimes_A k \cong \operatorname{H}_n(K(A) \otimes_A k) = \operatorname{H}_n(\widetilde{K}(A)) = 0.$$

Now for each n, we consider $\operatorname{H}_n(K(A))$ to be a graded right A-module, with grading induced by that on A. Note that for any nonzero graded A-module M, $A_+ M$ is a proper submodule of M. As a right A-module, note that $\operatorname{H}_n(K(A)) \otimes_A k \cong \operatorname{H}_n(K(A))/A_+ \operatorname{H}_n(K(A))$. It follows that $\operatorname{H}_n(K(A)) = 0$ for all $n \geq 1$. A calculation shows that $\operatorname{H}_0(K(A)) \cong A$. So $K(A)$ is exact, and thus is a free resolution of A as an A^e-module. We have shown that (iv) implies (v). $\qquad\square$

We next give some standard examples of Koszul algebras.

Example 3.4.11. Let $A = k[x_1, \ldots, x_n] \cong T(V)/(R)$, where V is the vector space with basis $x_1, \ldots, x_n$ and

$$R = \operatorname{Span}_k\{v \otimes w - w \otimes v \mid v, w \in V\}.$$

We claim that $K(A)$, given by (3.4.4), for this algebra A is equivalent to the resolution (3.1.4) given in Example 3.1.3. To see this, define the map

$\phi : A \otimes \bigwedge^{\bullet}(V) \otimes A \to B_{\bullet}(A)$ by

$$(3.4.12) \quad \phi(1 \otimes v_1 \wedge \cdots \wedge v_n \otimes 1) = \sum_{\sigma \in S_n} \mathrm{sgn}(\sigma) \otimes v_{\sigma(1)} \otimes \cdots \otimes v_{\sigma(n)} \otimes 1$$

for all $v_1, \ldots, v_n \in V$, where S_n is the symmetric group on n letters. A calculation shows that ϕ is a chain map, ϕ is injective, and that the image of ϕ is precisely $K(A)$. We showed in Example 3.1.3 that $A \otimes \bigwedge^{\bullet}(V) \otimes A$ is a free resolution of A as an A^e-module, so $K(A)$ is a free resolution of A as an A^e-module. By Theorem 3.4.6(i) and (v), A is a Koszul algebra. Its Hochschild cohomology ring is computed in Example 3.1.3.

Similarly, the skew polynomial ring $A = k_{\mathbf{q}}[x_1, \ldots, x_m]$ of Example 3.2.1 is a Koszul algebra: the resolution discussed in Example 3.2.5 can be shown to be equivalent to the complex $K(A)$ given by (3.4.4) for this algebra. The Hochschild cohomology ring may be found by using this resolution; see, e.g., Exercises 3.2.9 and 3.2.10.

Example 3.4.13. Let $\mathbf{q} = \{q_{ij}\}$ be a set of scalars as in Example 3.2.1, so that $q_{ii} = 1$ and $q_{ji} = q_{ij}^{-1}$ $(1 \le i, j \le m)$. Let

$$A = k_{\mathbf{q}}[x_1, \ldots, x_m]/(x_1^2, \ldots, x_m^2) \cong T(V)/(R),$$

where V is the vector space with basis $x_1, \ldots, x_m$ and

$$R = \mathrm{Span}_k\{x_i \otimes x_j - q_{ij} x_j \otimes x_i \mid 1 \le i, j \le m\} \oplus \mathrm{Span}_k\{x_1 \otimes x_1, \ldots, x_m \otimes x_m\}.$$

The Koszul complex $K(A)$ given by (3.4.4) in case $m = 2$ is equivalent to the resolution constructed in Example 3.2.6, as may be shown by an argument similar to that in Example 3.4.11. Thus A is Koszul. The Hochschild cohomology ring is discussed in Example 3.2.6.

There is a close relationship between Hochschild cohomology and the Ext algebra $\mathrm{Ext}_A^*(k, k)$ when A is a Koszul algebra: let $\phi_k : \mathrm{HH}^*(A) \to \mathrm{Ext}_A^*(k, k)$ be the map given by $- \otimes_A k$, as defined by equation (2.5.3). By Corollary 2.5.9, for any augmented algebra A, the image of ϕ_k is contained in the graded center $Z_{\mathrm{gr}}(\mathrm{Ext}_A^*(k, k))$ (see Definition 2.5.8). In fact, for Koszul algebras, even more is true.

Theorem 3.4.14. *Let A be a Koszul algebra. Then the image of the map $\phi_k : \mathrm{HH}^*(A) \to \mathrm{Ext}_A^*(k, k)$ defined by equation (2.5.3) is precisely the graded center, $Z_{\mathrm{gr}}(\mathrm{Ext}_A^*(k, k))$.*

For a proof of the theorem, see [**39**, Theorem 4.1].

Exercise 3.4.15. In Example 3.4.1, verify that $R^{\perp}$ is as stated there.

Exercise 3.4.16. Verify that the definition of minimal free resolution used in this section agrees with the definition of minimal resolution given in Section A.2.

Exercise 3.4.17. Verify that $K(A)$ given by the sequence (3.4.4) is indeed a subcomplex of $B(A)$, that is, the differential on $B(A)$ takes $K_n(A)$ to $K_{n-1}(A)$ for each n.

Exercise 3.4.18. Verify that ϕ in Example 3.4.11 is a chain map and its image is $K(A)$.

Exercise 3.4.19. Let $A = k_{\mathbf{q}}[x_1, \ldots, x_m]$. Define a map analogous to ϕ, of Example 3.4.11, from the resolution described in Exercise 3.2.10 to the bar resolution. Conclude that $k_{\mathbf{q}}[x_1, \ldots, x_m]$ is a Koszul algebra.

Exercise 3.4.20. Verify the claim made in Example 3.4.13 via an argument similar to that in Example 3.4.11. Constructing the map ϕ may take some work. See [**38**].

3.5. Skew group algebras

When a group acts on an algebra by automorphisms, there is a larger algebra called a skew group algebra that encodes both structures. When a group acts on a geometric space such as a manifold or an algebraic variety, it correspondingly acts on a suitable ring of functions on the space, and the associated skew group algebra is studied in noncommutative geometry. We will look at the special case of a finite group action here, and some techniques for studying the Hochschild homology and cohomology of the resulting skew group algebras, finishing with the special case of a group action on a polynomial ring.

Let G be a finite group acting by automorphisms on an algebra A. We use a left superscript to denote the action in order to distinguish it from multiplication in an algebra, that is, $^g a$ is the result of applying $g \in G$ to $a \in A$. The *skew group algebra* $A \rtimes G$ (also denoted $A\#G$ or $A\#kG$ or $A * G$) is $A \otimes kG$ as a vector space, with multiplication given by

$$(a \otimes g)(b \otimes h) = a(^g b) \otimes gh$$

for all $a, b \in A$ and $g, h \in G$. Note that A is isomorphic to the subalgebra $A \otimes k$ of $A \rtimes G$ and that the group algebra kG is isomorphic to the subalgebra $k \otimes kG$ of $A \rtimes G$. For simplicity of notation, we will abbreviate $a \otimes g$ by ag when it will cause no confusion. In this notation, the action of G on A is by conjugation in $A \rtimes G$: $gag^{-1} = {}^g a$.

Under some conditions, the skew group algebra $A \rtimes G$ is Morita equivalent to the subalgebra of A consisting of G-invariant elements, denoted A^G in equation (3.5.1) below. See, for example, [**37**, Proposition 6.10]. This fact was used in [**4**] to find the Hochschild cohomology rings of some invariant subrings of Weyl algebras under finite group actions.

There is a spectral sequence describing the Hochschild (co)homology of $A \rtimes G$ in terms of that of A and of G. This is a special case of a construction given in Section 9.6 for smash products with Hopf algebras. For now, we will work in a more specialized setting: assume the characteristic of k does not divide the order of G, so that the group algebra kG is semisimple by Maschke's Theorem. In this case, we will find the Hochschild (co)homology space of $A \rtimes G$ using more elementary methods.

For any set X on which G acts, we use a superscript G to denote the set of *invariants*, that is,

$$(3.5.1) \qquad X^G = \{x \in X \mid {}^g x = x \text{ for all } g \in G\},$$

and a subscript G to denote the set of *coinvariants*,

$$X_G = X/\sim, \quad \text{where } x \sim y \text{ if } y = {}^g x \text{ for some } g \in G.$$

In the following, we take the algebra structure on $\mathrm{HH}^n(A, A \rtimes G)$ to be that described in Remark 1.3.5.

Theorem 3.5.2. *Assume the characteristic of the field k does not divide the order of the finite group G. Let A be a k-algebra on which G acts by algebra automorphisms. There are actions of G for which*

$$\mathrm{HH}^*(A \rtimes G) \cong \mathrm{HH}^*(A, A \rtimes G)^G \quad \text{and} \quad \mathrm{HH}_*(A \rtimes G) \cong \mathrm{HH}_*(A, A \rtimes G)_G$$

as graded algebras and graded vector spaces, respectively.

Proof. One proof relies on a spectral sequence described for cohomology in Section 9.6 in a more general setting. (See Corollary 9.6.6 and Exercise 9.6.9.) We will give a more elementary proof here. We will see in the course of the proof that the action of G is that induced by its action on the bar resolution of A (diagonal on each tensor factor) and by conjugation on $A \rtimes G$. Let

$$(3.5.3) \qquad \mathcal{D} = \bigoplus_{g \in G} (Ag) \otimes (A^{\mathrm{op}} g^{-1}),$$

a subspace of $(A \rtimes G)^e$. A calculation shows that in fact $\mathcal{D}$ is a subalgebra of $(A \rtimes G)^e$, and further that $(A \rtimes G)^e$ is free as a right $\mathcal{D}$-module under multiplication, with free basis $\{1 \otimes g \mid g \in G\}$.

We claim that there is an isomorphism of $(A \rtimes G)^e$-modules,

$$(3.5.4) \qquad A \rtimes G \xrightarrow{\sim} A{\uparrow}_{\mathcal{D}}^{(A \rtimes G)^e},$$

where the arrow denotes tensor induction: $A{\uparrow}_{\mathcal{D}}^{(A \rtimes G)^e} = (A \rtimes G)^e \otimes_{\mathcal{D}} A$, on which $(A \rtimes G)^e$ acts by multiplication on the leftmost factor. This isomorphism is given by sending $a \otimes g$ to $(1 \otimes g) \otimes a$ for all $a \in A$ and $g \in G$. In light of this isomorphism (3.5.4), by the Eckmann-Shapiro Lemma (Lemma A.6.2),

$$\mathrm{Ext}^*_{(A\rtimes G)^e}(A\rtimes G, A\rtimes G) \cong \mathrm{Ext}^*_{\mathcal{D}}(A, A\rtimes G).$$

Now, any $\mathcal{D}$-projective resolution of A may be viewed, on restriction to A^e, as an A^e-projective resolution having an action of G (through the elements $g\otimes g^{-1}$ for g in G) that commutes with the differentials. For any pair of $\mathcal{D}$-modules U, V, note that $\mathrm{Hom}_{\mathcal{D}}(U,V)\cong \mathrm{Hom}_{A^e}(U,V)^G$, where the action of G on such functions is given by $({}^g f)(u) = {}^g(f({}^{g^{-1}}u))$ for $g\in G$, $f\in \mathrm{Hom}_{A^e}(U,V)$, $u\in U$. Since the characteristic of k does not divide the order of G, the space of G-invariants of any kG-module V is the image of $\frac{1}{|G|}\sum_{g\in G}g$ as an operator on V. It follows that taking G-invariants commutes with taking (co)homology, and so $\mathrm{Ext}^*_{\mathcal{D}}(A, A\rtimes G)\cong (\mathrm{Ext}^*_{A^e}(A, A\rtimes G))^G$. A calculation shows that this is an algebra isomorphism.

Similar to (3.5.4) is an isomorphism $A\rtimes G \cong A\otimes_{\mathcal{D}}(A\rtimes G)^e$ of *right* $(A\rtimes G)^e$-modules. We use this isomorphism to see that

$$\mathrm{HH}_n(A\rtimes G)\cong \mathrm{Tor}^{\mathcal{D}}_n(A, A\rtimes G),$$

as follows. Let $P_{\textstyle\cdot}$ be the bar resolution (1.1.4) of A as an A^e-module. The resolution admits an action of G commuting with the differentials, and this action may be used to extend the A^e-module structure on each P_i to a $\mathcal{D}$-module structure. Tensoring with $(A\rtimes G)^e$ over $\mathcal{D}$, we obtain $P_{\textstyle\cdot}\otimes_{\mathcal{D}}(A\rtimes G)^e$. Let k be the trivial kG-module, that is, each group element acts as the identity. There is an isomorphism $P_i\otimes_{\mathcal{D}}(A\rtimes G)\xrightarrow{\sim} k\otimes_{kG}(P_i\otimes_{A^e}(A\rtimes G))$ for each i, given by sending $x\otimes b$ to $1\otimes x\otimes b$ for $x\in P_i$, $b\in A\rtimes G$. The inverse map is given by $1\otimes x\otimes b\mapsto x\otimes b$. The action of kG on the tensor product of two modules is standard: g acts as $g\otimes g$ for all $g\in G$. It may be checked directly that these maps are well-defined. Finally, we see that $k\otimes_{kG}(P_i\otimes_{A^e}(A\rtimes G))\xrightarrow{\sim}(P_i\otimes_{A^e}(A\rtimes G))_G$ by sending $1\otimes x\otimes b$ to $G\cdot(x\otimes b)$, the orbit of $x\otimes b$. The inverse map sends $G\cdot(x\otimes b)$ to $1\otimes x\otimes b$ (a well-defined map due to the tensor product over kG). $\qquad\square$

We may further rewrite the expressions in the theorem. As an A^e-module, $A\rtimes G\cong \bigoplus_{g\in G}Ag$, which yields an isomorphism of graded vector spaces,

$$(3.5.5)\qquad\qquad \mathrm{HH}^*(A, A\rtimes G)\cong \bigoplus_{g\in G}\mathrm{HH}^*(A, Ag).$$

The action of G permutes the components of the direct sum via conjugation: letting $h\in G$, we have ${}^h(ag) = ({}^h a)hgh^{-1}$ for all $a\in A$ and $g\in G$, and so h takes $\mathrm{HH}^*(A, Ag)$ to $\mathrm{HH}^*(A, Ahgh^{-1})$. We may then apply h^{-1} to see that these two components are isomorphic as vector spaces, that is, the components of (3.5.5) are permuted according to the conjugation action of G

on itself. The G-invariant subspace is the image of the operator $\frac{1}{|G|} \sum_{g \in G} g$ since $|G|$ is invertible in k. We choose one representative element g in each conjugacy class to rewrite the sum. By Theorem 3.5.2,

$$(3.5.6) \qquad \mathrm{HH}^*(A \rtimes G) \cong \bigoplus_{g \in \overline{G}} (\mathrm{HH}^*(A, Ag))^{C(g)},$$

where $\overline{G}$ is a set of conjugacy class representatives in G, and $C(g)$ is the centralizer in G of g. We will use this isomorphism in the example of a polynomial ring next.

Example 3.5.7. For this example, we take k to be algebraically closed, as we will need to use eigenvalues of operators. Let V be a finite-dimensional kG-module, and let $A = S(V)$, the symmetric algebra on V (see Section 3.1). Recall that $S(V)$ is isomorphic to the polynomial ring on a basis of V. The action of G on V may be extended to an action on A by algebra automorphisms. We use techniques similar to those of Farinati [**77**] and of Ginzburg and Kaledin [**90**] to find the structure of the Hochschild cohomology ring of A via Theorem 3.5.2. Let $g \in G$. We wish to find an expression for the component $\mathrm{HH}^*(S(V), S(V)g)$ of (3.5.5). We will use the Koszul resolution of $S(V)$ as an $S(V)$-bimodule, from Example 3.1.3, 3.3.5, or 3.4.11. Since the element g of G has finite order, we may choose a basis $x_1, \ldots, x_n$ of V consisting of eigenvectors of g. Let $\lambda_1, \ldots, \lambda_n \in k$ be the corresponding eigenvalues. Assume they are ordered so that $\lambda_1 = 1, \ldots, \lambda_r = 1$ and $\lambda_{r+1} \neq 1, \ldots, \lambda_n \neq 1$. The invariant subspace V^g is then the k-linear span of $x_1, \ldots, x_r$. Let

$$V_g = \mathrm{Span}_k\{x_{r+1}, \ldots, x_n\} = \mathrm{Im}(1 - g) \cong V/V^g.$$

Consider the Koszul complex $K(x_1 \otimes 1 - 1 \otimes x_1, \ldots, x_n \otimes 1 - 1 \otimes x_n)$ defined in Example 3.3.5, and note that it is the tensor product over $S(V)^e$ of the two Koszul complexes $K(x_1 \otimes 1 - 1 \otimes x_1, \ldots, x_r \otimes 1 - 1 \otimes x_r)$ and $K(x_{r+1} \otimes 1 - 1 \otimes x_{r+1}, \ldots, x_n \otimes 1 - 1 \otimes x_n)$. Applying $\mathrm{Hom}_{S(V)^e}(-, S(V)g)$ and writing $S(V)g = S(V^g) \otimes S(V_g)g$, we obtain

$$\textstyle\bigwedge((V^g)^*) \otimes S(V^g) \otimes \bigwedge((V_g)^*) \otimes S(V_g)g.$$

To find the differentials, note that for each index i, for $s \in S(V)$ and $g \in G$,

$$(x_i \otimes 1 - 1 \otimes x_i) \cdot sg = x_i sg - sg x_i = x_i sg - s^g x_i g = (1 - \lambda_i)(x_i sg).$$

When $\lambda_i \neq 1$, the differential is thus just multiplication by a nonzero scalar multiple of x_i. It follows that the complex $\bigwedge((V_g)^*) \otimes S(V_g)g$ is exact other than in degree $n - r$, where it has homology $S(V_g)/(x_{r+1}, \ldots, x_n)S(V_g)g \cong k$ by Theorem 3.3.4 (applying Hom reverses the arrows). We will identify this with the top exterior power $\bigwedge^{n-r}((V_g)^*)$. The complex $\bigwedge((V^g)^*) \otimes S(V^g)$, by contrast, has differentials all 0. By the Künneth Theorem (Theorem A.5.2), since the tensor product is now over the field k, the homology of the complex is $\bigwedge((V^g)^*) \otimes S(V^g) \otimes \bigwedge^{n-r}((V_g)^*)g$. Since $n - r = \operatorname{codim} V^g$, applying (3.5.6), we have

$$\mathrm{HH}^n(S(V) \rtimes G) \cong \bigoplus_{g \in \overline{G}} \left(S(V^g)g \otimes \bigwedge^{n-\operatorname{codim} V^g}((V^g)^*) \otimes \bigwedge^{\operatorname{codim} V^g}((V_g)^*) \right)^{C(g)}.$$

In this expression, the factor $\bigwedge^{\operatorname{codim} V^g}((V_g)^*)$ is isomorphic to k as a vector space, but it potentially has a nontrivial $C(g)$-action, so we retain the factor in the notation.

A related example is given by the action of a group on a Weyl algebra (see Example 5.1.10) in papers by Alev, Farinati, Lambre, and Solotar [4] and Suárez-Álvarez [**210**].

In this chapter, our examples have included many algebras built from other algebras whose Hochschild cohomology rings may be understood from knowledge of the components, such as the tensor product and twisted tensor product algebras of Sections 3.1 and 3.2 and the skew group algebras of this section. There are other constructions of algebras from component parts that we do not consider here; some examples and results for their Hochschild cohomology rings are the split algebras and trivial extensions [**52**].

Exercise 3.5.8. Verify the following claims in the proof of Theorem 3.5.2:

(a) $(A \rtimes G)^e$ is free as a right $\mathcal{D}$-module.

(b) The map (3.5.4) is an $(A \rtimes G)$-bimodule isomorphism. What is the inverse map?

(c) $\operatorname{Hom}_{\mathcal{D}}(U, V) \cong \operatorname{Hom}_{A^e}(U, V)^G$ as vector spaces.

(d) $\operatorname{Ext}^*_{\mathcal{D}}(A, A \rtimes G) \cong (\operatorname{Ext}^*_{A^e}(A, A \rtimes G))^G$ as algebras.

Exercise 3.5.9. Let $k = \mathbb{C}$, let $G = S_3$, and let V be the 3-dimensional vector space that is a $\mathbb{C}G$-module by permutations of a chosen basis. Describe the graded vector space structure of $\mathrm{HH}^*(S(V) \rtimes G)$ by using the isomorphism given at the end of Example 3.5.7.

3.6. Path algebras and monomial algebras

In this section we define monomial algebras and present a construction due to Bardzell [**18**] of a bimodule resolution of a monomial algebra. For a generalization of this construction to all algebras defined by quivers and relations, see the paper of Chouhy and Solotar [**49**]. See also the paper of Kobayashi [**131**]. These resolutions are bimodule analogues of the Anick resolution for left modules [**8**, **100**].

A *quiver* Q is a directed graph, that is, Q consists of a set Q_0 of *vertices*, a set Q_1 of *arrows*, and two maps $s : Q_1 \to Q_0$, $t : Q_1 \to Q_0$, associating to each arrow α its *source* $s(\alpha)$ and *target* $t(\alpha)$. The quiver is *finite* if the sets Q_0 and Q_1 are both finite. A *path* in Q is a sequence of arrows $(\alpha_1, \ldots, \alpha_l)$ for which $t(\alpha_i) = s(\alpha_{i+1})$ for all i. We denote the associated path by $\alpha_1 \cdots \alpha_l$. Its *length* is l. There is a path of length 0 associated to each vertex a of Q_0, denoted e_a. The *path algebra* of Q, denoted kQ, is the associative k-algebra whose underlying vector space is the set of all paths $\alpha_1 \cdots \alpha_l$ of length $l \geq 0$ with multiplication determined by

$$(\alpha_1 \cdots \alpha_l) \cdot (\beta_1 \cdots \beta_m) = \begin{cases} \alpha_1 \cdots \alpha_l \beta_1 \cdots \beta_m, & \text{if } t(\alpha_l) = s(\beta_1), \\ 0, & \text{otherwise,} \end{cases}$$

and $e_a e_b = \delta_{a,b} e_a$, $e_{s(\alpha)} \alpha = \alpha$, and $\alpha e_{t(\alpha)} = \alpha$ for all vertices $a, b \in Q_0$ and all arrows $\alpha, \alpha_1 \ldots, \alpha_l, \beta, \beta_1, \ldots, \beta_m \in Q_1$.

Example 3.6.1. The path algebra of the leftmost quiver below has dimension 3 as a vector space, with basis e_1, e_2, α (note that $\alpha^2 = 0$ since the source and target of α are different). It is isomorphic to the algebra of upper triangular 2×2 matrices with entries in k. The path algebra of the rightmost quiver is the polynomial ring $k[x]$.

Remark 3.6.2. Any finite-dimensional algebra A is Morita equivalent to a quotient kQ/I for some quiver Q and ideal I. That is, the category of A-modules is equivalent to the category of kQ/I-modules [10, Corollary I.6.10 and Theorem II.3.7]. Thus quiver techniques are very important in the representation theory of finite-dimensional algebras in particular.

Definition 3.6.3. A *monomial algebra* is an algebra of the form $A = kQ/I$, where Q is a finite quiver and I is the ideal generated by a finite set of paths of length at least 2.

Example 3.6.4. The algebra $k[x]/(x^n)$ is a monomial algebra with quiver Q given by the rightmost quiver in Example 3.6.1.

We need further notation and terminology for the resolution we will construct next. Let $A = kQ/I$ be a monomial algebra. Let R be a minimal set of paths, of minimal length, for which $I = (R)$. Let u, w be paths in Q. We say that y is a *subpath* of w if $w = xyz$ for some paths x, z. We write $x = l(y)$ and $z = r(y)$ (for brevity, in the notation l and r, the dependence on the expression xyz is suppressed).

We define the *Bardzell resolution* $P_\bullet$ of A in a similar manner to the presentation of Redondo and Román [**183**]. See also [**18, 199**]. Let

$$
\begin{aligned}
P_0 &= A \otimes_{kQ_0} kQ_0 \otimes_{kQ_0} A \cong A \otimes_{kQ_0} A, \\
P_1 &= A \otimes_{kQ_0} kQ_1 \otimes_{kQ_0} A, \\
P_2 &= A \otimes_{kQ_0} kR \otimes_{kQ_0} A,
\end{aligned}
$$

where kQ_0 denotes the vector space with basis Q_0 (a subalgebra of kQ), and similarly for kQ_1, kR. For $n > 2$, let $P_n = A \otimes_{kQ_0} P'_n \otimes_{kQ_0} A$, where P'_n is as follows. Let $w = \alpha_1 \cdots \alpha_l$ be any path in Q and order all vertices occurring in the path according to its layout, with multiplicities as needed:

$$
s(\alpha_1) < s(\alpha_2) < \cdots < s(\alpha_l) < t(\alpha_l).
$$

Let $R(w)$ be the set of paths in R that are subpaths of w. We construct sets $L_1, L_2, \ldots$ recursively: choose $p_1 \in R(w)$. Let

$$
L_1 = \{p \in R(w) \mid s(p_1) < s(p) < t(p_1)\}.
$$

If $L_1 \neq \varnothing$, let $p_2 \in R(w)$ be a path for which $s(p_2)$ is minimal for paths in L_1. In general, let $j \geq 1$, assume $L_1, \ldots, L_j$ have been defined, and let $p_{j+1} \in R(w)$ be a path for which $s(p_{j+1})$ is minimal for paths in L_j. Let

$$
L_{j+1} = \{p \in R(w) \mid t(p_j) \leq s(p) < t(p_{j+1})\}.
$$

For each n, write $w(p_1, \ldots, p_{n-1})$ for the *support* of the sequence $p_1, \ldots, p_{n-1}$, that is, the path from $s(p_1)$ to $t(p_{n-1})$ that contains $p_1, \ldots, p_{n-1}$ as subpaths (note that this is a subpath of w by construction). Let P'_n be the set of all supports of all such sequences $p_1, \ldots, p_{n-1}$ for all paths w in Q, called n-*concatenations*. For each path w in P'_n, let

$$
\mathrm{Sub}(w) = \{w' \in P'_{n-1} \mid w' \text{ is a subpath of } w\}.
$$

Next we describe the differentials on the Bardzell resolution $P_\bullet$. The map $\pi : P_0 \to A$ is given by multiplication on $A \otimes_{kQ_0} A$. The map $d_1 : P_1 \to P_0$ is defined by

$$
d_1(1 \otimes \alpha \otimes 1) = \alpha \otimes 1 - 1 \otimes \alpha
$$

for all $\alpha \in Q_1$, where we have suppressed the subscript kQ_0 on the tensor symbol in expressions for elements in order to reduce clutter. The map $d_2 : P_2 \to P_1$ is given by

$$
d_2(1 \otimes w \otimes 1) = \sum_{w' \in \mathrm{Sub}(w)} l(w') \otimes w' \otimes r(w').
$$

More generally, in even degrees, the differential is similar:

$$
d_{2m}(1 \otimes w \otimes 1) = \sum_{w' \in \mathrm{Sub}(w)} l(w') \otimes w' \otimes r(w'),
$$

while in odd degrees we set

$$d_{2m+1}(1 \otimes w \otimes 1) = l(w_2) \otimes w_2 \otimes 1 - 1 \otimes w_1 \otimes r(w_1),$$

where $w = l(w_2)w_2 = w_1 r(w_1)$ with w_1, w_2 the supports of the corresponding (unique) $2m$-concatenations. See [18, 183, 199] for more details.

Exactness of $P_\bullet$ may be concluded from existence of the following contracting homotopy defined by Sköldberg [199]: $s_{-1}(a) = 1 \otimes 1 \otimes a$ for all $a \in A$ and $s_n : P_n \to P_{n+1}$ is the $kQ_0 \otimes A^{\mathrm{op}}$-module homomorphism defined by

$$s_n(u \otimes v \otimes 1) = \sum_{\substack{w \in P'_{n+1} \\ w \text{ a subpath of } uv}} l(w) \otimes w \otimes r(w)$$

for all $v \in P'_n$ and images u in A of paths. The proof that $s_\bullet$ is indeed a contracting homotopy is lengthy, and we refer to [199] for the details. A different approach and some examples are given in [18, Section 7]. Redondo and Román [183] provided chain maps between the Bardzell resolution and the bar resolution with the goal of computing some of the structure of Hochschild cohomology of monomial algebras.

Example 3.6.5. Let $A = k[x]/(x^n)$ as in Example 1.1.21. We show that the resolution (1.1.22) is equivalent to the Bardzell resolution. Let Q be the quiver with one vertex and one arrow:

$$\overset{\bullet}{\underset{x}{\circlearrowright}}$$

Then $A \cong kQ/I$, where $I = (x^n)$. The paths in Q are all nonnegative integer powers of x, and we may take $R = \{x^n\}$. For each m, there is a unique m-concatenation: in even degrees m it is $x^{\frac{mn}{2}}$, and in odd degrees m it is $x^{\frac{(m-1)n}{2}+1}$. The above formula for the differentials on the Bardzell resolution agrees with the differentials in the sequence (1.1.22) once we identify $1 \otimes 1$ in each degree in (1.1.22) with these free generators in the Bardzell resolution. This resolution was used to compute the Hochschild cohomology ring $\mathrm{HH}^*(A)$ in Examples 2.2.2 and 2.2.3.

Some other algebras whose Hochschild cohomology rings are known as a consequence of these and other quiver techniques are finite-dimensional monomial algebras [97, 99], radical square zero algebras [51], quotients of path algebras for which there is at most one path between each pair of vertices [50], preprojective algebras [68, 69], and other algebras defined by quivers and relations [17, 66, 67, 144]. These techniques apply more generally than may appear at first since a finite-dimensional algebra is Morita equivalent to an algebra defined by a quiver and relations [10] and Hochschild cohomology is invariant under Morita equivalence [223, §9.5].

Exercise 3.6.6. Verify the claims in Example 3.6.5, that is:

(a) Check that the m-concatenations are as stated.

(b) Check that the formula for the differentials on the Bardzell resolution agrees with that for (1.1.22) under suitable identifications.

Exercise 3.6.7. Find the Bardzell resolution for the path algebra kQ/I, where Q is the following quiver and $I = (\alpha\beta\alpha)$:

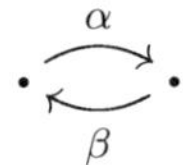

Chapter 4

Smooth Algebras and Van den Bergh Duality

In this chapter we look at noncommutative analogs of some commutative concepts, beginning with dimension, smoothness, differential forms, and square-zero extensions. These are defined using Hochschild cohomology. We present Van den Bergh's algebraic version of Poincaré duality, that is, a duality between Hochschild homology and cohomology for some types of smooth algebras. We define Calabi-Yau algebras and take a closer look at skew group algebras in this light. For Calabi-Yau algebras and for symmetric algebras we further define Batalin-Vilkovisky structures on Hochschild cohomology using different types of duality.

Throughout, k will be a field and A will be a k-algebra.

4.1. Dimension and smoothness

We first use the notion of projective dimension (see Section A.2) in a definition of dimension of the algebra A.

Definition 4.1.1. The *Hochschild dimension* of A is its projective dimension as an A^e-module:

$$\dim(A) = \operatorname{pdim}_{A^e}(A).$$

Some authors simply refer to this as the *dimension* of A. There are however many other types of dimension for algebras, depending on context, such as global dimension, Krull dimension, Gelfand-Kirillov dimension, or vector space dimension. See, for example, [**158**] for a discussion of dimension for noncommutative rings. Note that the global dimension of A is always less than or equal to its Hochschild dimension, since any A^e-projective resolution of A may be tensored over A with any module M to yield an A-projective resolution of M, as we saw in Section 2.5. For some algebras, more is known, for example if A is $\mathbb{N}$-graded and connected (i.e., $A_0 = k$), the Hochschild dimension is equal to the (left or right) global dimension, and both are equal to the projective dimension of the A-module k given by projection onto A_0 [26].

Example 4.1.2. By our work in Example 3.1.3, $\dim(k[x_1, \ldots, x_m]) = m$. Specifically, we found there a projective resolution of A as an A^e-module of length m. There cannot exist a shorter resolution since $\mathrm{HH}^m(A) \neq 0$.

Our first result describes a relationship among the Hochschild dimensions of two algebras and that of their tensor product algebra, or of their twisted tensor product algebra when the twisting is given by a bicharacter as defined in Section 3.2.

Theorem 4.1.3. *Let A and B be two algebras. Then*

(i) $\dim(A \otimes B) \leq \dim(A) + \dim(B)$, *and more generally*

(ii) *if A, B are graded by abelian groups Γ, Γ' and $t : \Gamma \times \Gamma' \to k^\times$ is a bicharacter, then $\dim(A \otimes^t B) \leq \dim(A) + \dim(B)$, where $A \otimes^t B$ is the twisted tensor product defined in Section 3.2.*

Proof. (i) Let $P_\bullet$ (respectively, $Q_\bullet$) be a projective resolution of A as an A^e-module (respectively, of B as a B^e-module). In Section 3.1 we showed that the total complex of the tensor product complex $P_\bullet \otimes Q_\bullet$ is a projective resolution of $A \otimes B$ as an $(A \otimes B)^e$-module. The length of this total complex is the sum of the lengths of $P_\bullet$ and $Q_\bullet$. Therefore there is a projective resolution of $A \otimes B$ as an $(A \otimes B)^e$-module of length $\dim(A) + \dim(B)$. It follows that $\dim(A \otimes B)$ is at most this number.

(ii) The proof of (i) applies more generally to the twisted tensor product algebra $A \otimes^t B$, using Theorem 3.2.3. $\qquad\qquad\square$

By contrast, there is no such theorem for the more general twisted tensor product algebra $A \otimes_\tau B$, arising from a twisting map $\tau : B \otimes A \to A \otimes B$, defined at the end of Section 3.2. There are conditions under which resolutions $P_\bullet$ and $Q_\bullet$ may be combined to form a complex of bimodules for this more general twisted tensor product algebra, and there are conditions under which the bimodules in such a complex are projective, but there are no general guarantees [196]. Indeed, there exist algebras of Hochschild dimension 0 for which a twisted tensor product algebra has infinite Hochschild dimension. See, e.g., [148, Proposition 3.3.6].

The following definition is due to Van den Bergh [216].

Definition 4.1.4. The algebra A is *smooth* if its Hochschild dimension is finite and it has a finite projective resolution as an A^e-module consisting of finitely generated projective modules.

Some authors use the term *homologically smooth* to distinguish this from other notions of smoothness. Note that if A and B are smooth, then so is $A \otimes B$ (and more generally $A \otimes^t B$): this follows from Theorem 4.1.3 and its proof, since the (twisted) tensor product of finitely generated modules is also finitely generated. If A is a finitely generated commutative algebra over k, this notion of smoothness is equivalent to more standard definitions as mentioned already in Section 3.3 (see [**216**]).

Example 4.1.5. A polynomial ring $A = k[x_1, \ldots, x_m]$ is smooth by our work in Example 3.1.3 (the resolution used there consists of finitely generated free modules). A skew polynomial ring $k_{\mathbf{q}}[x_1, \ldots, x_m]$ is smooth by our work in Example 3.2.5. If G is a finite group acting on $A = k[x_1, \ldots, x_m]$ by degree-preserving automorphisms and $\mathrm{char}(k)$ does not divide the order of G, then the skew group algebra $k[x_1, \ldots, x_m] \rtimes G$ of Section 3.5 is smooth: the group G acts on the Koszul resolution of $k[x_1, \ldots, x_m]$, as a subcomplex of the bar resolution via the map ϕ of (3.4.12). (The action of G on the bar resolution of A is diagonal, that is, $g \cdot (a_0 \otimes \cdots \otimes a_{n+1}) = {}^g a_0 \otimes \cdots \otimes {}^g a_{n+1}$ for all $g \in G$, $a_0, \ldots, a_{n+1} \in A$.) Our work in Section 3.5 shows that the Koszul resolution may be induced to a projective resolution of $k[x_1, \ldots, x_m] \rtimes G$ as a $(k[x_1, \ldots, x_m] \rtimes G)^e$-module, and this resolution consists of finitely generated modules. More specifically, under the action of G, the Koszul resolution may be viewed as a resolution of $A = k[x_1, \ldots, x_m]$ as a $\mathcal{D}$-module, where $\mathcal{D}$ is defined by equation (3.5.3). Induce from $\mathcal{D}$ to $(A \rtimes G)^e$. The map (3.5.4) is an isomorphism between $A \rtimes G$ and the $\mathcal{D}$-module A induced to an $(A \rtimes G)^e$-module.

Example 4.1.6. In contrast, the quantum complete intersections of Example 3.2.6, $A = k\langle x_1, x_2 \rangle (x_1 x_2 - q x_2 x_1, \ x_1^2, \ x_2^2)$, are not smooth. We explained there that A has infinite global dimension, forcing its Hochschild dimension to be infinite as well.

We will look closely at some algebras with Hochschild dimension 0 or 1.

Definition 4.1.7. An algebra A is *separable* if $\dim(A) = 0$. It is *quasi-free* if $\dim(A) \leq 1$.

This notion of quasi-free algebras is due to Cuntz and Quillen [**55**]. Quasi-free algebras have also been called *Cuntz-Quillen smooth* or *formally smooth*.

By definition, A is separable if and only if it is projective as an A^e-module. Another equivalent condition to separability is that any derivation from A to an A-bimodule is inner. Indeed, if A is separable, then $\mathrm{HH}^1(A, M) = 0$ for all A-bimodules M. By our work in Section 1.2, the vanishing of $\mathrm{HH}^1(A, M)$ is equivalent to the statement that any derivation from A to M is inner. Conversely, suppose that A is not projective as an A^e-module. Let K_1 be the first syzygy module of A in a given projective resolution $P_\bullet$ of A as an A^e-module. We claim that $\mathrm{HH}^1(A, K_1) \neq 0$. To see this, note that by the definitions, $\mathrm{HH}^1(A, K_1) \cong \mathrm{Hom}_{A^e}(K_1, K_1)/\mathrm{Im}(i_1^*)$, where i_1 is the embedding of K_1 into P_0 as in diagram (A.2.4). This quotient is nonzero, as if not, then the identity map from K_1 to K_1 is in the image of i_1^*, which implies that the short exact sequence $0 \to K_1 \to P_0 \to A \to 0$ splits and A is projective as an A^e-module, a contradiction.

We will look at some equivalent conditions to quasi-freeness in the next section. For now, we consider some examples and implications.

Example 4.1.8. $A = k[x]$ has Hochschild dimension 1 by our work in Example 1.1.18, and so is quasi-free. However, $k[x, y]$ has Hochschild dimension 2 by our work in Example 3.1.3, and so is not quasi-free. Thus the tensor product of two quasi-free algebras is not always quasi-free, and similarly for twisted tensor products. However, the free product of two quasi-free algebras is always quasi-free [**55**, Proposition 5.3]. It follows that the tensor algebra $T(V)$ of a finite-dimensional vector space V is quasi-free.

A quasi-free algebra is hereditary: if A is quasi-free, then there is a projective resolution $0 \to P_1 \to P_0 \to A \to 0$ of A as an A^e-module. For any A-module M, we may apply $-\otimes_A M$ to this sequence to obtain a projective resolution of M as an A-module, as explained in Section 2.5. Therefore the projective dimension $\mathrm{pdim}_A(M)$ of M is at most 1, and so $\mathrm{gldim}_l A \leq 1$. By definition then, A is hereditary.

We define the *Jacobson radical* (or simply *radical*) of A, denoted $\mathrm{rad}(A)$, to be the intersection of all the maximal left ideals of A. This can be shown to be the same as the intersection of all the maximal right ideals, and thus is a two-sided ideal. If A is finite dimensional as a vector space over k, we say that A is *semisimple* if $\mathrm{rad}(A) = 0$. Equivalently, every A-module is a direct sum of simple A-modules. The Wedderburn-Artin Theorem states that every semisimple algebra over k is a direct sum of matrix algebras over division rings. See, e.g., [**121**, Theorem IX.3.3].

Any semisimple algebra is separable, since if A is semisimple, then so is A^e, and so all A^e-modules are projective. For example, if G is a finite group whose order is not divisible by the characteristic of k, then the group algebra kG is semisimple by Maschke's Theorem, and so kG is separable.

Exercise 4.1.9. Find the Hochschild dimensions of the following algebras:

(a) A skew polynomial ring $k_{\mathbf{q}}[x_1, \ldots, x_m]$.

(b) A skew group algebra $k_{\mathbf{q}}[x_1, \ldots, x_m] \rtimes G$ in case the characteristic of k does not divide the order of G.

Exercise 4.1.10. Find the Hochschild dimension of $k[x]/(x^n)$. (See Example 2.5.10.)

Exercise 4.1.11. Find the Hochschild dimension of each of the algebras $k\langle x_1, x_2 \rangle / (x_1 x_2 - q x_2 x_1, \ x_1^{n_1}, \ x_2^{n_2})$ discussed in Example 3.2.2. (See Exercise 3.2.12 and Examples 3.2.6 and 4.1.6.)

4.2. Noncommutative differential forms

We study the quasi-free algebras of Definition 4.1.7 in more detail in this section. To this end, we introduce a noncommutative version of Kähler differentials. This material and notation are from Cuntz and Quillen [55] and Ginzburg [88].

Let A be a k-algebra and for each $n \geq 0$, let

$$(4.2.1) \qquad \Omega_{nc}^n A = A \otimes (\overline{A})^{\otimes n},$$

where $\overline{A} = A/k$ is the vector space quotient (the field k is identified with $k \cdot 1$ as a vector subspace of A). We write elements of $\overline{A}$ via notation from A, viewing $\overline{A}$ noncanonically as a vector space direct summand of A, when this will not cause confusion. We will identify the vector space $\Omega_{nc}^1 A$ with the kernel of the multiplication map $\pi : A \otimes A \to A$, at the same time giving it the structure of an A-bimodule, as follows. This will indicate a connection with the Heller operator of similar notation Ω (see Section A.2) and also a comparison to Kähler differentials [223] of similar notation Ω_{com}^1 in the special case that A is commutative.

Consider $\Omega_{nc}^1 A$ to be an A-bimodule under the following actions:

$$(4.2.2) \qquad c(a \otimes b) = ca \otimes b \quad \text{and} \quad (a \otimes b)c = a \otimes bc - ab \otimes c$$

for all $a, b, c \in A$. Let $j : A \otimes \overline{A} \to A \otimes A$ be given by $j(a \otimes b) = ab \otimes 1 - a \otimes b$ for $a, b \in A$. Then the sequence

$$(4.2.3) \qquad 0 \to \Omega_{nc}^1 A \xrightarrow{j} A \otimes A \xrightarrow{\pi} A \to 0$$

is an exact sequence of A-bimodules. To see this, note that j maps $\Omega_{nc}^1 A$ isomorphically onto $\mathrm{Ker}(\pi)$. Due to exactness of the bar resolution $B(A)$ of the A^e-module A given in (1.1.4), $\mathrm{Ker}(\pi) = \mathrm{Im}(d_1)$, but this is precisely the image of j. Further, j is injective. Assume that $j(\sum_i a_i \otimes b_i) = 0$ for some elements a_i, b_i. Then, since $b_i \in \overline{A}$ and

$$j\left(\sum_i a_i \otimes b_i\right) = \left(\sum_i a_i b_i\right) \otimes 1 - \sum_i a_i \otimes b_i,$$

we have $\sum_i a_i b_i = 0$ (the tensor product is over the field k). It follows that $\sum_i a_i \otimes b_i = j(\sum_i a_i \otimes b_i) = 0$. Thus the sequence (4.2.3) is exact as claimed, and $\Omega^1_{nc} A$ is a first syzygy module of A as an A-bimodule.

As one important property of the A-bimodule $\Omega^1_{nc} A$, we claim that for all A-bimodules M,

$$(4.2.4) \qquad \mathrm{Der}(A, M) \cong \mathrm{Hom}_{A^e}(\Omega^1_{nc} A, M),$$

where $\mathrm{Der}(A, M)$ is the space of k-derivations from A to M, defined in Section 1.2. This isomorphism follows immediately from our work in Section 1.2 interpreting Hochschild cohomology in degree 1, as this is precisely the space of Hochschild 1-cocycles, that is, the 1-cochains on the bar resolution that factor through the first syzygy module. The isomorphism (4.2.4) can be interpreted as saying that $\Omega^1_{nc} A$ represents the functor $\mathrm{Der}(A, -)$ on the category of A-bimodules.

In comparison, for commutative algebras A, we consider A-modules rather than A-bimodules. The Kähler differentials, defined next, represent the functor $\mathrm{Der}(A, -)$ on the category of A-modules [223].

Definition 4.2.5. Let A be a commutative algebra. The A-module of *Kähler differentials* $\Omega^1_{com} A$ is the A-module with one generator da for each $a \in A$ and $dc = 0$ for all $c \in k$. Relations are

$$d(a + b) = da + db \quad \text{and} \quad d(ab) = adb + bda$$

for all $a, b \in A$.

We claim that $\Omega^1_{com} A \cong \mathrm{Ker}\, \pi / (\mathrm{Ker}\, \pi)^2 \cong (\Omega^1_{nc} A)/(\mathrm{Ker}\, \pi)^2$ and that $\Omega^1_{com} A \cong \mathrm{HH}_1(A)$. See Exercises 4.2.14 and 4.2.15.

Returning to the general case of a not necessarily commutative algebra A, recall the definition (4.2.1) of $\Omega^n_{nc} A$.

Definition 4.2.6. Let A be an algebra. The space of *noncommutative differential forms* on A is

$$\Omega_{nc} A = \bigoplus_{n \geq 0} \Omega^n_{nc} A.$$

We will see that $\Omega_{nc} A$ is a differential graded algebra with

$$d(a_0 \otimes \cdots \otimes a_n) = 1 \otimes a_0 \otimes \cdots \otimes a_n,$$

$$(a_0 \otimes \cdots \otimes a_n)(a_{n+1} \otimes \cdots \otimes a_r) = \sum_{i=0}^{n} (-1)^{n-i} a_0 \otimes \cdots \otimes a_i a_{i+1} \otimes \cdots \otimes a_r$$

for all $a_0, \ldots, a_r \in A$, a theorem of Cuntz and Quillen [**55**, Proposition 1.1]. Moreover, $\Omega_{nc}A$ is universal with respect to differential graded algebras whose degree 0 term is the target of an algebra homomorphism from A.

Theorem 4.2.7. *The space $\Omega_{nc}A$ of noncommutative differential forms on an algebra A is a differential graded algebra with differential and multiplication given above, unique such that*

$$a_0(da_1) \cdots (da_n) = a_0 \otimes \cdots \otimes a_n$$

for all $a_0, \ldots, a_n \in A$. Moreover, for any differential graded algebra Γ and algebra homomorphism $u : A \to \Gamma^0$, there is a unique differential graded algebra homomorphism $u_* : \Omega_{nc}A \to \Gamma$ that extends u.

Proof. It may be checked directly that $\Omega_{nc}A$ is indeed a differential graded algebra. (See Exercise 4.2.16 or [**55**, Proposition 1.1].)

For the second statement, let Γ be a differential graded algebra and $u : A \to \Gamma^0$ an algebra homomorphism. Define $u_* : \Omega_{nc}A \to \Gamma$ by

$$u_*(a_0 \otimes a_1 \otimes \cdots \otimes a_n) = u(a_0)du(a_1) \cdots du(a_n).$$

It may be checked that u_* is a homomorphism of differential graded algebras, and it is uniquely determined. $\qquad\square$

Definition 4.2.8. A *square-zero extension* of A is an algebra R such that $A \cong R/I$ for some ideal I of R with $I^2 = 0$. Two square-zero extensions R, R' of A are *equivalent* if they are isomorphic as algebras via an isomorphism that induces the identity map on A.

The ideal I in the definition is necessarily an A-bimodule since $I^2 = 0$: given an element $a = r + I \in A$ for some $r \in R$, and given $x \in I$, define $ax = rx$ and $xa = xr$. Conversely, every A-bimodule M determines a square-zero extension: let $R = A \oplus M$ and define $(a_1, m_1) \cdot (a_2, m_2) = (a_1 a_2, a_1 m_2 + m_1 a_2)$ for all $a_1, a_2 \in A$ and $m_1, m_2 \in M$. This is called the *trivial extension*. More generally, let $f : A \otimes A \to M$ be a Hochschild 2-cocycle, that is,

$$af(b \otimes c) + f(a \otimes bc) = f(ab \otimes c) + f(a \otimes b)c$$

for all $a, b, c \in A$ as in (1.2.2). Then $R = A \oplus M$ is a ring with multiplication

$$(4.2.9) \qquad (a_1, m_1)(a_2, m_2) = (a_1 a_2, a_1 m_2 + m_1 a_2 + f(a_1 \otimes a_2))$$

for all $a_1, a_2 \in A$ and $m_1, m_2 \in M$. (Associativity is equivalent to the above Hochschild 2-cocycle condition.) This connection between Hochschild 2-cocycles and extensions was refined by Hochschild [**114**] to the following theorem.

Theorem 4.2.10. *Let M be an A-bimodule. Then $\mathrm{HH}^2(A, M)$ is in one-to-one correspondence with equivalence classes of square-zero extensions of A by M.*

Proof. We have already seen that a representative element of $\mathrm{HH}^2(A, M)$, at the chain level, determines a square-zero extension of A by M. A calculation shows that cohomologous cocycles correspond to equivalent square-zero extensions. Conversely, given a square-zero extension R of A by $I = M$, choose an A-bimodule isomorphism $A \oplus I \to R$ that sends each element of I to itself in R and when composed with the quotient map $R \to A$ is the identity on A. This is possible due to the A-bimodule structure of I described earlier. Let $a, b \in A$ and $x, y \in I$, and identify (a, x) and (b, y) with their images in R. Then

$$
\begin{aligned}
(a, x)(b, y) &= (a, 0)(b, 0) + (a, 0)(0, y) + (0, x)(b, 0) + (0, x)(0, y) \\
&= (a, 0)(b, 0) + (1, 0)a(0, y) + (0, x)b(1, 0) \\
&= (a, 0)(b, 0) + (0, ay) + (0, xb),
\end{aligned}
$$

since $I^2 = 0$. Necessarily $(a, 0)(b, 0) = (ab, f(a \otimes b))$ for some function $f : A \otimes A \to I$ that is a Hochschild 2-cocycle. A calculation shows that a different choice of map $A \oplus I \to R$ yields a cohomologous cocycle. $\qquad\square$

Square-zero extensions, noncommutative differential forms, and quasi-free algebras are all related as stated in the following theorem. By a *lifting* of a square-zero extension R of A, we mean an A-bimodule structure on R extending that on I and an A-bimodule homomorphism $A \to R$ that is a section of the quotient map $R \to A$.

Theorem 4.2.11. *The following are equivalent for an algebra A:*

    (i) *A is quasi-free.*

    (ii) *$\Omega^1_{nc}A$ is a projective A^e-module.*

    (iii) *$\mathrm{HH}^2(A, M) = 0$ for all A-bimodules M.*

    (iv) *For any square-zero extension R of A, there is a lifting $A \to R$.*

Proof. We have identified $\Omega^1_{nc}A$ with the first syzygy of A as an A^e-module. If A is quasi-free, there exists a projective resolution $P_{\bullet}$ of A as an A^e-module of length 1, that is, $P_i = 0$ for all $i \geq 2$. Thus the first syzygy module is P_1, a projective module. By Schanuel's Lemma (Lemma A.2.5), any other first syzygy module is projective as well, so in particular $\Omega^1_{nc}A$ is projective. Thus (i) implies (ii). Conversely, if $\Omega^1_{nc}A$ is a projective A^e-module, then the Hochschild dimension of A is at most 1, so A is quasi-free, that is, (ii) implies (i).

By dimension shifting (Theorem A.3.3),

$$
\mathrm{HH}^2(A, M) \cong \mathrm{Ext}^1_{A^e}(\Omega^1_{nc}A, M).
$$

So (ii) implies (iii). Conversely, assume that $\mathrm{Ext}^1_{A^e}(\Omega^1_{nc}A, M) = 0$ for all A-bimodules M, so that every A^e-extension of $\Omega^1_{nc}A$ splits. Map a projective

A^e-module onto $\Omega^1_{nc}A$, and consider the extension of $\Omega^1_{nc}A$ by the kernel of this map. It splits, forcing $\Omega^1_{nc}A$ itself to be projective. So (iii) implies (ii).

Finally, if $\mathrm{HH}^2(A, M) = 0$ for all A-bimodules M, then every square-zero extension splits by Theorem 4.2.10. So if R is a square-zero extension of A, there is a lifting $A \to R$. That is, (iii) implies (iv). Conversely, let R be a square-zero extension of A by an ideal I. Assume there is a lifting $A \to R$. The lifting is a splitting of the sequence of A-bimodules, $0 \to I \to R \to A \to 0$. So if any square-zero extension lifts, then in particular the square-zero extension $A \oplus M$ lifts for any A-bimodule M, and so the sequence $0 \to M \to A \oplus M \to A \to 0$ splits as a sequence of A-bimodules. By Theorem 4.2.10, $\mathrm{HH}^2(A, M) = 0$ for all A-bimodules M. Thus (iv) implies (iii). $\qquad\square$

Schelter [190] proposed a condition similar to (iv) in the theorem as a definition of smoothness for noncommutative algebras.

In case A is commutative, consider the related condition that for every *commutative* square-zero extension R of A, there is a lifting $A \to R$. This is equivalent to smoothness for commutative algebras [223, Section 9.3.1] but is a weaker condition than being quasi-free. In fact, a commutative algebra is smooth in the sense of Definition 4.1.4 if and only if it is smooth in this classical sense [216].

Exercise 4.2.12. Verify that equations (4.2.2) do indeed give $\Omega^1_{nc}A$ the structure of an A-bimodule, that is, (a) the action is well-defined, and (b) it is an A-bimodule action.

Exercise 4.2.13. Verify that the map j in the sequence (4.2.3) is a well-defined A-bimodule map.

Exercise 4.2.14. Show that if A is a commutative algebra, then the A-module $\Omega^1_{com}A$ of Definition 4.2.5 may equivalently be defined as

$$A \otimes A / (ab \otimes c - a \otimes bc + ac \otimes b \mid a, b, c \in A)$$

by considering the left A-module map from $A \otimes A$ to $\Omega^1_{com}A$ that sends $1 \otimes b$ to db for all $b \in A$. This can also be identified with Hochschild homology $\mathrm{HH}_1(A)$.

Exercise 4.2.15. Verify the isomorphism $\Omega^1_{com}A \cong \operatorname{Ker}\pi/(\operatorname{Ker}\pi)^2$ for a commutative algebra A by using Exercise 4.2.14 and the map $A \otimes A \to I$ that sends $a \otimes b$ to $ab \otimes 1 - a \otimes b$ for all $a, b \in A$.

Exercise 4.2.16. Verify the claims from the proof of Theorem 4.2.7:

   (a) $\Omega_{nc}A$ is a differential graded algebra. In particular, check that $d^2 = 0$, the multiplication is associative, and the differential is a graded derivation.

(b) The map u_* is a homomorphism of differential graded algebras, and is unique.

Exercise 4.2.17. Verify that formula (4.2.9) defines an associative multiplication on $A \oplus M$.

4.3. Van den Bergh duality and Calabi-Yau algebras

For some smooth algebras, both commutative and noncommutative, there exists a duality between Hochschild homology and cohomology, an analog of Poincaré duality in geometry. We state this duality in Theorem 4.3.2. A special case is Corollary 4.3.8 for Calabi-Yau algebras, which are defined in this section.

In general, if P is a left A^e-module, then $\mathrm{Hom}_{A^e}(P, A^e)$ is a right A^e-module by setting $(f \cdot (a \otimes b))(p) = bf(p)a$ for all $a, b \in A$, $p \in P$, and $f \in \mathrm{Hom}_{A^e}(P, A^e)$. This gives the structure of a right A^e-module to the Hochschild cohomology space of A with coefficients in A^e, that is, to $\mathrm{HH}^i(A, A^e)$ for each i. This action appears in the statement of Theorem 4.3.2 below.

Definition 4.3.1. An A-bimodule U is *invertible* if there is an A-bimodule V such that $U \otimes_A V \cong A$ and $V \otimes_A U \cong A$ as A-bimodules.

The invertible A-bimodules correspond one-to-one with autoequivalences of the category A-mod, that is, equivalences between the category of A-modules and itself, given by tensor products with the invertible A-bimodules. The identity autoequivalence is given by the invertible A-bimodule A.

An invertible A-bimodule gives a duality between Hochschild homology $\mathrm{HH}_*(A)$ and cohomology $\mathrm{HH}^*(A)$ under some conditions, as stated in the following theorem of Van den Bergh [216, Theorem 1].

Theorem 4.3.2. *Let A be a smooth algebra. Assume that there is a positive integer d for which $\mathrm{HH}^i(A, A^e) = 0$ for all $i \neq d$ and that $U = \mathrm{HH}^d(A, A^e)$ is an invertible A-bimodule. There is an isomorphism of vector spaces*

$$\mathrm{HH}^n(A, M) \cong \mathrm{HH}_{d-n}(A, U \otimes_A M)$$

for all A-bimodules M and $n \in \{0, \ldots, d\}$, and $\mathrm{HH}^n(A, M) = 0$ for $n > d$.

As a consequence of the statement, the integer d in the theorem is equal to the Hochschild dimension $\dim(A)$ of A: by hypothesis, $\mathrm{HH}^d(A, A^e) \neq 0$, implying $d \leq \dim(A)$. By equation (A.2.10), there exists an A^e-module M for which $\mathrm{HH}^{\dim(A)}(A, M) \neq 0$, and so by the last statement of the theorem, $d \geq \dim(A)$.

Definition 4.3.3. If the hypotheses of the theorem are satisfied, we call U the *dualizing bimodule* of A and we say that A has *Van den Bergh duality*.

Proof of Theorem 4.3.2. The proof is a special case of a proof in [**136**], and we choose the same indexing for ease of comparison. Since A is smooth, there is an A^e-projective resolution $(P_\bullet, \delta_\bullet)$ of A such that each P_i is finitely generated and $P_i = 0$ for all $i > n = \dim(A)$. (Note that $d \le n$ since $U = \mathrm{HH}^d(A, A^e) \ne 0$ by hypothesis, and as explained above, it will follow from the proof and equation (A.2.10) that $d = n$.) Let $Q_\bullet \xrightarrow{\varepsilon} M$ be an A^e-projective resolution of M. Since U is invertible, the functor $U \otimes_A -$ is a category equivalence, and so $U \otimes_A Q_\bullet$ is an A^e-projective resolution of $U \otimes_A M$. Let

$$C_{p,q} = \mathrm{Hom}_{A^e}(P_{-p}, Q_q)$$

for all $p \le 0$, $q \ge 0$. We claim that since P_{-p} is finitely generated and projective, for each p, q, there is an isomorphism of vector spaces,

$$(4.3.4) \qquad \mathrm{Hom}_{A^e}(P_{-p}, A^e) \otimes_{A^e} Q_q \xrightarrow{\sim} C_{p,q}$$

given by $f \otimes y \mapsto (x \mapsto (-1)^{pq} f(x)y)$ for all $f \in \mathrm{Hom}_{A^e}(P_{-p}, A^e)$ and $y \in Q_q$. Indeed, in case $P_{-p} = A^e$, this is clearly an isomorphism, as it is if P_{-p} is a free module of finite rank. The claim then follows for any finitely generated projective A^e-module, and thus for each P_{-p}. Note that the tensor product here is taken over A^e instead of over A.

The rest of the proof uses a comparison of two spectral sequences for a bicomplex (Section A.7). Consider the columns in the following diagram, where $n = \dim(A)$:

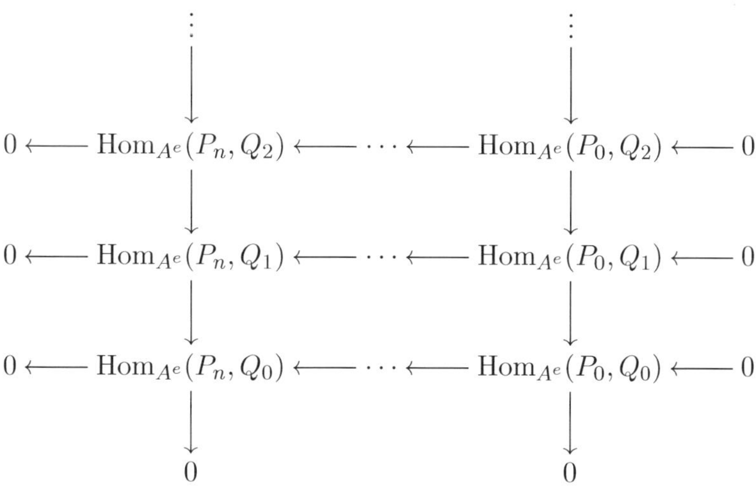

We will apply the two spectral sequences described in Section A.7 to the double complex $C_{\bullet,\bullet}$.

Let E'' denote the first spectral sequence described in Section A.7 for a double complex, given specifically by equation (A.7.5). That is, E'' begins with vertical differentials only, and we write $E''_1 \cong \mathrm{H}''(C)$ and

$E_2'' \cong \mathrm{H}'(\mathrm{H}''(C))$. Since each P_i is projective and $Q_\bullet \xrightarrow{\varepsilon} M$ is exact, the ith column of the above diagram is exact when augmented with $\mathrm{Hom}_{A^e}(P_i, M)$. Thus we find that E_1'' consists of $\mathrm{Hom}_{A^e}(P_\bullet, M)$ in the bottom row, with 0 in all other positions. It follows that E_2'' consists of $\mathrm{HH}^*(A, M)$ in the bottom row, with zero differentials. So the spectral sequence collapses and this is the cohomology of $C_{\bullet,\bullet}$.

For comparison, let E' denote the second spectral sequence described in Section A.7 for a double complex, given on page 233. That is, E' begins with horizontal differentials only, and we write $E_1' \cong \mathrm{H}'(C)$ and $E_2' \cong \mathrm{H}''(\mathrm{H}'(C))$. By hypothesis, E_1' consists of $U \otimes_{A^e} Q_\bullet$ as the $-d$th column and 0 in all other positions. It follows that E_2' is $\mathrm{Tor}^{A^e}_{d-*}(U, M)$, with zero differentials. We claim that $U \otimes_{A^e} Q_j \cong A \otimes_{A^e} (U \otimes_A Q_j)$ as A^e-modules for all j. To see this, first note that it is true for a free A^e-module since $U \xrightarrow{\sim} A \otimes_{A^e} (U \otimes_A A^e)$ via the map $u \mapsto 1 \otimes (u \otimes (1 \otimes 1))$ which has inverse $a \otimes (u \otimes (b \otimes c)) \mapsto caub$. Each Q_j is projective, so is a direct summand of a free module, and this isomorphism preserves such a direct sum. Therefore $\mathrm{Tor}^{A^e}_{d-*}(U, M) \cong \mathrm{HH}_{d-\bullet}(A, U \otimes_A M)$, and this is the cohomology of $C_{\bullet,\bullet}$, completing the proof.
$\qquad\square$

We next show that polynomial rings have Van den Bergh duality.

Example 4.3.5. Let $A = k[x]$. Consider the Koszul resolution (1.1.19) of A as an A^e-module. Apply $\mathrm{Hom}_{k[x]^e}(-, k[x]^e)$ to obtain

$$0 \longleftarrow \mathrm{Hom}_{k[x]^e}(k[x]^e, k[x]^e) \longleftarrow \mathrm{Hom}_{k[x]^e}(k[x]^e, k[x]^e) \longleftarrow 0.$$

Under the isomorphism $\mathrm{Hom}_{k[x]^e}(k[x]^e, k[x]^e) \cong \mathrm{Hom}_k(k, k[x]^e) \cong k[x]^e$, this sequence is equivalent to

$$0 \longleftarrow k[x]^e \longleftarrow k[x]^e \longleftarrow 0,$$

the nonzero map being given by multiplication by $x \otimes 1 - 1 \otimes x$. So $\mathrm{HH}^0(k[x], k[x]^e) = 0$ and $\mathrm{HH}^1(k[x], k[x]^e) \cong k[x]$, an invertible $k[x]$-bimodule. The hypotheses of Theorem 4.3.2 are satisfied and so

$$\mathrm{HH}^n(k[x], M) \cong \mathrm{HH}_{1-n}(k[x], M)$$

for all $k[x]$-bimodules M and $n = 0, 1$. A similar argument applies to a polynomial ring in more indeterminates: in Example 4.1.5 we explained that $k[x_1, \ldots, x_m]$ is smooth. We find that

$$\mathrm{HH}^m(k[x_1, \ldots, x_m], k[x_1, \ldots, x_m]^e) \cong k[x_1, \ldots, x_m],$$

while $\mathrm{HH}^i(k[x_1, \ldots, x_m], k[x_1, \ldots, x_m]^e) = 0$ for $i \neq m$, and so by Theorem 4.3.2,

$$\mathrm{HH}^n(k[x_1, \ldots, x_m], M) \cong \mathrm{HH}_{m-n}(k[x_1, \ldots, x_m], M)$$

for all $k[x_1, \ldots, x_m]$-bimodules M and $n = 0, \ldots, m$.

Definition 4.3.6. A smooth algebra A is *Calabi-Yau* if it has Van den Bergh duality with dualizing bimodule $U \cong A$.

Example 4.3.7. As a consequence of our work in Example 4.3.5, polynomial rings $k[x_1, \ldots, x_m]$ are Calabi-Yau.

Calabi-Yau algebras were first defined by Ginzburg [89] as an analog, in the noncommutative setting, of rings of functions on Calabi-Yau varieties. There is also a notion of a *twisted Calabi-Yau algebra*: in the definition of Calabi-Yau algebra, allow more generally an isomorphism $U \cong A_\sigma$, where σ is an algebra automorphism of A and $A_\sigma = A$ with A^e-module structure twisted by σ on the right, that is, the right action of A on A_σ is given by $a \cdot b = a\sigma(b)$ for all $a, b \in A$. See, e.g., [94].

When A is a Calabi-Yau algebra, we may replace U by A in Theorem 4.3.2 and apply the isomorphism $A \otimes_A M \cong M$ to obtain the following corollary.

Corollary 4.3.8. *If A is a Calabi-Yau algebra of Hochschild dimension d, then*

$$\mathrm{HH}^n(A, M) \cong \mathrm{HH}_{d-n}(A, M)$$

for all A-bimodules M and integers n.

Examples of noncommutative Calabi-Yau algebras include some Sklyanin algebras [163] and some deformed preprojective algebras [7]. Skew group algebras can be Calabi-Yau, and we give details for some of these examples in the next section.

Exercise 4.3.9. Prove that invertible bimodules correspond with autoequivalences of the category A-mod.

Exercise 4.3.10. Let $A = k[x_1, \ldots, x_m]$. Verify the claimed structure of $\mathrm{HH}^i(A, A^e)$ stated in Example 4.3.5.

Exercise 4.3.11. Let $A = k_q[x_1, x_2]$, as defined in Example 3.2.1. Find $\mathrm{HH}^i(A, A^e)$ for each i. Is A Calabi-Yau?

Exercise 4.3.12. Let $A = k\langle x_1, x_2 \rangle / (x_1 x_2 - q x_2 x_1, \ x_1^{n_1}, \ x_2^{n_2})$, as in Example 3.2.6. Find $\mathrm{HH}^i(A, A^e)$ for each i. Is A Calabi-Yau? Cf. Exercise 3.2.12.

4.4. Skew group algebras

Let G be a finite group, let k be a field of characteristic not dividing $|G|$, and let V be a kG-module of finite dimension d as a vector space. In this section we show that the skew group algebra $A = S(V) \rtimes G$ has Van den Bergh duality and determine conditions under which it is Calabi-Yau. In Example 3.5.7, we found expressions for the Hochschild cohomology spaces of A. Here, Van den Bergh duality leads to an alternate computation of the Hochschild cohomology spaces in the case that k is algebraically closed, through knowledge of Hochschild homology. We present this computation, due to Farinati [**77**], in this section.

By our work in Example 4.3.5 (see also Example 3.1.3),

$$\mathrm{HH}^d(S(V), S(V)^e) \cong S(V) \otimes \textstyle\bigwedge^d(V^*)$$

and $\mathrm{HH}^n(S(V), S(V)^e) = 0$ for $n \neq d = \dim_k(V)$. We retain the tensor factor $\bigwedge^d(V^*)$ in this expression even though it is a one-dimensional vector space, as it may have a nontrivial group action. The dual vector space $V^* = \mathrm{Hom}_k(V, k)$ is a kG-module via $({}^g f)(v) = f({}^{g^{-1}} v)$ for all $g \in G$, $f \in V^*$, and $v \in V$, and G acts factorwise on the tensor product $S(V) \otimes \bigwedge^d(V^*)$.

Note that as an $S(V)^e$-module, $A^e \cong S(V)^e \otimes k(G \times G)$, where $S(V)^e$ acts only on the left tensor factor $S(V)^e$. (Map $sg \otimes g's' \mapsto (s \otimes s') \otimes (g, (g')^{-1})$ for all $s, s' \in S(V)$ and $g, g' \in G$.) Applying first the techniques in the proof of Theorem 3.5.2, and then Van den Bergh duality for $S(V)$ as in Example 4.3.5, we obtain isomorphisms for each n:

$$\begin{aligned}
\mathrm{HH}^n(A, A^e) \ &\cong\ \mathrm{HH}^n(S(V), A^e)^G \\
&\cong\ \mathrm{HH}^n(S(V), S(V)^e \otimes k(G \times G))^G \\
&\cong\ \mathrm{HH}_{d-n}(S(V), S(V)^e \otimes k(G \times G))^G.
\end{aligned}$$

This is Hochschild homology, obtained from the tensor product of an $S(V)^e$-projective resolution of $S(V)$, over $S(V)^e$, with $S(V)^e \otimes k(G \times G)$. By definition, the differential is the identity map on the factor $k(G \times G)$ and so tensoring with $k(G \times G)$ commutes with taking homology. That is, paying close attention to placement of parentheses,

$$\mathrm{HH}^n(A, A^e) \cong (\mathrm{HH}_{d-n}(S(V), S(V)^e) \otimes k(G \times G))^G.$$

Now apply Van den Bergh duality again to obtain

$$\mathrm{HH}^n(A, A^e) \cong (\mathrm{HH}^n(S(V), S(V)^e) \otimes k(G \times G))^G.$$

If $n \neq d$, this is 0, while if $n = d$, this is isomorphic to

$$(S(V) \otimes \textstyle\bigwedge^d(V^*) \otimes k(G \times G))^G$$

as a vector space. We claim that the A-bimodule structure on the above space is as it is on $A \otimes \bigwedge^d(V^*)$: if G acts by linear transformations of determinant 1, the isomorphism is given by

$$\begin{aligned}
S(V) \rtimes G \ &\longrightarrow\ (S(V) \otimes k(G \times G))^G \\
s \otimes g \ &\mapsto\ \frac{1}{|G|} \sum_{h \in G} {}^h s \otimes (hg, h^{-1})
\end{aligned}$$

for all $s \in S(V)$ and $g \in G$. In the more general case where G does not act by linear transformations of determinant 1, a similar isomorphism also yields $\mathrm{HH}^d(A, A^e) \cong A \otimes \bigwedge^d(V^*)$. Thus in either case, A has Van den Bergh duality with dualizing bimodule $A \otimes \bigwedge^d(V^*)$, so that the following theorem holds.

Theorem 4.4.1. *Let G be a finite group, let k be a field of characteristic not dividing the order of G, and let V be a kG-module of finite dimension d as a vector space. The skew group ring $S(V) \rtimes G$ has Van den Bergh duality with dualizing bimodule $(S(V) \rtimes G) \otimes \bigwedge^d(V^*)$. If G acts on V via linear transformations of determinant 1, then $S(V) \rtimes G$ is Calabi-Yau.*

Now assume further that k *is algebraically closed*. We use Van den Bergh duality to compute Hochschild cohomology from Hochschild homology. We begin with a computation of Hochschild homology. By Theorem 3.5.2 and further analysis similar to that leading to expression (3.5.6) for Hochschild cohomology, the Hochschild homology space of A in degree n is

$$
\begin{aligned}
\mathrm{HH}_n(S(V) \rtimes G) &\cong \mathrm{HH}_n(S(V), S(V) \rtimes G)_G \\
&\cong \bigoplus_{g \in \overline{G}} \mathrm{HH}_n(S(V), S(V)g)_{C(g)},
\end{aligned}
$$

where $\overline{G}$ is a set of representatives of conjugacy classes of G and $C(g)$ is the centralizer in G of g. Letting V^g be the subspace of V invariant under g and $V_g = \mathrm{Im}(1 - g)$, we have $V = V^g \oplus V_g$ and an $S(V)$-bimodule isomorphism $S(V)g \cong S(V^g) \otimes S(V_g)g$. By Exercise 3.1.10, it now follows that

$$
\begin{aligned}
\mathrm{HH}_n(S(V), S(V)g)_{C(g)} &\cong \mathrm{HH}_n(S(V^g) \otimes S(V_g), S(V^g) \otimes S(V_g)g)_{C(g)} \\
&\cong \left(\bigoplus_{p+q=n} \mathrm{HH}_p(S(V^g)) \otimes \mathrm{HH}_q(S(V_g), S(V_g)g) \right)_{C(g)}.
\end{aligned}
$$

By Exercise 3.1.11 or Theorem 3.3.6, $\mathrm{HH}_*(S(V^g)) \cong S(V^g) \otimes \bigwedge(V^g)$. We compute $\mathrm{HH}_*(S(V_g), S(V_g)g)$. Diagonalize the action of g on V_g, so that V_g has a basis of eigenvectors for g, namely $x_1, \ldots, x_r$ with eigenvalues $\lambda_1, \ldots, \lambda_r$. By the Künneth Theorem (Theorem A.5.2),

$$
\mathrm{HH}_*(S(V_g), S(V_g)g) \cong \bigotimes_{i=1}^{r} \mathrm{HH}_*(k[x_i], k[x_i]g).
$$

We claim that $\mathrm{HH}_0(k[x_i], k[x_i]g) \cong k$ and $\mathrm{HH}_1(k[x_i], k[x_i]g) = 0$ since $\lambda_i \neq 1$ (as $x_i \in V_g$): if we consider the $k[x]^e$-module $k[x]g$, where ${}^g x = \lambda x$ for some scalar λ, then applying $- \otimes_{k[x]^e} k[x]g$ to (1.1.19), we obtain

$$
0 \to k[x]g \to k[x]g \to 0,
$$

where the nonzero map is given by applying $x \otimes 1 - 1 \otimes x$, and thus becomes multiplication by $(1-\lambda)x$. So $\mathrm{HH}_0(k[x], k[x]g) \cong k$ and $\mathrm{HH}_1(k[x], k[x]g) = 0$, as claimed. Note also that $h \in C(g)$ may be simultaneously diagonalized with g. So we now have

$$\mathrm{HH}_n(S(V), S(V)g)_{C(g)} \cong (S(V^g) \otimes \textstyle\bigwedge^n(V^g))_{C(g)}.$$

We are ready to compute Hochschild cohomology via Van den Bergh duality. From our earlier work, using dualizing bimodule $A \otimes \bigwedge^d(V^*)$, we have

$$
\begin{aligned}
\mathrm{HH}^n(S(V) \rtimes G) \;&\cong\; \mathrm{HH}_{d-n}(S(V) \rtimes G, (S(V) \rtimes G) \otimes \textstyle\bigwedge^d(V^*)) \\
&\cong\; \bigoplus_{g \in \overline{G}} (S(V^g) \otimes \textstyle\bigwedge^{d-n}(V^g) \otimes \textstyle\bigwedge^d(V^*))_{C(g)}.
\end{aligned}
$$

Note that

$$
\begin{aligned}
\textstyle\bigwedge^{d-n}(V^g) \otimes \textstyle\bigwedge^d(V^*) \;&\cong\; \textstyle\bigwedge^{d-n}(V^g) \otimes \textstyle\bigwedge^{\dim V^g}((V^g)^*) \wedge \textstyle\bigwedge^{\mathrm{codim}\, V^g}((V_g)^*) \\
&\cong\; \textstyle\bigwedge^{n-\mathrm{codim}\, V^g}((V^g)^*) \otimes \textstyle\bigwedge^{\mathrm{codim}\, V^g}((V_g)^*)
\end{aligned}
$$

as $kC(g)$-modules via the evaluation map pairing V and V^*, noting also that the vector space dimension of $\bigwedge^{n-\mathrm{codim}\, V^g}((V^g)^*)$ is the same as that of $\bigwedge^{d-n}((V^g)^*)$ since $d = \dim_k V$, the vector space dimension of V. In comparison to our calculation of Example 3.5.7, the duality makes the degree shift due to the factor $\bigwedge^{\mathrm{codim}\, V^g}((V^g)^*)$ appear very naturally. For this comparison, we identify the space of orbits under the action of $C(g)$ with its image under the map $s \mapsto \frac{1}{|C(g)|} \sum_{h \in C(g)} {}^h s$.

Exercise 4.4.2. Under the hypotheses at the start of the section, what is the inverse of the dualizing bimodule $(S(V) \rtimes G) \otimes \bigwedge^d(V^*)$?

Exercise 4.4.3. Let $k = \mathbb{C}$, let $G = S_3$, and let V be the 3-dimensional vector space that is a $\mathbb{C}G$-module by permutations of a chosen basis. Use Van den Bergh duality to determine $\mathrm{HH}^*(S(V) \rtimes G)$ and compare to the results of Exercise 3.5.9.

4.5. Connes differential and Batalin-Vilkovisky structure

We introduce the Connes differential on Hochschild homology. For Calabi-Yau algebras, we use this differential in combination with Van den Bergh duality to define a new operation on Hochschild cohomology, called a Batalin-Vilkovisky operator. For finite-dimensional symmetric algebras, we use it in combination with another duality relation to define a Batalin-Vilkovisky operator. The former method can be generalized to some twisted Calabi-Yau algebras, and the latter to some Frobenius algebras. The Connes differential arises in cyclic homology [146].

Recall the contracting homotopy s. for the bar resolution, given by maps $s : A^{\otimes(n+1)} \to A^{\otimes(n+2)}$ for each $n \geq 0$ in (1.1.3) as:

$$s(a_0 \otimes \cdots \otimes a_n) = 1 \otimes a_0 \otimes \cdots \otimes a_n$$

for all $a_0, \ldots, a_n \in A$. (We suppress indices on maps here for legibility in formulas, generically writing s in place of s_{n-1}.) For each n, define a map $t : A^{\otimes(n+1)} \to A^{\otimes(n+1)}$ to be the signed cyclic permutation of tensor factors given by

$$t(a_0 \otimes \cdots \otimes a_n) = (-1)^n a_n \otimes a_0 \otimes \cdots \otimes a_{n-1}$$

for all $a_0, \ldots, a_n \in A$, and define $N : A^{\otimes(n+1)} \to A^{\otimes(n+1)}$ by

$$N = 1 + t + t^2 + \cdots + t^n.$$

Definition 4.5.1. The *Connes differential* $\mathcal{B} : A^{\otimes(n+1)} \to A^{\otimes(n+2)}$ is the map defined for each $n \geq 0$ by

$$\mathcal{B} = (1 - t)sN.$$

Calculations show that $\mathcal{B}$ is a chain map of degree 1, on the bar complex, and so it induces a map on Hochschild homology.

Calabi-Yau algebras. Assume A is a Calabi-Yau algebra of Hochschild dimension d so that Corollary 4.3.8 implies $\mathrm{HH}^n(A) \cong \mathrm{HH}_{d-n}(A)$ for all n. We define an operator

$$\Delta : \mathrm{HH}^n(A) \to \mathrm{HH}^{n-1}(A)$$

to be that induced by the Connes differential $\mathcal{B}$ under this Van den Bergh duality isomorphism. That is, define Δ by the following commuting diagram for all n:

$$
\begin{array}{ccc}
\mathrm{HH}^n(A) & \xrightarrow{\ \Delta\ } & \mathrm{HH}^{n-1}(A) \\
\Big\downarrow{\scriptstyle\cong} & & \Big\downarrow{\scriptstyle\cong} \\
\mathrm{HH}_{d-n}(A) & \xrightarrow{\ \mathcal{B}\ } & \mathrm{HH}_{d-n+1}(A)
\end{array}
$$

Definition 4.5.2. The map Δ defined by the above diagram is the *Batalin-Vilkovisky operator* on the Hochschild cohomology ring of the Calabi-Yau algebra A.

Remark 4.5.3. There is a relationship between the Batalin-Vilkovisky operator Δ and the Gerstenhaber bracket. Let A be a Calabi-Yau algebra and let α, β be homogeneous elements in $\mathrm{HH}^*(A)$. Then

$$[\alpha, \beta] = \Delta(\alpha \smile \beta) - \Delta(\alpha) \smile \beta - (-1)^{|\alpha|}\alpha \smile \Delta(\beta).$$

For a proof, see, e.g., [**89**, Section 9.3], or [**1**] for a more general context.

For a Calabi-Yau algebra A, we say that $(\mathrm{HH}^*(A), \smile, [\ ,\ ], \Delta)$ is a *Batalin-Vilkovisky algebra*. Such a structure on Hochschild cohomology has also been defined similarly for some twisted Calabi-Yau algebras [135].

Symmetric algebras. A different duality can be used to define a Batalin-Vilkovisky structure in the case of a finite-dimensional symmetric algebra (see, e.g., [215]) and for some Frobenius algebras (see [139, 219]). We give a description of this structure for symmetric algebras based on ideas in [139].

Let A be a finite-dimensional *symmetric algebra*, that is, there is a non-degenerate symmetric associative bilinear form $\langle\ ,\ \rangle : A \times A \to k$. Give the dual vector space $D(A) = \mathrm{Hom}_k(A, k)$ the structure of an A^e-module via $(afb)(c) = f(bca)$ for all $a, b, c \in A$ and $f \in D(A)$. (We use the notation $D(A)$ here rather than A^* to avoid confusion.) The form $\langle\ ,\ \rangle$ gives an isomorphism $A \cong D(A)$ as A^e-modules. Let $B(A)$ denote the bar resolution (1.1.4) of A as an A^e-module. For Hochschild homology, we consider the complex $A \otimes_{A^e} B(A)$ with terms $A \otimes_{A^e} B_n(A)$, which are left A-modules by multiplication on the leftmost tensor factor. Note that the corresponding tensor induced A^e-module is

$$A^e \otimes_A (A \otimes_{A^e} B_n(A)) \cong B_n(A).$$

Applying the Nakayama relations (Lemma A.6.1) twice (the first time involving coinduction from k to A and the second time involving induction from A to A^e), we find

$$
\begin{aligned}
\mathrm{Hom}_k(A \otimes_{A^e} B_n(A), k) &\cong \mathrm{Hom}_A(A \otimes_{A^e} B_n(A), D(A)) \\
&\cong \mathrm{Hom}_{A^e}(B_n(A), D(A)) \\
&\cong \mathrm{Hom}_{A^e}(B_n(A), A),
\end{aligned}
$$

the last isomorphism due to the isomorphism $D(A) \cong A$ of A^e-modules. Taking cohomology of the corresponding complexes, we obtain

$$D(\mathrm{HH}_n(A)) \cong \mathrm{HH}^n(A).$$

We use this duality to define a Batalin-Vilkovisky structure on $\mathrm{HH}^*(A)$ just as in the Calabi-Yau setting. Let $\mathcal{B}^t$ denote the transpose of the Connes differential $\mathcal{B}$, that is, $\mathcal{B}^t(f) = f\mathcal{B}$ for all $f \in D(\mathrm{HH}_*(A))$. Then

$$\Delta : \mathrm{HH}^n(A) \to \mathrm{HH}^{n-1}(A)$$

is defined by the following commuting diagram:

$$
\begin{array}{ccc}
\mathrm{HH}^n(A) & \xrightarrow{\ \Delta\ } & \mathrm{HH}^{n-1}(A) \\
\cong \downarrow & & \downarrow \cong \\
D(\mathrm{HH}_n(A)) & \xrightarrow{\ \mathcal{B}^t\ } & D(\mathrm{HH}_{n-1}(A))
\end{array}
$$

Just as in the case of a Calabi-Yau algebra, the relation between the Batalin-Vilkovisky operator and the Gerstenhaber bracket given in Remark 4.5.3 holds. A generalization to Frobenius algebras with diagonalizable Nakayama automorphism may be found in the two papers [**139, 219**] using different methods. More details and examples may be found in these papers.

Exercise 4.5.4. Find a formula for the Connes differential $\mathcal{B}$ of Definition 4.5.1 in cases $n = 1, 2$.

Exercise 4.5.5. Verify that the Connes differential $\mathcal{B}$ is indeed a chain map.

Chapter 5

Algebraic Deformation Theory

In this chapter we will define and examine some types of deformations of associative algebras, focusing on the role played by Hochschild cohomology and its Gerstenhaber bracket. Surveys of further aspects of the theory beyond what we will develop here include that of Schedler [189] on deformations arising in noncommutative geometry and that of Giaquinto [87] for deformation formulas arising from bialgebra actions and for deformations of bialgebras. Here we generally discuss formal deformations, infinitesimal deformations, deformation quantization, and graded deformations. We summarize the theory of Braverman and Gaitsgory [33] for graded deformations of Koszul algebras in particular, and give their proof of the classical Poincaré-Birkhoff-Witt Theorem for Lie algebras as an application. Their theory makes heavy use of Hochschild cohomology. A different homological treatment was given by Polishchuk and Positselski [175, 176].

Let A be an algebra over a field k.

5.1. Formal deformations

Let t be an indeterminate. Denote by $A[[t]]$ the algebra whose elements are formal power series $\sum_{i \geq 0} a_i t^i$ with coefficients $a_i \in A$. Multiplication is given by the Cauchy product:

$$(5.1.1) \qquad \left(\sum_{i \geq 0} a_i t^i\right)\left(\sum_{j \geq 0} b_j t^j\right) = \sum_{l \geq 0} \left(\sum_{i+j=l} a_i b_j\right) t^l.$$

The algebra $A[[t]]$ is a $k[[t]]$-module under multiplication via the identification of k with the subalgebra $k \cdot 1$ of A. We are interested in new associative algebra structures on this $k[[t]]$-module $A[[t]]$ for which the quotient by the ideal generated by t is isomorphic to A. Precisely, we have the following definition.

Definition 5.1.2. A *formal deformation* $(A_t, *)$ of A (also called a *deformation of A over $k[[t]]$*) is an associative k-bilinear multiplication $*$ on the $k[[t]]$-module $A[[t]]$, such that in the quotient by the ideal generated by t, the multiplication corresponds to that on A; this multiplication is required to be determined by a multiplication on elements of A and extended to $A[[t]]$ by the Cauchy product rule (5.1.1). Define similarly a *deformation of A over $k[t]$* or *over $k[t]/(t^n)$*.

To be clear, we intend in the definition a *new* multiplication on elements of A, taking values in the $k[[t]]$-module $A[[t]]$, and extended to a multiplication on $A[[t]]$ by Cauchy products. We give an explicit description via a multiplication formula (5.1.4) below.

Remark 5.1.3. There are more general types of deformations. Let R be any commutative augmented k-algebra (such as $R = k[[t_1, \ldots, t_m]]$), with augmentation map $\varepsilon : R \to k$, that is complete with respect to the $(\mathrm{Ker}\,\varepsilon)$-adic topology. A *deformation of A over R* is an associative R-algebra A_R that is isomorphic to the completed tensor product of A with R as an R-module and for which there is a k-algebra isomorphism $A_R/(\mathrm{Ker}\,\varepsilon) \xrightarrow{\sim} A$. Often it is assumed that A_R is free as an R-module, or more generally that A_R is flat as an R-module (for a *flat deformation*). In this book we will only consider deformations over $k[[t]]$, $k[t]$, or $k[t]/(t^n)$ as in Definition 5.1.2, and we will not need this more general definition.

Any multiplication $*$ as in Definition 5.1.2 is determined by products of pairs of elements of A: for $a, b \in A$, write

$$(5.1.4) \qquad a * b = ab + \mu_1(a \otimes b)t + \mu_2(a \otimes b)t^2 + \mu_3(a \otimes b)t^3 + \cdots ,$$

where ab is the usual product in A and $\mu_1, \mu_2, \mu_3, \ldots$ are functions from $A \otimes A$ to A giving the coefficients of $t, t^2, t^3, \ldots$ as indicated. Sometimes we write $\mu_0(a \otimes b) = ab$ so that the formula becomes

$$a * b = \sum_{i \geq 0} \mu_i(a \otimes b)t^i.$$

The functions μ_i are necessarily k-linear. We call μ_i the *ith multiplication map* of the deformation. We sometimes denote the deformation $(A_t, *)$ by (A_t, μ_t), writing

$$\mu_t = \mu_0 + \mu_1 t + \mu_2 t^2 + \cdots$$

as a function from $A \otimes A$ to A_t. When needed, we extend μ_t to be a function on the tensor product of the $k[[t]]$-module $A[[t]]$ with itself, completed so that expressions with formal power series as tensor factors make sense:

$$\left(\sum_{i\geq 0} a_i t^i\right) \otimes_{k[[t]]} \left(\sum_{j\geq 0} b_j t^j\right) = \sum_{n\geq 0}\left(\sum_{i+j=n} a_i \otimes b_j\right) t^n.$$

In this case, the value of μ_t on an element $\sum_{n\geq 0} c_n t^n$, where $c_n \in A \otimes A$ for all n, is taken to be $\sum_{n\geq 0}(\sum_{i+j=n} \mu_i(c_j))t^n$. Some authors write $\widehat{\otimes}$ in place of $\otimes$ to denote a completed tensor product.

We will next derive some properties of the multiplication maps μ_i. Assume that $(A, *)$ is a formal deformation of A as in Definition 5.1.2, so that the multiplication $*$ is associative and given by maps μ_i as above. Calculating each side of the equation $(a * b) * c = a * (b * c)$ for all $a, b, c \in A$ by using formula (5.1.4), we find that

$$\begin{aligned}
(a * b) * c \;=\;& abc + (\mu_1(ab \otimes c) + \mu_1(a \otimes b)c)t \\
&+ (\mu_2(ab \otimes c) + \mu_1(\mu_1(a \otimes b) \otimes c) + \mu_2(a \otimes b)c)t^2 + \cdots
\end{aligned}$$

while

$$\begin{aligned}
a * (b * c) \;=\;& abc + (\mu_1(a \otimes bc) + a\mu_1(b \otimes c))t \\
&+ (\mu_2(a \otimes bc) + \mu_1(a \otimes \mu_1(b \otimes c)) + a\mu_2(b \otimes c))t^2 + \cdots .
\end{aligned}$$

Equating coefficients of t, it follows that

$$(5.1.5) \qquad \mu_1(ab \otimes c) + \mu_1(a \otimes b)c = \mu_1(a \otimes bc) + a\mu_1(b \otimes c)$$

for all $a, b, c \in A$. Comparing with equation (1.2.2), we see that μ_1 may be identified with a Hochschild 2-cocycle on the bar resolution of A. That is, $d_3^*(\mu_1) = 0$, where d_3 is the differential on the bar resolution (1.1.4), under the identification $\mathrm{Hom}_{A^e}(A^{\otimes 5}, A) \cong \mathrm{Hom}_k(A^{\otimes 3}, A)$ given by (1.1.11) with $M = A$ and $n = 3$. Equating coefficients of t^2, we have

$$\begin{aligned}
&\mu_1(\mu_1(a \otimes b) \otimes c) - \mu_1(a \otimes \mu_1(b \otimes c)) \\
&= a\mu_2(b \otimes c) - \mu_2(ab \otimes c) + \mu_2(a \otimes bc) - \mu_2(a \otimes b)c
\end{aligned}$$

for all $a, b, c \in A$. Comparing with Definition 1.4.1, the left side of the above equation is the circle product $\mu_1 \circ \mu_1$, which in characteristic not 2 is half of the Gerstenhaber bracket $[\mu_1, \mu_1]$, applied to $a \otimes b \otimes c$. The right side is $d_3^*(\mu_2)$ applied to $a \otimes b \otimes c$, where d_3 again denotes the differential on the bar resolution. Thus associativity of $*$ implies that, at the chain level, if $\mathrm{char}(k) \neq 2$,

$$(5.1.6) \qquad\qquad\qquad [\mu_1, \mu_1] = 2d_3^*(\mu_2).$$

If $\mathrm{char}(k) = 2$, we must express the condition as $\mu_1 \circ \mu_1 = d_3^*(\mu_2)$, or $\mathrm{Sq}(\mu_1) = d_3^*(\mu_2)$, where Sq is the divided square operation of Definition 1.4.2. A similar analysis shows that

$$(5.1.7) \qquad\qquad\qquad [\mu_1, \mu_2] = d_3^*(\mu_3)$$

and more generally that

$$(5.1.8) \quad \sum_{j=1}^{i-1}(\mu_j(\mu_{i-j}(a \otimes b) \otimes c) - \mu_j(a \otimes \mu_{i-j}(b \otimes c))) = d_3^*(\mu_i)(a \otimes b \otimes c)$$

for all $a, b, c \in A$ and $i \geq 2$. Alternatively we may apply Lemma 1.4.5(ii), with $\pi = \mu_0$, to rewrite the differential on $\mathrm{Hom}_k(A^{\otimes *}, A)$ as $(-1)[-, \mu_0]$ and change the indexing of the above sum to include $j = 0$ and $j = i$. Then equation (5.1.8) becomes

$$(5.1.9) \quad \sum_{j=0}^{i}(\mu_j(\mu_{i-j}(a \otimes b) \otimes c) - \mu_j(a \otimes \mu_{i-j}(b \otimes c))) = 0.$$

We have thus found that there are infinitely many conditions that must be satisfied, one for each i, given by equation (5.1.8) or (5.1.9), in order that $*$ be an associative multiplication on $A[[t]]$. We call the left side of equation (5.1.8), viewed as a function on $A^{\otimes 3}$, the $(i-1)st$ *obstruction*. Considering the right side of equation (5.1.8), as a consequence of associativity of $*$, we find that the $(i-1)$st obstruction must represent the element 0 in the Hochschild cohomology space $\mathrm{HH}^3(A)$. It follows that a set of k-linear functions $\mu_i : A \otimes A \to A$ potentially defining an associative product $*$ on $A[[t]]$ via formula (5.1.4) is prevented from doing so if any of the obstructions is nonzero in $\mathrm{HH}^3(A)$.

We give several examples next. Each example is a deformation over $k[[t]]$ or $k[t]$ which is then specialized to a particular value of the parameter t where possible. Such specialization is a common source of new algebras arising from deformations of given algebras.

Example 5.1.10. Let $A = k[x, y]$. Define a multiplication $*$ on $A[t]$ or on $A[[t]]$ by

$$x^i * x^j = x^{i+j}, \quad y^i * y^j = y^{i+j}, \quad x^i * y^j = x^i y^j, \quad y * x = xy + t$$

for all i, j, and extend by requiring $*$ to be associative. This gives rise to a deformation of A over $k[[t]]$ or $k[t]$. Over $k[t]$, we may substitute $t = 1$ (that is, take the quotient by the ideal $(t - 1)$) to obtain the *Weyl algebra*

$$A_1 = k\langle x, y\rangle/(yx - xy - 1).$$

This may be done more generally to obtain the Weyl algebra A_m on $2m$ generators, that is, the algebra with generators $x_1, \ldots, x_m, y_1, \ldots, y_m$ and relations $x_i x_j = x_j x_i$, $y_i y_j = y_j y_i$, and $y_i x_j = x_j y_i + \delta_{i,j}$ for all i, j.

Example 5.1.11. Let $k = \mathbb{C}$ for this example, so that we may take advantage of convergence of the exponential function. Let $A = \mathbb{C}[x, y]$. Define a multiplication $*$ on $A[[t]]$ by

$$x^i * x^j = x^{i+j}, \quad y^i * y^j = y^{i+j}, \quad x^i * y^j = x^i y^j,$$

$$y * x = xy\left(1 + t + \frac{1}{2!}t^2 + \frac{1}{3!}t^3 + \cdots\right) = xy \cdot \exp(t)$$

for all i, j, and extend by requiring $*$ to be associative. This gives rise to a formal deformation of A. Let $t_0 \in \mathbb{C}$ and substitute $t = t_0$ in the subalgebra of $A[[t]]$ generated by x and y. Let $q = \exp(t_0)$. The resulting algebra is the quantum plane

$$\mathbb{C}_q[x, y] = \mathbb{C}\langle x, y\rangle / (yx - qxy).$$

This is done more generally to obtain a skew polynomial ring $\mathbb{C}_{\mathbf{q}}[x_1, \ldots, x_n]$ as defined in Example 3.2.1.

For the next example, recall that a *Lie algebra* $\mathfrak{g}$ is a vector space with a linear map $[-, -] : \mathfrak{g} \otimes \mathfrak{g} \to \mathfrak{g}$ for which $[x, x] = 0$ for all $x \in \mathfrak{g}$ and the *Jacobi identity* holds:

$$(5.1.12) \qquad [x, [y, z]] + [y, [z, x]] + [z, [x, y]] = 0$$

for all $x, y, z \in \mathfrak{g}$. If the characteristic of k is not 2, the condition $[x, x] = 0$ for all x is equivalent to the condition that $[x, y] = -[y, x]$ for all x, y, and in any case implies it, that is, the bracket is antisymmetric. The *universal enveloping algebra* of $\mathfrak{g}$ is

$$U(\mathfrak{g}) = T(\mathfrak{g})/(x \otimes y - y \otimes x - [x, y] \mid x, y \in \mathfrak{g}).$$

Example 5.1.13. Let k be a field of characteristic not 2. Let $A = k[e, f, h]$. Define a multiplication on $A[t]$ by

$$f * e = ef - ht, \quad h * e = eh + 2et, \quad h * f = fh - 2ft,$$

and all products of monomials in alphabetical order are as in A, for example, $e^i f^j * h^m = e^i f^j h^m$ for all i, j, m. Extend by requiring $*$ to be associative. Substitute $t = 1$ to obtain the universal enveloping algebra of the Lie algebra $\mathfrak{sl}_2$:

$$U(\mathfrak{sl}_2) = k\langle e, f, h\rangle / (fe - ef + h, \ he - eh - 2e, \ hf - fh + 2f).$$

The last example above is essentially generalized below in Section 5.5 in a restatement of the classical Poincaré-Birkhoff-Witt (PBW) Theorem. In the proof of Theorem 5.5.5 we will see that the universal enveloping algebra $U(\mathfrak{g})$ of a Lie algebra $\mathfrak{g}$ is a particular type of deformation of a polynomial ring, termed a PBW deformation due to its appearance in that theorem.

Exercise 5.1.14. Derive the formula (5.1.7) by equating coefficients of t^3 in the equation $(a * b) * c = a * (b * c)$.

Exercise 5.1.15. Identify the Hochschild cocycle μ_1 inherent in Example 5.1.10. What is μ_2?

Exercise 5.1.16. Identify the Hochschild cocycle μ_1 inherent in Example 5.1.11. What is μ_2?

Exercise 5.1.17. Let B be the *Jordan plane*:

$$B = k\langle x, y\rangle/(yx - xy - x^2).$$

One may realize B as a specialization to $t = 1$ of a deformation of the polynomial ring $k[x, y]$ over $k[t]$. What is the corresponding Hochschild 2-cocycle μ_1?

Exercise 5.1.18. Let A_1 be the Weyl algebra of Example 5.1.10. Show that $\mathrm{HH}^0(A_1) \cong k$ and $\mathrm{HH}^n(A) = 0$ for all $n > 0$. (Use the Koszul resolution K for A_1: set $K = A_1 \otimes \bigwedge(V) \otimes A_1$, where V is a vector space with basis $\{x, y\}$. The differentials are as in the Koszul resolution for $k[x, y]$. Check that K is indeed a free resolution of A_1 as an A_1^e-module.)

Exercise 5.1.19. Let A_m be the Weyl algebra on $2m$ generators of Example 5.1.10. Use Exercise 5.1.18 and Theorem 3.1.2 to show that $\mathrm{HH}^0(A_m) \cong k$ and $\mathrm{HH}^n(A_m) = 0$ for all $n > 0$.

5.2. Infinitesimal deformations and rigidity

An algebra is rigid if it cannot be deformed, and we make this notion precise in this section. We saw in the last section that every formal deformation has associated to it a Hochschild 2-cocycle. If Hochschild cohomology vanishes in degree 2, the algebra is necessarily rigid, as stated in Proposition 5.2.8 below. Otherwise we examine more closely the cocycles that can arise, starting with the next definition.

Definition 5.2.1. A k-linear function $\mu_1 : A \otimes A \to A$ is an *infinitesimal deformation* if (5.1.5) holds, that is, μ_1 is a Hochschild 2-cocycle. Its *primary* (or *first*) *obstruction vanishes* if $[\mu_1, \mu_1]$ is a coboundary in the space $\mathrm{Hom}_k(A^{\otimes 3}, A)$. (In characteristic 2, instead require $\mathrm{Sq}(\mu_1)$ to be a coboundary, where Sq is given in Definition 1.4.2.) The function μ_1 is *integrable* if there is a formal deformation (A_t, μ_t) for which μ_1 is the first multiplication map.

If μ_1 is an infinitesimal deformation, then it defines an associative algebra structure on $A[t]/(t^2)$, that is, it defines a deformation of A over $k[t]/(t^2)$, the ring of dual numbers: let

$$a * b = ab + \mu_1(a \otimes b)t$$

for all $a, b \in A$ and extend $k[t]/(t^2)$-bilinearly to $A[t]/(t^2)$. (For notational convenience, we identify t with its image under the quotient map from $k[t]$ to $k[t]/(t^2)$.) Conversely, a deformation of A over $k[t]/(t^2)$ is determined by the coefficient of t in the above equation, which necessarily satisfies equation (5.1.5). Sometimes when we refer to an infinitesimal deformation, we mean this corresponding algebra structure on $A[t]/(t^2)$.

In Section 7.1, we will more generally define infinitesimal n-deformations for any $n \geq 2$; our infinitesimal deformations here will be called infinitesimal 2-deformations there. We will show in Theorem 7.1.8 that a Hochschild n-cocycle corresponds to an infinitesimal n-deformation, generalizing the connection we have already seen among infinitesimal deformations, Hochschild 2-cocycles, and deformations over $k[t]/(t^2)$.

We will next define a notion of equivalence of formal deformations. We will see that the Hochschild 2-cocycles corresponding to equivalent deformations are cohomologous.

Definition 5.2.2. Two formal deformations (A_t, μ_t), (A'_t, μ'_t) are *equivalent* if there is a $k[[t]]$-linear function $\phi_t : A_t \to A'_t$ determined by its values on A given in the form

$$(5.2.3) \qquad \phi_t(a) = a + \phi_1(a)t + \phi_2(a)t^2 + \cdots$$

for k-linear functions $\phi_i : A \to A$ such that

$$(5.2.4) \qquad \phi_t \mu_t(a \otimes b) = \mu'_t(\phi_t(a) \otimes \phi_t(b))$$

for all $a, b \in A$, and extended to A_t. A formal deformation (A_t, μ_t) is *trivial* if it is equivalent to $A[[t]]$.

In the definition, we extend (5.2.3) to all elements of A_t by defining $\phi_t(\sum_{i \geq 0} a_i t^i)$ to be $\sum_{n \geq 0}(\sum_{i+j=n} \phi_i(a_j))t^n$, where ϕ_0 is the identity map on A.

Note that the function ϕ_t is an isomorphism of algebras by its definition above. (Any function of the given form is necessarily invertible as a formal power series; see for example the proof of Lemma 5.2.6 below.) We view each function $\phi_i \in \mathrm{Hom}_k(A, A)$ as a 1-cochain on the Hochschild complex via the identification $\mathrm{Hom}_k(A, A) \cong \mathrm{Hom}_{A^e}(A^{\otimes 3}, A)$. This view gives meaning to the expression $d_2^*(\phi_1)$ in the statement of the next lemma.

Lemma 5.2.5. *If (A_t, μ_t), (A'_t, μ'_t) are equivalent via a function ϕ_t as in Definition 5.2.2, then $\mu'_1 = \mu_1 - d_2^*(\phi_1)$. In particular, if (A_t, μ_t) is trivial, then μ_1 is a coboundary.*

Proof. Expanding equation (5.2.4), we have

$$ab + (\phi_1(ab) + \mu_1(a \otimes b))t + \cdots = ab + (\mu'_1(a \otimes b) + \phi_1(a)b + a\phi_1(b))t + \cdots$$

for all $a, b \in A$. Equating coefficients of t, and using the techniques of Section 1.2 to evaluate $(d_2^*(\phi_1))(a \otimes b)$, we see that $\mu'_1 = \mu_1 - d_2^*(\phi_1)$ as claimed. If (A_t, μ_t) is trivial, then it is equivalent to (A'_t, μ'_t) with $\mu'_0 = \mu_0$ and $\mu'_i = 0$ for all $i > 0$ via some function ϕ_t. Thus $\mu_1 = d_2^*(\phi_1)$. $\square$

In the next lemma, we will view functions $\mu'_n \in \mathrm{Hom}_k(A \otimes A, A)$ as 2-cochains on the Hochschild complex under the standard identification of $\mathrm{Hom}_k(A \otimes A, A)$ with $\mathrm{Hom}_{A^e}(A^{\otimes 4}, A)$. Given a sequence $\mu'_1, \mu'_2, \ldots$ of such cochains, we will refer in the next lemma to the first nonvanishing cochain, by which we mean μ'_n such that $\mu'_n \neq 0$ and $\mu'_i = 0$ for all $i < n$.

Lemma 5.2.6. *A nontrivial formal deformation (A_t, μ_t) of A is equivalent to a formal deformation (A'_t, μ'_t) with the property that the first nonvanishing cochain μ'_n is a Hochschild 2-cocycle that is not a coboundary.*

Proof. Suppose (A_t, μ_t) is a formal deformation of A that is not equivalent to a formal deformation (A'_t, μ'_t) for which the first nonvanishing cochain is not a coboundary. We will show that (A_t, μ_t) is necessarily trivial.

It follows from our assumption that the first nonvanishing cochain of (A_t, μ_t) is a coboundary. Write

$$\mu_t(a \otimes b) = ab + \mu_n(a \otimes b)t^n + \mu_{n+1}(a \otimes b)t^{n+1} + \cdots$$

for all $a, b \in A$, where $\mu_n = d_2^*(\beta)$ for some $\beta \in \mathrm{Hom}_k(A, A)$. Let

$$\phi_t(a) = a + \beta(a)t^n$$

for all $a \in A$, and note that

$$\phi_t^{-1}(a) = a - \beta(a)t^n + \beta^2(a)t^{2n} - \beta^3(a)t^{3n} + \cdots,$$

where $\beta^2(a) = \beta(\beta(a))$, $\beta^3(a) = \beta(\beta(\beta(a)))$, etc. Set $\mu'_t = \phi_t\mu_t(\phi_t^{-1} \otimes \phi_t^{-1})$. A calculation shows that μ'_t is the multiplication map for a deformation (A'_t, μ'_t) which by definition is equivalent to (A_t, μ_t). Since $\mu_n = d_2^*(\beta)$, there is some function μ'_{n+1} such that

$$\mu'_t(a \otimes b)$$
$$= \phi_t\mu_t((a - \beta(a)t^n + \cdots) \otimes (b - \beta(b)t^n + \cdots))$$
$$= \phi_t(ab + (\mu_n(a \otimes b) - a\beta(b) - \beta(a)b)t^n + \cdots)$$
$$= ab + (\beta(ab) + \mu_n(a \otimes b) - a\beta(b) - \beta(a)b)t^n + \mu'_{n+1}(a \otimes b)t^{n+1} + \cdots$$
$$= ab + \mu'_{n+1}(a \otimes b)t^{n+1} + \cdots.$$

By assumption, μ'_{n+1} must be a coboundary, and so similarly (A'_t, μ'_t) is equivalent to (A''_t, μ''_t), where

$$\mu''_t(a \otimes b) = ab + \mu''_{n+2}(a \otimes b)t^{n+2} + \cdots$$

via a function ϕ'_t with $\phi'_t(a) = a + \beta'(a)t^{n+1}$. Continuing in this fashion, we may let Φ_t be the function

$$\Phi_t(a) = a + \Phi_n(a)t^n + \Phi_{n+1}(a)t^{n+1} + \cdots$$

defined as the composition of all $\ldots, \phi_t'', \phi_t', \phi_t$. This composition is well-defined: the coefficient function of each power of t in Φ_t is a finite polynomial in $\beta, \beta', \beta'', \ldots$. For example, if $n = 1$, then

$$\Phi_t(a) = a + \beta(a)t + \beta'(a)t^2 + (\beta''(a) + \beta'(\beta(a)))t^3 + \cdots .$$

Applying Φ_t, we see that (A_t, μ_t) is trivial (see Definition 5.2.2) as claimed. Note that for any given power of t, its coefficient function in the resulting equivalent deformation only involves a composition of finitely many such equivalences, and we find that the coefficient function is indeed 0. $\qquad\square$

We now focus on algebras having no such deformations.

Definition 5.2.7. An algebra A is *rigid* if it has no nontrivial formal deformations.

As an immediate consequence of Lemma 5.2.6, we have the following proposition.

Proposition 5.2.8. *If* $\mathrm{HH}^2(A) = 0$, *then* A *is rigid.*

Examples of algebras A for which $\mathrm{HH}^2(A) = 0$ include separable algebras (as a consequence of their definition), universal enveloping algebras of complex semisimple Lie algebras [**125**, Section XVIII.3], Weyl algebras (see Exercise 5.1.19), and tensor algebras over a field (see Example 4.1.8). Thus all of these algebras are rigid. We point out however that universal enveloping algebras of semisimple Lie algebras do have bialgebra deformations [**72**], yielding one approach to some types of quantum groups. It is bialgebra cohomology that governs these deformations in an analogous theory [**87**].

There is a related statement for deformations of smooth finitely generated commutative algebras. These algebras have no nontrivial *commutative* formal deformations since the Harrison cohomology in degree 2 is 0. Indeed, see Remark 3.3.7; a calculation shows that any nontrivial commutative deformation would give rise to a Harrison 2-cocycle that is not a coboundary, and no such cocycle exists. These algebras can however have *noncommutative* deformations, as we saw in the examples of Section 5.1.

The converse of Proposition 5.2.8 is false. Some counterexamples were found by Gerstenhaber and Schack [**83**].

Exercise 5.2.9. For Examples 5.1.10 and 5.1.11, determine the structure of the corresponding deformations over $k[t]/(t^2)$.

Exercise 5.2.10. Justify the claim that a function ϕ_t defined as in (5.2.3) is always an isomorphism.

Exercise 5.2.11. Let $A = k\langle x, y\rangle/(yx - xy - 1)$, a Weyl algebra. Verify directly the claim that A is rigid. (Use Proposition 5.2.8 and the result of Exercise 5.1.18.)

Exercise 5.2.12. Verify the claim that smooth finitely generated commutative algebras have no nontrivial *commutative* formal deformations. First show that for any commutative algebra, a nontrivial commutative formal deformation gives rise to a Harrison 2-cocycle that is not a coboundary (see Definition 1.6.1 of Harrison cohomology). Then apply Remark 3.3.7.

5.3. Maurer-Cartan equation and Poisson bracket

For this section, we assume the characteristic of the field k is not 2. We will give another interpretation of the conditions (5.1.9) for a deformation, deriving the Maurer-Cartan equation (5.3.1) below. We will in particular examine the first obstruction in the case of commutative algebras.

Let (A_t, μ_t) be a formal deformation of A. Write $\mu_t = \sum_{i \geq 0} \mu_i t^i$ as before. The conditions (5.1.9) for all $i \geq 2$, together with associativity of μ_0 and the assumption that μ_1 is a Hochschild 2-cocycle, may be combined and reinterpreted as stating that

$$[\mu_t, \mu_t] = 0.$$

Set

$$\mu_t = \mu_0 + \mu'.$$

Note that associativity of A is the condition that $[\mu_0, \mu_0] = 0$. Since μ_0 and μ' may be viewed as 2-cochains, the graded commutativity of the bracket gives $[\mu', \mu_0] = [\mu_0, \mu']$. So the equation $[\mu_t, \mu_t] = 0$ is equivalent to

$$2[\mu_0, \mu'] + [\mu', \mu'] = 0.$$

By Lemma 1.4.5(ii) with $\pi = \mu_0$, the differential on the Hochschild complex is $(-1)[-, \mu_0]$, so the above equation is equivalent to

$$(5.3.1) \qquad\qquad -d^*(\mu') + \frac{1}{2}[\mu', \mu'] = 0.$$

With appropriate sign conventions, this is the *Maurer-Cartan equation* (also called the *Berikashvili equation*). We have shown that the deformed multiplication μ_t is associative if and only if μ' satisfies the Maurer-Cartan equation. Focusing on μ_1 and μ_2 (giving rise to coefficients of t and t^2 in expression (5.1.4)), this equation implies μ_1 is a Hochschild 2-cocycle and $d_3^*(\mu_2) = \frac{1}{2}[\mu_1, \mu_1]$, as we saw before in equation (5.1.6).

We next look more closely at commutative algebras and their potentially noncommutative deformations.

Definition 5.3.2. A *Poisson algebra* is a commutative associative algebra A that is also a Lie algebra under a binary operation $\{\ ,\ \}$ for which

$$\{a, bc\} = \{a, b\}c + b\{a, c\}$$

for all $a, b, c \in A$. We call $\{\ ,\ \}$ a *Poisson bracket*.

Note that a Poisson bracket $\{\ ,\ \}$ is by definition a Lie bracket, that is, it is antisymmetric and satisfies the Jacobi identity (5.1.12). One source of Poisson brackets on commutative algebras is Hochschild cohomology: let $\mu_1 : A \otimes A \to A$ be a Hochschild 2-cocycle on A whose primary obstruction vanishes *at the chain level*, that is, $[\mu_1, \mu_1] = 0$ as a cochain on the bar resolution. Let

$$(5.3.3) \qquad \{a, b\} = \frac{1}{2}(\mu_1(a \otimes b) - \mu_1(b \otimes a))$$

for all $a, b \in A$. A calculation using commutativity of A and the Hochschild 2-cocycle condition (5.1.5) shows that $\{\ ,\ \}$ is a Poisson bracket on A. Conversely, a calculation shows that a Poisson bracket $\{\ ,\ \}$ is a Hochschild 2-cocycle.

Definition 5.3.4. Let A be a Poisson algebra. A *deformation quantization* of A is a formal deformation $(A_t, *)$ for which

$$a * b - b * a = \{a, b\}t \pmod{t^2}$$

for all $a, b \in A$.

Example 5.3.5. The Weyl algebra A_1 of Example 5.1.10 is the specialization of a deformation quantization of the polynomial ring $k[x, y]$ with Poisson bracket $\{x, y\} = -1$. Similar statements hold for Weyl algebras in more indeterminates, and for the quantum plane and other skew polynomial rings described in Example 5.1.11.

We will discuss in Section 7.6 some general conditions under which it is known that a Poisson algebra has a deformation quantization (see Theorem 7.6.3). In the noncommutative setting, an infinitesimal deformation μ_1 of a not necessarily commutative algebra A is sometimes regarded as a noncommutative Poisson structure on A when its primary obstruction vanishes.

Exercise 5.3.6. Verify the claim that the condition $[\mu_t, \mu_t] = 0$ is equivalent to conditions (5.1.9) for all $i \geq 2$ together with associativity of μ_0 and the property that μ_1 is a Hochschild 2-cocycle.

Exercise 5.3.7. Verify that a Poisson bracket on a commutative algebra is a Hochschild 2-cocycle.

Exercise 5.3.8. Verify that formula (5.3.3) defines a Poisson bracket on a commutative algebra A.

Exercise 5.3.9. Find the Poisson bracket on the polynomial ring $\mathbb{C}[x, y]$ arising from viewing Example 5.1.11 as a deformation quantization.

5.4. Graded deformations

Let A be an $\mathbb{N}$-graded algebra. Let t be an indeterminate, and assign t the degree 1. Consider the resulting $\mathbb{N}$-graded algebra $A[t]$, in which $|at^i| = |a|+i$ for all homogeneous $a \in A$ and $i \in \mathbb{N}$. A *graded deformation* of A over $k[t]$ is a deformation A_t of A over $k[t]$ that is also a graded algebra. We will translate this condition to one on the multiplication maps μ_i of equation (5.1.4), viewed as elements of $\mathrm{Hom}_k(A \otimes A, A)$. First note that the grading on A induces a grading on this space of homomorphisms. For $m \in \mathbb{Z}$,

$$\mathrm{Hom}_k(A \otimes A, A)_m = \{f \in \mathrm{Hom}_k(A \otimes A, A) \mid f((A \otimes A)_i) \subseteq A_{i+m} \text{ for all } i\}.$$

For a graded deformation, since $|t| = 1$, each function μ_j of equation (5.1.4) must be homogeneous of degree $-j$, that is, $\mu_j \in \mathrm{Hom}_k(A \otimes A, A)_{-j}$. An *$i$th-level graded deformation of A* is a deformation A_i of A over $k[t]/(t^{i+1})$ that is also a graded algebra. It has underlying $k[t]/(t^{i+1})$-module $A[t]/(t^{i+1})$ and multiplication given on elements $a, b \in A$ by

$$a * b = ab + \mu_1(a \otimes b)t + \mu_2(a \otimes b)t^2 + \cdots + \mu_i(a \otimes b)t^i$$

for k-linear functions $\mu_j : A \otimes A \to A$ that are homogeneous of degree $-j$. (For notational convenience, we identify t with its image under the quotient map from $k[t]$ to $k[t]/(t^{i+1})$.) A first-level graded deformation of A then corresponds to an infinitesimal deformation of A that is graded. We say that an $(i-1)$st-level graded deformation A_{i-1} of A *lifts* to an ith-level graded deformation A_i of A if the jth multiplication maps of A_{i-1} and A_i agree for all $j \leq i-1$. Just as in Section 5.1, a calculation shows that if A_{i-1} lifts to A_i, then equation (5.1.9) holds. Accordingly, just as in Section 5.1, we say that the obstruction to existence of such a lifting is the left side of equation (5.1.8), which must define a coboundary on the Hochschild complex as a condition for lifting.

Braverman and Gaitsgory [**33**] proved the following proposition regarding lifting graded deformations. We will need to use a grading on Hochschild cohomology induced by that on the Hochschild complex which generalizes the grading on $\mathrm{Hom}_k(A \otimes A, A)$ discussed above. We define this grading next.

The bar resolution of the $\mathbb{N}$-graded algebra A is itself graded, with

$$|a_0 \otimes \cdots \otimes a_{n+1}| = |a_0| + \cdots + |a_{n+1}|$$

for all homogeneous $a_0, \ldots, a_{n+1} \in A$. The A^e-module $A^{\otimes(n+2)}$ is a graded A^e-module in this way, and the differentials are graded maps, that is, they preserve the grading. A grading on $\mathrm{Hom}_{A^e}(A^{\otimes(n+2)}, A) \cong \mathrm{Hom}_k(A^{\otimes n}, A)$ for each n is defined by

$$\mathrm{Hom}_k(A^{\otimes n}, A)_m = \{f \in \mathrm{Hom}_k(A^{\otimes n}, A) \mid f((A^{\otimes n})_i) \subseteq A_{i+m} \text{ for all } i\},$$

and again the differentials are graded maps. Thus the cohomology inherits the grading. Hochschild cohomology $\mathrm{HH}^*(A)$ is therefore graded both by this grading from A and by homological degree. We call the grading on $\mathrm{HH}^*(A)$ coming from A its *internal grading* and that of homological degree its *homological grading*. Correspondingly, we use the terms *internal degree* and *homological degree* for homogeneous elements. Denote by $\mathrm{HH}^{i,j}(A)$ the subspace of $\mathrm{HH}^i(A)$ consisting of elements of internal degree j (not to be confused with a component of the Hodge decomposition (1.6.2) for which this same notation is sometimes used in the literature).

Proposition 5.4.1. *A first-level graded deformation of an $\mathbb{N}$-graded algebra A corresponds to an element of $\mathrm{HH}^{2,-1}(A)$. An obstruction to lifting an $(i-1)$st-level graded deformation of A to an ith-level graded deformation of A is in $\mathrm{HH}^{3,-i}(A)$. Specifically, an $(i-1)$st-level graded deformation lifts to an ith-level graded deformation if and only if the $(i-1)$st obstruction defined by the left side of equation (5.1.8) becomes 0 in cohomology.*

Proof. We have noted before that a first-level deformation corresponds to an infinitesimal deformation, that is, a Hochschild 2-cocycle. If, in addition, it is graded, then necessarily the internal degree of the Hochschild 2-cocycle is -1, as noted before. Two first-level graded deformations are isomorphic if and only if their corresponding cocycles are cohomologous. Therefore the first statement of the proposition holds.

A graded deformation of A necessarily involves functions μ_j that are homogeneous of degree $-j$, as was noted at the beginning of this section, and the same is true of an ith-level graded deformation. We have already noted the obstructions (5.1.8), and in this graded setting, we see that equation (5.1.8) indeed involves functions of internal degree $-i$. Thus the second statement holds. $\qquad\square$

Exercise 5.4.2. Let $P_\bullet$ be a free resolution of an $\mathbb{N}$-graded algebra A as an A^e-module consisting of graded modules and graded differentials. Show that the internal grading on $\mathrm{HH}^*(A)$ is induced from that on $P_\bullet$.

Exercise 5.4.3. Let $P_\bullet$ be the resolution (3.1.4) of a polynomial ring $A = k[x_1, \dots, x_m]$. Define a grading on each module P_i under which $P_\bullet$ becomes a graded resolution, that is, the differentials are graded maps. Use this grading on $P_\bullet$ to describe each space $\mathrm{HH}^{i,j}(A)$.

Exercise 5.4.4. Let A be a Koszul algebra and let $K(A)$ be the Koszul resolution given by (3.4.4). Put a grading on each $K_i(A)$ for which the differentials are graded maps, that is, they preserve the grading.

Exercise 5.4.5. Verify, by direct calculation, the claim in the proof of Proposition 5.4.1 that two first-level graded deformations are isomorphic if and only if their corresponding cocycles are cohomologous.

Exercise 5.4.6. Identify a first-level graded deformation of the polynomial ring $k[e, f, h]$ corresponding to the deformation of Example 5.1.13.

5.5. Braverman-Gaitsgory theory and the PBW Theorem

We present the theory of Braverman and Gaitsgory [**33**] on particular types of deformations and its application to the classical Poincaré-Birkhoff-Witt (PBW) Theorem on the structure of universal enveloping algebras of Lie algebras. We only consider here the original setting of [**33**] where A is a connected graded Koszul algebra, although this theory has been generalized in a number of directions (see, for example, the survey [**195**]).

Let A be a Koszul algebra, as defined in Section 3.4. Write $A = T(V)/(R)$, where V is a finite-dimensional vector space whose elements are given degree 1, and R is a subspace of $V \otimes V$. We will construct a particular type of deformation of A from choices of k-linear maps $\alpha : R \to V$ and $\beta : R \to k$ as follows. Consider the subspace of $T_0(V) \oplus T_1(V) \oplus T_2(V)$ given by

$$(5.5.1) \qquad P = P_{\alpha,\beta} = \{x - \alpha(x) - \beta(x) \mid x \in R\}.$$

Set $A' = T(V)/(P)$, where (P) denotes the ideal of $T(V)$ generated by $P = P_{\alpha,\beta}$. If α and β are both zero maps, then $P = R$ and $A' = A$ is a graded algebra. In general, A' is not graded but is a *filtered algebra* in the following way. Let $F_n A'$ be the image in A' of the subspace $\bigoplus_{0 \leq i \leq n} T_i(V)$ of $T(V)$. Then

$$F_0 A' \subseteq F_1 A' \subseteq F_2 A' \subseteq \cdots$$

and $(F_i A')(F_j A') \subseteq F_{i+j} A'$. Denote the *associated graded algebra* to A' by $\operatorname{gr} A'$, that is, as a vector space,

$$\operatorname{gr} A' = F_0 A' \oplus (F_1 A'/F_0 A') \oplus (F_2 A'/F_1 A') \oplus \cdots,$$

and the product of an element in $F_i A'/F_{i-1} A'$ with one in $F_j A'/F_{j-1} A'$ is defined by multiplying their inverse images in $F_i A'$ and $F_j A'$, and taking the resulting image in $F_{i+j} A'/F_{i+j-1} A'$. By definition of P, since $|\alpha| = -1$ and $|\beta| = -2$ as maps, the relations R hold in $\operatorname{gr} A'$. So there is a surjective algebra homomorphism

$$(5.5.2) \qquad A \to \operatorname{gr} A'$$

induced by mapping the generating space V to its image in $\operatorname{gr} A'$. In general the map is not injective, and the cases where it is are given the name PBW deformations.

Definition 5.5.3. The algebra $A' = T(V)/(P)$ is a *PBW deformation* (also called a *filtered deformation*) of A if the surjective algebra homomorphism (5.5.2) is an isomorphism.

A PBW deformation turns out to be a specialization of a graded deformation to a value of the parameter t. To see this, given P as in (5.5.1), let

$$P[t] = \{x - \alpha(x)t - \beta(x)t^2 \mid x \in R\},$$

a homogeneous subspace of $T(V)[t]$ of degree 2. The specializations of $T(V)[t]/P[t]$ to $t = 0$ and to $t = 1$ are isomorphic to A and to A', respectively. The condition that A' is a PBW deformation of A implies that $T(V)[t]/P[t]$ is a graded deformation of A over $k[t]$.

Braverman and Gaitsgory gave necessary and sufficient conditions for $A' = T(V)/(P)$ to be a PBW deformation of A. The following theorem is essentially a combination of [**33**, Lemma 0.4, Theorem 0.5, and Lemma 3.3].

Theorem 5.5.4. *Let $A = T(V)/(R)$ be a Koszul algebra as defined in Section 3.4, and let P be as in (5.5.1). The algebra $A' = T(V)/(P)$ is a PBW deformation of A if and only if the following conditions hold:*

(i) $P \cap F_1(T(V)) = \{0\}$,

(ii) $\mathrm{Im}(\alpha \otimes 1 - 1 \otimes \alpha) \subseteq R$,

(iii) $\alpha(\alpha \otimes 1 - 1 \otimes \alpha) = -(\beta \otimes 1 - 1 \otimes \beta)$,

(iv) $\beta(\alpha \otimes 1 - 1 \otimes \alpha) = 0$,

where the maps $\alpha \otimes 1 - 1 \otimes \alpha$ and $\beta \otimes 1 - 1 \otimes \beta$ are defined on the subspace $(R \otimes V) \cap (V \otimes R)$ of $T(V)$.

Proof. Assume that A' is a PBW deformation of A. Then necessarily condition (i) holds, in order that the subspace $k \oplus V$ of $T(V)/(R)$ maps isomorphically onto a copy of itself in $\mathrm{gr}\, A'$ under the map (5.5.2). We will show that the other three conditions also hold. These can be interpreted in terms of the Hochschild cohomology $\mathrm{HH}^*(A)$ as follows.

Let $K = K(A)$ be the Koszul resolution (3.4.4) of A and let $\iota : K \hookrightarrow B$ be its inclusion into the bar resolution $B = B(A)$ of A, as described in Section 3.4. Let $\psi : B \to K$ be a chain map lifting the identity map on A. Note that ψ may be chosen so that $\psi\iota = 1_K$ (the identity map on K) since $\iota_j(K_j)$ is a direct summand of B_j as an A^e-module for each j (a free basis of a complementary module is given by all $1 \otimes x \otimes 1$, where x runs over a vector space basis of a vector space complement to $K'_j = K'_j(A)$ in $A^{\otimes j}$). (See also [**39**].)

Now α and β are functions on R, and the degree 2 term of the Koszul resolution is $K_2 = A \otimes R \otimes A$. We identify α and β with functions on K_2 by setting $\alpha(1 \otimes x \otimes 1) = \alpha(x)$ and $\beta(1 \otimes x \otimes 1) = \beta(x)$ for all $x \in R$ and extending to A-bimodule homomorphisms. Set

$$\mu_1 = \alpha\psi_2, \qquad \mu_2 = \beta\psi_2$$

to obtain maps in $\mathrm{Hom}_{A^e}(A^{\otimes 4}, A) \cong \mathrm{Hom}_k(A^{\otimes 2}, A)$.

If A' is a PBW deformation of A, then

$$T(V)[t]/(P[t],\ t^2) = T(V)[t]/(x - \alpha(x)t,\ t^2 \mid x \in R)$$

corresponds to an infinitesimal deformation of A. Considering the elements of $(R \otimes V) \cap (V \otimes R)$, associativity implies condition (ii). Further, the quotient $T(V)[t]/(P[t],\ t^3)$ is a second-level graded deformation of A, which implies that

$$\mu_1(\mu_1 \otimes 1 - 1 \otimes \mu_1) = d_3^*(\mu_2),$$

considered as functions on $A \otimes A \otimes A$. Applying both sides of this equation to $(R \otimes V) \cap (V \otimes R)$, since $\psi\iota$ is the identity map, this is equivalent to condition (iii). Similarly, there is a μ_3 such that

$$\mu_1(\mu_2 \otimes 1 - 1 \otimes \mu_2) + \mu_2(\mu_1 \otimes 1 - 1 \otimes \mu_1) = d_3^*(\mu_3).$$

Applying both sides of this equation to elements of $(R \otimes V) \cap (V \otimes R)$, for degree reasons (since $|\mu_3| = -3$, $|\mu_2| = -2$, $|\mu_1| = -1$), the terms $\mu_1(\mu_2 \otimes 1 - 1 \otimes \mu_2)$ have image zero, and we obtain $\beta(\alpha \otimes 1 - 1 \otimes \alpha) = 0$, which is condition (iv).

Conversely, suppose conditions (i)–(iv) hold. Then $\mu_1\iota_2 = \alpha$ since $\psi_2\iota_2 = 1_{K_2}$. It follows that μ_1 is a Hochschild 2-cocycle on the bar resolution, and so defines an infinitesimal deformation of A. Similarly $\mu_2\iota_2 = \beta$. We modify μ_2 so that it satisfies condition (5.1.8) with $i = 2$ as a function on the bar resolution. Let

$$\gamma = -\mu_2 d_3 + \mu_1(\mu_1 \otimes 1 - 1 \otimes \mu_1).$$

Then $\gamma\iota_3 = -\beta d_3 + \alpha(\alpha \otimes 1 - 1 \otimes \alpha) = 0$ on K_3, which implies γ is a coboundary on the bar resolution, that is, $\gamma = \mu d_3$ for some μ. Now

$$\mu\iota_2 d_3 = \mu d_3\iota_3 = \gamma\iota_3 = 0,$$

so $\mu\iota_2$ is a cocycle on K. Consequently there is a cocycle μ', of internal degree -2, on the bar resolution with $\mu'\iota_2 = \mu\iota_2$. Then $(\mu_2 - \mu + \mu')\iota_2 = \beta$ and

$$(\mu_2 - \mu + \mu')d_3 + \mu_1(\mu_1 \otimes 1 - 1 \otimes \mu_1)$$

is zero on the bar resolution since $\mu'd_3 = 0$ and

$$-(\mu_2 - \mu)d_3 = \mu_1(\mu_1 \otimes 1 - 1 \otimes \mu_1).$$

We replace μ_2 by $\mu_2 - \mu + \mu'$.

Now, the map μ_1 and the new map μ_2 satisfy (5.1.8) with $i = 1$, $i = 2$. Thus there is a second-level graded deformation of A defined by μ_1, μ_2. By condition (iv), considering internal degree, the obstruction

$$\mu_2(\mu_1 \otimes 1 - 1 \otimes \mu_1) + \mu_1(\mu_2 \otimes 1 - 1 \otimes \mu_2)$$

to lifting to a third-level deformation of A becomes 0 as a cochain on K, on applying ι_3. So this is a coboundary on the bar resolution, and there is a μ_3 of internal degree -3 satisfying (5.1.8) with $i = 3$.

Now by Proposition 5.4.1, the obstruction to lifting to a fourth-level graded deformation lies in $\mathrm{HH}^{3,-4}(A)$. However, this is 0: $K_3 = A \otimes K_3' \otimes A$ and elements of K_3' all have degree 3, so $\mathrm{Hom}_{A^e}(K_3, A) \cong \mathrm{Hom}_k(K_3', A)$ consists of elements of internal degree at least -3. It follows that there is a μ_4 defining a fourth-level deformation. The same argument shows that this may be lifted to a fifth-level deformation and so on. Letting A_t denote the graded deformation obtained in this manner, we see that A' is isomorphic to $A_t|_{t=1}$: a map of vector spaces from V to $A_t|_{t=1}$ induces a map $A' \to A_t|_{t=1}$, which may be seen to be an isomorphism by counting dimensions in each degree. $\qquad\square$

As an application, we obtain the classical Poincaré-Birkhoff-Witt (PBW) Theorem for Lie algebras next. (Recall the definitions of Lie algebra and universal enveloping algebra in Section 5.1.) The proof of the next theorem shows that the universal enveloping algebra $U(\mathfrak{g})$ of a Lie algebra $\mathfrak{g}$ is a PBW deformation of a polynomial ring. Specifically, the polynomial ring may be identified as $S(\mathfrak{g})$, the symmetric algebra on the underlying vector space of $\mathfrak{g}$, that is,

$$S(\mathfrak{g}) = T(\mathfrak{g})/(x \otimes y - y \otimes x \mid x, y \in \mathfrak{g}).$$

Theorem 5.5.5 (Poincaré-Birkhoff-Witt Theorem). *Let $\mathfrak{g}$ be a finite-dimensional Lie algebra, and let $U(\mathfrak{g})$ be its universal enveloping algebra. The associated graded algebra of $U(\mathfrak{g})$ is isomorphic to $S(\mathfrak{g})$.*

Proof. Let $V = \mathfrak{g}$ and $A = S(\mathfrak{g}) = T(V)/(R)$, where R is the vector subspace of $V \otimes V$ spanned by all $x \otimes y - y \otimes x$ for $x, y \in \mathfrak{g}$. Let

$$P = \{x \otimes y - y \otimes x - [x, y] \mid x, y \in \mathfrak{g}\},$$

so that $U(\mathfrak{g}) = T(\mathfrak{g})/(P)$ by definition. Set $\alpha(x \otimes y - y \otimes x) = [x, y]$ and $\beta(x \otimes y - y \otimes x) = 0$. By antisymmetry of the Lie bracket $[\ ,\ ]$, condition (i) of Theorem 5.5.4 holds. A calculation shows that $(R \otimes V) \cap (V \otimes R)$ consists of linear combinations of elements of the form

$$x \otimes y \otimes z - y \otimes x \otimes z + y \otimes z \otimes x - z \otimes y \otimes x + z \otimes x \otimes y - x \otimes z \otimes y$$

for $x, y, z \in \mathfrak{g}$, and that condition (ii) of Theorem 5.5.4 holds. Condition (iii) is equivalent to the Jacobi identity, and condition (iv) automatically holds since β is 0. By Theorem 5.5.4, $U(\mathfrak{g})$ is a PBW deformation of $S(\mathfrak{g})$, and in particular, $\mathrm{gr}\, U(\mathfrak{g}) \cong S(\mathfrak{g})$. $\qquad\square$

For a survey of some of the many generalizations of the classical Poincaré-Birkhoff-Witt Theorem, as well as other methods of proof, see [**195**].

Exercise 5.5.6. Let $A = k\langle x, y\rangle/(yx - xy - 1)$, the Weyl algebra. What is its associated graded algebra $\operatorname{gr} A$? Show that A is a PBW deformation of $\operatorname{gr} A$.

Exercise 5.5.7. Consider $U(\mathfrak{sl}_2)$ as defined in Example 5.1.13. Show directly that $\operatorname{gr} U(\mathfrak{sl}_2)$ is isomorphic to a polynomial ring in three indeterminates.

Exercise 5.5.8. Look up Sridharan enveloping algebras (e.g., in [204]) and express them in the notation of this section. That is, each is a PBW deformation of a polynomial ring. What are α and β as in (5.5.1)?

Exercise 5.5.9. Verify the claims in the proof of Theorem 5.5.5, in particular that conditions (ii) and (iii) of Theorem 5.5.4 hold under the hypotheses of Theorem 5.5.5.

Chapter 6

Gerstenhaber Bracket

In Section 1.4, we introduced the Gerstenhaber bracket on Hochschild cohomology under which it becomes a graded Lie algebra. In Chapter 5 we saw applications of this Lie structure in algebraic deformation theory. In this chapter we give some equivalent definitions of the bracket, partially paralleling Chapter 2 where we saw equivalent definitions of the cup product. The definition of Gerstenhaber bracket that is most suitable in any given setting will vary. Some definitions we consider here lead to computational techniques, and we illustrate with some small examples.

We begin in Section 6.1 with a realization of Hochschild cohomology as the homology of a complex of coderivations on the tensor coalgebra of A, related to the bar resolution. The Gerstenhaber bracket is a graded commutator of coderivations [207]. We next derive a formula in Section 6.2 for brackets of elements in degree 1 with those in arbitrary degree n [212]. The degree 1 elements are identified with derivations and thus with functions on the bar resolution, while degree n elements and the bracket formula are given on an arbitrary resolution. Thus our tour of Gerstenhaber bracket techniques begins to depart from the historical setting of the bar resolution, and the remainder of this chapter involves techniques for other resolutions and exact sequences. We present in Section 6.3 the notion of homotopy liftings that allow Gerstenhaber brackets to be expressed on an arbitrary resolution as essentially graded commutators [220]. We discuss related computational techniques in Section 6.4 for resolutions that are differential graded coalgebras, and these techniques apply in particular to Koszul algebras [165]. For a topological approach, we outline in Section 6.5 a construction of brackets as loops in the classifying space of an extension category [194].

Recall the standard sign convention (2.3.1) that we make heavy use of in this chapter.

6.1. Coderivations

In this section, we present Stasheff's realization [**207**] of the Gerstenhaber bracket on Hochschild cohomology of A as a graded commutator of coderivations on the tensor coalgebra of A. See also Quillen [**180, 181**] for related constructions and Schlessinger and Stasheff [**191**] for the case of Harrison cohomology. We start by defining the tensor coalgebra.

Let $T = T(A) = \bigoplus_{n \geq 0} A^{\otimes n}$, where we set $A^{\otimes 0} = k$, considered as a complex with differential d_T given by

$$d_T(a_1 \otimes \cdots \otimes a_n) = \sum_{i=1}^{n-1} (-1)^i a_1 \otimes \cdots \otimes a_i a_{i+1} \otimes \cdots \otimes a_n$$

for all $a_1, \ldots, a_n \in A$. Then $T(A)$ is a graded vector space with $T_n(A) = A^{\otimes n}$. Here we will use the notation $\mathrm{Hom}_k(T(A), T(A))$ for the graded vector space whose mth component is

$$\mathrm{Hom}_k(T(A), T(A))_m = \{f \mid f(A^{\otimes n}) \subseteq A^{\otimes (n+m)} \text{ for all } n\}.$$

Then $\mathrm{Hom}_k(T(A), T(A))$ is a graded algebra under composition of functions, and it is a bicomplex whose total complex has differential ∂ given by

$$(6.1.1) \qquad \partial(f) = d_T f - (-1)^{|f|} f d_T$$

for all homogeneous functions f (see Section A.5). By abuse of notation, as is customary, we use the same notation and terminology for the bicomplex and its total complex, and it will be clear from context which is meant. Note that $|d_T| = -1$. A calculation shows that

$$\partial(fg) = \partial(f)g + (-1)^{|f|} f \partial(g),$$

where fg denotes composition of these functions. That is, ∂ is a graded derivation on $\mathrm{Hom}_k(T(A), T(A))$. Thus $\mathrm{Hom}_k(T(A), T(A))$ is a differential graded algebra.

The complex $\mathrm{Hom}_k(T(A), T(A))$ has another binary operation given by the graded commutator:

$$(6.1.2) \qquad [f, g] = fg - (-1)^{|f||g|} gf$$

for all homogeneous $f, g \in \mathrm{Hom}_k(T(A), T(A))$. By virtue of being a graded commutator, it enjoys a graded Jacobi identity just as in Lemma 1.4.3(ii). Calculations show that

$$\partial([f, g]) = [\partial(f), g] + (-1)^{|f|} [f, \partial(g)]$$

and

$$(6.1.3) \qquad \partial(f) = [d_T, f].$$

Define a k-linear map $\Delta_T : T(A) \to T(A) \otimes T(A)$ by

$$\Delta_T(a_1 \otimes \cdots \otimes a_n) = 1 \otimes (a_1 \otimes \cdots \otimes a_n) + (a_1 \otimes \cdots \otimes a_n) \otimes 1$$
$$+ \sum_{i=1}^{n-1} (a_1 \otimes \cdots \otimes a_i) \otimes (a_{i+1} \otimes \cdots \otimes a_n)$$

for all $a_1, \ldots, a_n \in A$. Under this map, $T(A)$ is a differential graded coalgebra, that is, Δ_T is a chain map, $(\Delta_T \otimes 1)\Delta_T = (1 \otimes \Delta_T)\Delta_T$, and $(\varepsilon \otimes 1)\Delta_T = 1 = (1 \otimes \varepsilon)\Delta_T$, where $\varepsilon : T(A) \to k$ projects onto $T(A)_0 = k$. Some authors write $T^c(A)$ for this differential graded coalgebra in order to distinguish it from the algebra on the same underlying vector space. Let $\Delta_T^{(2)} : T(A) \to T(A) \otimes T(A) \otimes T(A)$ be defined by

$$\Delta_T^{(2)} = (\Delta_T \otimes 1_T)\Delta_T = (1_T \otimes \Delta_T)\Delta_T.$$

Definition 6.1.4. A *graded coderivation* on $T(A)$ is a graded k-linear map $f : T(A) \to T(A)$, of some degree j, for which

$$\Delta_T f = (f \otimes 1_T + 1_T \otimes f)\Delta_T$$

as functions from $T(A)$ to $T(A) \otimes T(A)$. Denote by $\mathrm{Coder}(T(A))$ the vector space spanned by the graded coderivations on $T(A)$.

A calculation shows that the space $\mathrm{Coder}(T(A))$ is closed under the graded commutator bracket (6.1.2) by its definition (recalling the sign convention (2.3.1)). Also note that the differential d_T is itself a coderivation since Δ_T is a chain map. $\mathrm{Coder}(T(A))$ is thus a subcomplex of the Hom complex $(\mathrm{Hom}_k(T(A), T(A)), \partial)$ since $\partial = [d_T, -]$ as noted in (6.1.3).

The following connection with Hochschild cohomology goes back to work of Stasheff [**207**] (see also Quillen [**180, 181**]). Let $B = B(A)$ be the bar resolution (1.1.4) of A as an A^e-module, so that $B_n = A^{\otimes(n+2)}$ for all $n \geq 0$. We take the differential d^* on $\mathrm{Hom}_k(T(A), A) \cong \mathrm{Hom}_{A^e}(B, A)$ to be that induced by the differential d on the bar resolution of A given by equation (1.1.2), which in turn is related to the differential d_T on $T(A)$:

$$d^*(f)(a_1 \otimes \cdots \otimes a_m)$$
$$= a_1 f(a_2 \otimes \cdots \otimes a_m) + f d_T(a_1 \otimes \cdots \otimes a_m) + (-1)^m f(a_1 \otimes \cdots \otimes a_{m-1})a_m$$

for all $a_1, \ldots, a_m \in A$ and $f \in \mathrm{Hom}_k(A^{\otimes m}, A)$. Note that the degree of such a function f is taken to be $m - 1$ in our context here (not m as in other contexts). A calculation shows that f may be extended uniquely to a coderivation $D_f : T(A) \to T(A)[1 - m]$ as follows:

$$(6.1.5) \qquad D_f = (1_T \otimes f \otimes 1_T)\Delta_T^{(2)},$$

where if $l < m$, we interpret D_f to be 0 on $A^{\otimes l}$. On elements then, applying the sign convention (2.3.1), we have

(6.1.6)
$$D_f(a_1 \otimes \cdots \otimes a_l)$$
$$= \sum_{i=1}^{l-m+1} (-1)^u a_1 \otimes \cdots \otimes a_{i-1} \otimes f(a_i \otimes \cdots \otimes a_{i+m-1}) \otimes a_{i+m} \otimes \cdots \otimes a_l$$

for all $a_1, \ldots, a_l \in A$, where $u = (m-1)(i-1)$. (Existence and uniqueness of D_f is due to the corresponding truncated complex being cofree in a certain category of coalgebras; see, for example, [**153**, Section II.3.7].)

Next we show that the complex of coderivations is isomorphic to the bar complex $\mathrm{Hom}_k(T(A), A) \cong \mathrm{Hom}_{A^e}(B, A)$ from which Hochschild cohomology is obtained.

Theorem 6.1.7. *The complex* $(\mathrm{Coder}(T(A)), \partial)$ *is a subcomplex of the complex* $(\mathrm{Hom}_k(T(A), T(A)), \partial)$ *that is isomorphic, as a differential graded vector space, to* $(\mathrm{Hom}_k(T(A), A), d^*)$.

Proof. We have already seen that the space $\mathrm{Coder}(T(A))$ is a subcomplex of $(\mathrm{Hom}_k(T(A), T(A)), \partial)$, since the differential d_T is a coderivation, $\partial = [d_T, -]$, and $\mathrm{Coder}(T(A))$ is closed under bracket. Given an element f of $\mathrm{Hom}_k(T(A), A)$, it extends uniquely to a coderivation D_f from $T(A)$ to $T(A)$ given by (6.1.5). On the other hand, given a coderivation from $T(A)$ to $T(A)$, its composition with projection onto $T_1(A) = A$ is an element of $\mathrm{Hom}_k(T(A), A)$. A calculation shows that the differential ∂ on $\mathrm{Coder}(T(A))$ corresponds to d^* on $\mathrm{Hom}_k(T(A), A)$. $\qquad\square$

As a consequence of the theorem, Hochschild cohomology $\mathrm{HH}^*(A)$ is the cohomology of the complex $(\mathrm{Coder}(T(A)), \partial)$. We may realize the Gerstenhaber bracket in a natural way on $\mathrm{Coder}(T(A))$ as follows. Recall the degree shift by 1 here in making comparisons to earlier sections.

Theorem 6.1.8. *The bracket* (6.1.2) *induces the Gerstenhaber bracket of Definition 1.4.1 on Hochschild cohomology* $\mathrm{HH}^*(A)$ *under the isomorphism of complexes given in Theorem 6.1.7.*

Proof. The isomorphism of Theorem 6.1.7 sends cochains f, g on the bar resolution $B = B(A)$ to their corresponding coderivations D_f, D_g on $T(A)$ given by formula (6.1.6). We claim that the formula (6.1.2) applied to D_f, D_g coincides with Definition 1.4.1 of Gerstenhaber bracket. To see this, note that projecting values of $[D_f, D_g] = D_f D_g - (-1)^{|D_f||D_g|} D_g D_f$ onto $T_1(A) = A$ yields the formula

(6.1.9)										$f D_g - (-1)^{|D_f||D_g|} g D_f$

for their bracket as an element of $\mathrm{Hom}_k(T(A), A)$. If $f \in \mathrm{Hom}_k(A^{\otimes m}, A)$ and $g \in \mathrm{Hom}_k(A^{\otimes n}, A)$, then $|D_f| = m - 1$ and $|D_g| = n - 1$. Applying the formula (6.1.6) for D_f and D_g in terms of f and g, and comparing to the formula for the Gerstenhaber bracket in Definition 1.4.1, we see that they are the same. $\qquad\square$

Exercise 6.1.10. Verify that Δ_T is a chain map.

Exercise 6.1.11. Verify that $\mathrm{Coder}(T(A))$ is closed under the graded commutator bracket (6.1.2).

Exercise 6.1.12. Verify that the differential d_T is a coderivation.

Exercise 6.1.13. Verify that the differential ∂ is a graded derivation with respect to the graded commutator (6.1.2).

Exercise 6.1.14. Verify that (6.1.5) (equivalently, (6.1.6)) defines a coderivation on $T(A)$.

6.2. Derivation operators

In this section, we present Suárez-Álvarez' methods [212] for computing Gerstenhaber brackets with elements of homological degree 1 on an arbitrary resolution. These methods may be used for example to find the Lie structure on degree 1 Hochschild cohomology $\mathrm{HH}^1(A)$ and the structure of its module $\mathrm{HH}^*(A)$. Suárez-Álvarez worked in a broader context of derivation operators and actions on Ext. Here we consider only that part of his theory that is directly relevant to the Gerstenhaber bracket on $\mathrm{HH}^*(A)$, and refer to [212] for more general results.

Let $P \xrightarrow{\mu_P} A$ be a projective resolution of the A^e-module A with differential d. Let $f : A \to A$ be a derivation, so that it represents an element of $\mathrm{HH}^1(A)$, as explained in Section 1.2. Let $f^e : A^e \to A^e$ be defined by

$$(6.2.1) \qquad\qquad f^e = f \otimes 1 + 1 \otimes f,$$

and note that f^e is a derivation on A^e. Functions satisfying equation (6.2.3) below are termed *derivation operators* (or more specifically f^e-*operators*). More generally, the notion of a δ-operator, for any derivation δ on an algebra, is defined in [212].

The following lemma is related to work of Gopalakrishnan and Sridharan [95].

Lemma 6.2.2. *Let $f : A \to A$ be a derivation. There is a k-linear chain map $\tilde{f}_\bullet : P_\bullet \to P_\bullet$ lifting f with the property that for each n,*

$$(6.2.3) \qquad \tilde{f}_n((a \otimes b) \cdot x) = f(a)xb + a\tilde{f}_n(x)b + axf(b)$$

for all $a, b \in A$ and $x \in P_n$. Moreover, $\tilde{f}_\bullet$ is unique up to A^e-module chain homotopy.

Proof. We wish to define each $\tilde{f}_i$ so that it satisfies equation (6.2.3), and so that the following diagram commutes:

$$
\begin{array}{ccccccccc}
\cdots \xrightarrow{d_2} & P_1 & \xrightarrow{d_1} & P_0 & \xrightarrow{\mu_P} & A & \longrightarrow & 0 \\
& \downarrow{\tilde{f}_1} & & \downarrow{\tilde{f}_0} & & \downarrow{f} & & \\
\cdots \xrightarrow{d_2} & P_1 & \xrightarrow{d_1} & P_0 & \xrightarrow{\mu_P} & A & \longrightarrow & 0
\end{array}
$$

If P_0 is free as an A^e-module, choose a free basis $\{x_j \mid j \in J\}$, where J is some indexing set. Since μ_P is surjective, for each $j \in J$, there exists a $y_j \in P_0$ such that $\mu_P(y_j) = f(\mu_P(x_j))$. Set $\tilde{f}_0(x_j) = y_j$. Extend to P_0 by requiring

$$\tilde{f}_0((a \otimes b) \cdot x_j) = f(a)x_j b + a y_j b + a x_j f(b)$$

for all $a, b \in A$ and $j \in J$. Note the rightmost square in the diagram indeed commutes since f is a derivation and the action of A^e on A is by left and right multiplication. If P_0 is not free, we may realize it as a direct summand of a free module and argue similarly.

Now $\tilde{f}_0 d_1$ has image contained in the image of d_1, since $\mu_P \tilde{f}_0 d_1 = f \mu_P d_1 = 0$. We may apply the same argument as above to define $\tilde{f}_1$, and so on. Thus we have a k-linear chain map $\tilde{f}_\bullet$ satisfying (6.2.3).

If $\tilde{f}_\bullet$ and $\tilde{f}'_\bullet$ are two such k-linear chain maps, then $\tilde{f}_\bullet - \tilde{f}'_\bullet$ is a chain map lifting the zero map from A to A. Since each of $\tilde{f}_\bullet, \tilde{f}'_\bullet$ satisfies (6.2.3), their difference $\tilde{f}_\bullet - \tilde{f}'_\bullet$ is A^e-linear, and so it is A^e-chain homotopic to 0. $\qquad \square$

A standard example is given by functions on the bar resolution, as we explain next.

Example 6.2.4. Let B be the bar resolution on A, and let $f : A \to A$ be a derivation. For each i, let

$$\tilde{f}_i(a_0 \otimes \cdots \otimes a_{i+1}) = \sum_{j=0}^{i+1} a_0 \otimes \cdots \otimes a_{j-1} \otimes f(a_j) \otimes a_{j+1} \otimes \cdots \otimes a_{i+1}$$

for all $a_0, \ldots, a_{i+1} \in A$ and extend k-linearly. Then $\tilde{f}_\bullet : B \to B$ is a derivation operator, that is, it satisfies equation (6.2.3).

The following theorem is due to Suárez-Álvarez [212]. For Hochschild cocycles defined on a resolution P other than the bar resolution B, their Gerstenhaber bracket is defined as a function on P via chain maps between P and B. (See Exercise 6.2.11.) This is the definition of Gerstenhaber bracket used in the theorem.

Theorem 6.2.5. *Let $f : A \to A$ be a derivation. Let P be a projective resolution of A as an A^e-module. Let $g \in \operatorname{Hom}_{A^e}(P_n, A)$ be a cocycle, and let $\tilde{f}_n : P_n \to P_n$ be a map satisfying (6.2.3). The Gerstenhaber bracket of f and g is represented by*

$$(6.2.6) \qquad\qquad [f, g] = fg - g\tilde{f}_n$$

as a cocycle on P_n.

Proof. First suppose that P is the bar resolution $B = B(A)$. Let $\tilde{f}_n$ be chosen as in Example 6.2.4. Then formula (6.2.6) agrees with the historical formula of Definition 1.4.1 for the Gerstenhaber bracket, since f is a 1-cocycle. Also, since $gd = 0$ and $\tilde{f}_\bullet$ is unique up to chain homotopy as stated in Lemma 6.2.2, the element of Hochschild cohomology given by formula (6.2.6) does not depend on the choice of $\tilde{f}_\bullet$.

If P is not the bar resolution, let $\theta : B \to P$ and $\iota : P \to B$ be comparison maps, that is, chain maps lifting the identity map on A. Identify the derivation f with a cocycle on B as described in Section 1.2. The Gerstenhaber bracket of f and g is by definition $[f, g\theta]\iota$, where $[f, g\theta]$ denotes the Gerstenhaber bracket defined as usual on B (see Exercise 6.2.11). Let $\tilde{f}'_\bullet : B \to B$ be a k-linear chain map satisfying (6.2.3) on B. A calculation shows that for each i, the function $\theta\tilde{f}'_i - \tilde{f}_i\theta$ is in fact an A^e-module homomorphism. Since $\theta\tilde{f}'_\bullet - \tilde{f}_\bullet\theta$ lifts the zero map from A to A, it must be A^e-chain homotopic to 0. By our arguments in the first paragraph above, $[f, g\theta] = fg\theta - g\theta\tilde{f}'_n$ represents the Gerstenhaber bracket of f and $g\theta$ at the chain level on B. Using the notation $\sim$ to indicate that cocycles are cohomologous, on P we have

$$[f, g\theta]\iota \ \sim \ fg\theta\iota - g\theta\tilde{f}'_n\iota \ \sim \ fg\theta\iota - g\tilde{f}_n\theta\iota \ \sim \ fg - g\tilde{f}_n,$$

since $\theta\iota$ is chain homotopic to the identity map and $gd = 0$. $\qquad\square$

The above proof of Theorem 6.2.5 via Lemma 6.2.2 is constructive, giving rise to a method for computing Gerstenhaber brackets with 1-cocycles. We illustrate this derivation operator method next with a small example. Other examples are in the literature; e.g., [**147, 159**]. Used in combination with the relation given in Lemma 1.4.7 between cup product and Gerstenhaber bracket, the derivation operator method sometimes suffices to compute the full Gerstenhaber algebra structure on Hochschild cohomology.

Example 6.2.7. Let $A = k[x, y]$. We will find a general formula for the Gerstenhaber bracket of a 1-cocycle with a 2-cocycle on the Koszul resolution P given by (3.1.4), using formula (6.2.6). Other brackets may be found similarly. Let $f = x^i y^j \frac{\partial}{\partial x}$, a derivation on A. Let $g = qx^* \wedge y^*$ for some

$q \in A$. We first find functions $\tilde{f}_0, \tilde{f}_1, \tilde{f}_2$ as in Lemma 6.2.2:

$$\tilde{f}_0(a \otimes b) = f(a) \otimes b + a \otimes f(b),$$

$$\tilde{f}_1(a \otimes x \otimes b) = f(a) \otimes x \otimes b + \sum_{l=1}^{j} ax^i y^{j-l} \otimes y \otimes y^{l-1} b$$

$$+ \sum_{l=1}^{i} ax^{i-l} \otimes x \otimes x^{l-1} y^j b + a \otimes x \otimes f(b),$$

$$\tilde{f}_1(a \otimes y \otimes b) = f(a) \otimes y \otimes b + a \otimes y \otimes f(b),$$

$$\tilde{f}_2(a \otimes x \wedge y \otimes b) = f(a) \otimes x \wedge y \otimes b + \sum_{l=1}^{i} ax^{i-l} \otimes x \wedge y \otimes x^{l-1} y^j b$$

$$+ a \otimes x \wedge y \otimes f(b)$$

for all $a, b \in A$. By Theorem 6.2.5, setting $p = x^i y^j$,

$$[f, g](x \wedge y) = (fg - g\tilde{f}_2)(x \wedge y)$$

$$= f(q) - g\left(\sum_{l=1}^{i} x^{i-l} \otimes x \wedge y \otimes x^{l-1} y^j\right)$$

$$= p\frac{\partial}{\partial x}(q) - q\frac{\partial}{\partial x}(p).$$

So $[f, g] = (p\frac{\partial}{\partial x}(q) - q\frac{\partial}{\partial x}(p))x^* \wedge y^*$.

Exercise 6.2.8. Verify that f^e, defined by (6.2.1), is a derivation on A^e.

Exercise 6.2.9. Verify that $\tilde{f}_\bullet$ as defined in Example 6.2.4 is indeed a k-linear chain map on the bar resolution $B = B(A)$ and that it satisfies (6.2.3).

Exercise 6.2.10. Verify the claim in the first sentence of the proof of Theorem 6.2.5, that is, formula (6.2.6) agrees with the historical definition of Gerstenhaber bracket as defined on the bar resolution.

Exercise 6.2.11. Let P be a projective resolution of A as an A^e-module, and let B be the bar resolution of A. Let $\theta : B \to P$ and $\iota : P \to B$ be comparison maps. Define a bilinear operation $[\ ,\ ]$ on $\mathrm{Hom}_{A^e}(P_\bullet, A)$ as follows. Let $f \in \mathrm{Hom}_{A^e}(P_m, A)$ and $g \in \mathrm{Hom}_{A^e}(P_n, A)$. Let $[f, g] = [f\theta, g\theta]\iota$, where the bracket on the right side is the Gerstenhaber bracket of Definition 1.4.1. Show that this induces a well-defined operation $[\ ,\ ]$ on Hochschild cohomology that agrees with the Gerstenhaber bracket under the isomorphism induced by the chain maps θ, ι.

Exercise 6.2.12. Let $A = k[x, y]$, notation as in Example 6.2.7. Find $[x^i y^j \frac{\partial}{\partial x}, qx^*]$ and $[x^i y^j \frac{\partial}{\partial x}, qy^*]$. What more must be computed to obtain all possible Gerstenhaber brackets among elements of $\mathrm{HH}^*(A)$?

6.3. Homotopy liftings

In this section we present Volkov's approach to brackets on Hochschild cohomology expressed directly on an arbitrary resolution, and explain how results of Sections 6.1 and 6.2 fit with this approach. More details and applications may be found in [**220**].

Let $P \xrightarrow{\mu_P} A$ be a projective resolution of A as an A^e-module with differential d. We work with the Hom complex $\operatorname{Hom}_{A^e}(P, P)$ in which the differential $\mathbf{d}$ is given by

$$\mathbf{d}(f) = df - (-1)^m fd$$

for all A^e-maps $f : P \to P[-m]$. (See Section A.1 for the degree shift notation $P[-m]$ and Section A.5 for the Hom complex.) The Hom complex is quasi-isomorphic to $\operatorname{Hom}_{A^e}(P, A)$ via the augmentation μ_P. We use the notation $\sim$ in this section to indicate that two functions are cohomologous in $\operatorname{Hom}_{A^e}(P, P)$. Equivalently, they are cohomologous in $\operatorname{Hom}_{A^e}(P, A)$ after application of μ_P, since μ_P induces a quasi-isomorphism between the complexes $\operatorname{Hom}_{A^e}(P, P)$ and $\operatorname{Hom}_{A^e}(P, A)$.

Let $f \in \operatorname{Hom}_{A^e}(P_m, A)$ and $g \in \operatorname{Hom}_{A^e}(P_n, A)$ be cocycles, that is, $fd = 0$ and $gd = 0$. Inspired by the expression (6.1.9) for the Gerstenhaber bracket obtained in Stasheff's coderivation theory, we aim to express the Gerstenhaber bracket $[f, g]$ analogously as a function on P similar to a graded commutator with respect to function composition. We will define functions $\psi_f : P \to P[1 - m]$ and $\psi_g : P \to P[1 - n]$ for which the Gerstenhaber bracket is represented at the chain level by

$$(6.3.1) \qquad [f, g] = f\psi_g - (-1)^{(m-1)(n-1)} g\psi_f.$$

We caution that we have chosen different sign conventions and notation from that of Volkov [**220**].

We will impose a condition on the functions ψ_f, ψ_g analogous to a property of the circle product stated in Lemma 1.4.5(i). For an m-cocycle f and an n-cocycle g on the bar resolution, this is:

$$(6.3.2) \qquad (-1)^m (g \circ f)d = (-1)^{mn} f \smile g - g \smile f.$$

We wish to define an analog of the circle product, as a function on an arbitrary resolution, that has such a relationship to the cup product. With this in mind, let $\Delta_P : P \to P \otimes_A P$ be a diagonal map, that is, an A^e-module chain map lifting the canonical isomorphism $A \to A \otimes_A A$ of A^e-modules. The cup product $f \smile g$ may be represented by the function $(f \otimes g)\Delta_P$ as in formula (2.3.2), where we have suppressed the subscript A on the tensor product symbol between functions since the subscript is clear from the domain $P \otimes_A P$. Accordingly, we require ψ_f to satisfy the following equation analogous to (6.3.2) for all cocycles g in $\mathrm{Hom}_{A^e}(P_n, A)$:

$$(6.3.3) \qquad (-1)^m g\psi_f d = ((-1)^{mn} f \otimes g - g \otimes f)\Delta_P.$$

We impose these two conditions (6.3.1) and (6.3.3) on functions ψ_f, ψ_g, and derive from these some further conditions, leading to Definition 6.3.6 below of homotopy lifting. We will show in Theorem 6.3.11 that the conditions are sufficient to define the Gerstenhaber bracket as (6.3.1), justifying this approach.

We consider the second imposed condition (6.3.3) first. Fixing f, since $gd = 0$ and $|\psi_f| = m - 1$, the condition is (recalling the Koszul sign convention (2.3.1)):

$$g\mathbf{d}(\psi_f) = (-1)^m g\psi_f d = ((-1)^{mn} f \otimes g - g \otimes f)\Delta_P = g(f \otimes 1_P - 1_P \otimes f)\Delta_P$$

for all $n \geq 0$ and all n-cocycles g. This will hold if

$$(6.3.4) \qquad \mathbf{d}(\psi_f) = (f \otimes 1_P - 1_P \otimes f)\Delta_P.$$

We consider the first imposed condition (6.3.1) in the case that g is the 0-cocycle μ_P, rewriting it as follows. Let B denote the bar resolution on A, and let $\theta : B \to P$ and $\iota : P \to B$ be comparison maps. Then $f\theta$ is a cocycle on B, and so the Gerstenhaber bracket $[f\theta, \mu_B]$ becomes 0 in cohomology by Lemma 1.4.5(ii) since μ_B is simply the multiplication map π on A. Using the historical definition of Gerstenhaber bracket and the comparison maps ι, θ to translate to cocycles on P, the Gerstenhaber bracket of f and μ_P is

$$[f, \mu_P] = [f\theta, \mu_P\theta]\iota = [f\theta, \mu_B]\iota \sim 0.$$

So, if $[f, \mu_P]$ may be expressed as in equation (6.3.1), then setting $\psi = \psi_{\mu_P}$, we have

$$(6.3.5) \qquad f\psi + (-1)^m \mu_P \psi_f \sim 0.$$

Note that by its definition, the function $f\psi + (-1)^m \mu_P \psi_f$ takes P_{m-1} to A and condition (6.3.5) is simply requiring ψ_f to take values in P_0 consistent with values of $f\psi$.

In fact these two conditions (6.3.4) and (6.3.5) are sufficient to define the bracket via formula (6.3.1), as we will see in Theorem 6.3.11. Next we will give a name to functions ψ_f having these properties, as in [220].

Definition 6.3.6. Let P be a projective resolution of A as an A^e-module, let $\Delta_P : P \to P \otimes_A P$ be a diagonal map, and let $f \in \mathrm{Hom}_{A^e}(P_m, A)$ be a cocycle. An A^e-module homomorphism $\psi_f : P \to P[1 - m]$ is a *homotopy* *lifting of f with respect to Δ_P* if

$$
\begin{aligned}
\mathbf{d}(\psi_f) &= (f \otimes 1_P - 1_P \otimes f)\Delta_P \quad \text{and} \\
\mu_P \psi_f &\sim (-1)^{m-1} f \psi
\end{aligned}
$$

for some $\psi : P \to P[1]$ for which $\mathbf{d}(\psi) = (\mu_P \otimes 1_P - 1_P \otimes \mu_P)\Delta_P$.

We will often use the term *homotopy lifting of f* without explicit reference to Δ_P if it is clear from the context which map Δ_P is intended, or in situations where the choice of Δ_P does not matter. We caution again that our homotopy lifting differs from that of Volkov [**220**] by signs.

It may be checked directly that if ψ_f, ψ_g are homotopy liftings for cocycles f, g with respect to Δ_P, then $[f, g]$ as defined in (6.3.1) is a cocycle. We check that if either f or g is a coboundary, then so is $[f, g]$ as defined in (6.3.1): if $f = hd$ for some cochain h, set

$$
(6.3.7) \qquad \psi_f = (-1)^m (h \otimes 1_P - 1_P \otimes h)\Delta_P.
$$

A calculation shows that ψ_f is a homotopy lifting for f. With this choice, $f\psi_g \sim (-1)^{(m-1)(n-1)} g\psi_f$, and so $[f, g]$ is a coboundary.

Example 6.3.8. Let $P = B$, the bar resolution of A, and let Δ_B be the standard diagonal map on B given by formula (2.3.3) (note this corresponds to the map Δ_T of Section 6.1 by writing $B = A \otimes T(A) \otimes A$ and extending Δ_T to an A^e-module homomorphism). Then $(\mu_B \otimes 1_B - 1_B \otimes \mu_B)\Delta_B = 0$, and we may let $\psi = 0$ in Definition 6.3.6. Let $f \in \mathrm{Hom}_{A^e}(B_m, A)$ be a cocycle. We may assume without loss of generality that $f(a_0 \otimes \cdots \otimes a_{m+1})$ is 0 whenever at least one of $a_1, \ldots, a_m$ is in the field k, since f is cohomologous to such a function (as may be seen by mapping to the reduced bar resolution). Let

$$
\psi_f(a_0 \otimes \cdots \otimes a_{l+1})
$$
$$
= \sum_{i=1}^{l-m+1} (-1)^u a_0 \otimes \cdots \otimes a_{i-1} \otimes f(a_i \otimes \cdots \otimes a_{i+m-1}) \otimes a_{i+m} \otimes \cdots \otimes a_{l+1},
$$

where $u = (m-1)(i-1)$, for all $l \geq m$ and $a_0, \ldots, a_{l+1} \in A$, and take ψ_f to be the zero map on B_l for $l \leq m-1$. Then ψ_f is a homotopy lifting of f with respect to Δ_B. A calculation shows that with this choice of ψ_f and a similar choice of ψ_g, the bracket $[f, g]$ as given by formula (6.3.1) is precisely the Gerstenhaber bracket as defined on the bar resolution in Definition 1.4.1.

We may view ψ_f defined by the formula in the example as a coderivation on B, or restrict to $T(A) \cong k \otimes T(A) \otimes k \hookrightarrow A \otimes T(A) \otimes A = B$ to obtain a coderivation $\psi_f|_{T(A)}$ on $T(A)$ as in Definition 6.1.4; see also formula (6.1.6). If f is a 1-cocycle, then $\psi_f|_{T(A)}$, viewed another way, may be extended to a derivation operator in the sense of Lemma 6.2.2 (see Example 6.2.4). Thus homotopy liftings encompass these two views—coderivations on the tensor coalgebra and derivation operators on the bar resolution—that were introduced in Sections 6.1 and 6.2.

We next state a needed existence and uniqueness result for homotopy liftings.

Lemma 6.3.9. *Let $f \in \mathrm{Hom}_{A^e}(P_m, A)$ be a cocycle, and let $\Delta_P : P \to P \otimes_A P$ be a diagonal map. There is a homotopy lifting $\psi_f : P \to P[1 - m]$ of f with respect to Δ_P. Moreover, it is unique up to chain homotopy.*

Proof. First we show existence of a homotopy lifting ψ of μ_P with respect to Δ_P. Consider the function $(\mu_P \otimes 1_P - 1_P \otimes \mu_P)\Delta_P$ in the Hom complex $\mathrm{Hom}_{A^e}(P, P)$. Apply the quasi-isomorphism μ_P to $\mathrm{Hom}_{A^e}(P, A)$. Note that $\mu_P(\mu_P \otimes 1_P - 1_P \otimes \mu_P) = \mu_P \otimes \mu_P - \mu_P \otimes \mu_P = 0$, and so under the quasi-isomorphism from $\mathrm{Hom}_{A^e}(P \otimes_A P, P)$ to $\mathrm{Hom}_{A^e}(P \otimes_A P, A)$, the map $\mu_P \otimes 1_P - 1_P \otimes \mu_P$ becomes 0. Further, a calculation shows that

$$\mathbf{d}(\mu_P \otimes 1_P - 1_P \otimes \mu_P) = 0,$$

and therefore $\mu_P \otimes 1_P - 1_P \otimes \mu_P$ is a boundary in $\mathrm{Hom}_{A^e}(P \otimes_A P, P)$. Precomposing with the chain map Δ_P, we see that $(\mu_P \otimes 1_P - 1_P \otimes \mu_P)\Delta_P = \mathbf{d}(\psi)$ for some $\psi : P \to P[1]$, as claimed.

Next we show existence of functions ψ_f satisfying the conditions (6.3.4) and (6.3.5). Now $(f \otimes 1_P - 1_P \otimes f)\Delta_P$ is a chain map from P to $P[-m]$ since $fd = 0$. Applying μ_P, since $|\mu_P| = 0$, we have

$$\begin{aligned}
\mu_P(f \otimes 1_P - 1_P \otimes f)\Delta_P &= f(1_P \otimes \mu_P - \mu_P \otimes 1_P)\Delta_P \\
&= -f\mathbf{d}(\psi) = -f\psi d,
\end{aligned}$$

that is, applying the quasi-isomorphism from $\mathrm{Hom}_{A^e}(P, P)$ to $\mathrm{Hom}_{A^e}(P, A)$ given by μ_P, we find that $\mu_P(f \otimes 1_P - 1_P \otimes f)\Delta_P$ is a coboundary. Consequently, in $\mathrm{Hom}_{A^e}(P, P)$, the function $(f \otimes 1_P - 1_P \otimes f)\Delta_P$ is a coboundary, so that

$$(f \otimes 1_P - 1_P \otimes f)\Delta_P = \mathbf{d}(\psi_f)$$

for some ψ_f, that is, condition (6.3.4) holds. We will show that some of the functions ψ_f satisfying (6.3.4) also satisfy condition (6.3.5). Combining the above two equations, we now have

$$\mu_P\mathbf{d}(\psi_f) = -f\psi d,$$

and since $\mathbf{d}(\psi_f) = d\psi_f + (-1)^m \psi_f d$ and $\mu_P d = 0$, this is equivalent to

$$((-1)^m \mu_P \psi_f + f\psi)d = 0.$$

However, we want $(-1)^m \mu_P \psi_f + f\psi \sim 0$. Set $g = (-1)^m \mu_P \psi_f + f\psi$, viewed as a map from P_{m-1} to A. We have seen that g is a cocycle, and thus it corresponds to a chain map $g_\bullet$ from P to $P[1-m]$. Define $\psi'_f = \psi_f - (-1)^m g_\bullet$.

Since $g_{\bullet}$ is a chain map, $\mathbf{d}(\psi'_f) = \mathbf{d}(\psi_f)$, and so ψ'_f also satisfies (6.3.4). Additionally we now have

$$(-1)^m \mu_P \psi'_f + f\psi = (-1)^m \mu_P \psi_f + f\psi - g = 0,$$

by definition of g, and so ψ'_f also satisfies (6.3.5). Replacing ψ_f by ψ'_f, we have shown that there exists a homotopy lifting of f with respect to Δ_P.

Finally, we show uniqueness up to chain homotopy. Let ψ_f and ψ'_f be two homotopy liftings of f with respect to Δ_P. Then $\mathbf{d}(\psi_f - \psi'_f) = 0$ and $\mu_P(\psi_f - \psi'_f) \sim 0$. Again, μ_P gives rise to the quasi-isomorphism from $\mathrm{Hom}_{A^e}(P, P)$ to $\mathrm{Hom}_{A^e}(P, A)$ and this implies $\psi_f - \psi'_f \sim 0$, as claimed. Note that this argument does not depend on choice of homotopy ψ for $(\mu_P \otimes 1_P - 1_P \otimes \mu_P)\Delta_P$, again as any two will be homotopic. $\square$

The following theorem and proof are [**220**, Theorem 4].

Theorem 6.3.10. *Let P be a projective resolution of A as an A^e-module with diagonal map $\Delta_P : P \to P \otimes_A P$. Let $f \in \mathrm{Hom}_{A^e}(P_m, A)$ and $g \in \mathrm{Hom}_{A^e}(P_n, A)$ be cocycles. The element of Hochschild cohomology $\mathrm{HH}^*(A)$ represented by $[f, g]$ as defined by formula (6.3.1) is independent of choice of resolution P, of diagonal map Δ_P, and of homotopy liftings ψ_f and ψ_g.*

Proof. We will prove independence of choices in the reverse order from what is stated. Independence of choice of ψ_f and ψ_g is immediate from the uniqueness of ψ_f and ψ_g up to chain homotopy stated in Lemma 6.3.9, since $fd = 0$ and $gd = 0$.

Let Δ_P and Δ'_P be two diagonal maps. Then $\Delta'_P - \Delta_P = \mathbf{d}(u)$ for some $u : P \to (P \otimes_A P)[1]$. Let ψ_f and ψ_g be homotopy liftings of f and g with respect to Δ_P. Let

$$\psi'_f = \psi_f + (-1)^m (f \otimes 1_P - 1_P \otimes f)u,$$

and similarly ψ'_g. A calculation shows that ψ'_f and ψ'_g are homotopy liftings of f and g with respect to Δ'_P, respectively. We find that

$$
\begin{aligned}
f\psi'_g &- (-1)^{(m-1)(n-1)} g\psi'_f \\
&= f\psi_g - (-1)^{(m-1)(n-1)} g\psi_f + (-1)^n f(g \otimes 1_P - 1_P \otimes g)u \\
&\quad - (-1)^{(m-1)(n-1)}(-1)^m g(f \otimes 1_P - 1_P \otimes f)u \\
&= f\psi_g - (-1)^{(m-1)(n-1)} g\psi_f,
\end{aligned}
$$

so these two expressions give the same bracket $[f, g]$ via formula (6.3.1). Thus the formula is independent of choice of diagonal map.

Let $Q \xrightarrow{\mu_Q} A$ be another projective resolution of A as an A^e-module, and let $\Delta_Q : Q \to Q \otimes_A Q$ be a diagonal map. Let $\iota : P \to Q$ and $\theta : Q \to P$ be chain maps lifting the identity map on A. Let $f \in \mathrm{Hom}_{A^e}(P_m, A)$ and

$g \in \mathrm{Hom}_{A^e}(P_n, A)$ be cocycles on P. Then $f\theta$ and $g\theta$ are cocycles on Q. Let $\psi_{f\theta}$ be a homotopy lifting for $f\theta$ with respect to Δ_Q. Set $\psi_f = \theta\psi_{f\theta}\iota$. We first check that ψ_f is a homotopy lifting for f with respect to $\Delta_P = (\theta \otimes \theta)\Delta_Q\iota$:

$$
\begin{aligned}
\mathbf{d}(\psi_f) &= \theta\mathbf{d}(\psi_{f\theta})\iota \\
&= \theta(f\theta \otimes 1_Q - 1_Q \otimes f\theta)\Delta_Q\iota \\
&= (f \otimes 1_P - 1_P \otimes f)(\theta \otimes \theta)\Delta_Q\iota \\
&= (f \otimes 1_P - 1_P \otimes f)\Delta_P,
\end{aligned}
$$

so ψ_f satisfies equation (6.3.4).

Set $\psi_P = \theta\psi_Q\iota$, where ψ_Q satisfies $\mathbf{d}(\psi_Q) = (\mu_Q \otimes 1_Q - 1_Q \otimes \mu_Q)\Delta_Q$ as well as $(-1)^m \mu_Q\psi_{f\theta} + f\theta\psi_Q \sim 0$. Then

$$
\begin{aligned}
\mathbf{d}(\psi_P) &= \theta\mathbf{d}(\psi_Q)\iota = \theta(\mu_Q \otimes 1_Q - 1_Q \otimes \mu_Q)\Delta_Q\iota \\
&= (\mu_P \otimes 1_P - 1_P \otimes \mu_P)(\theta \otimes \theta)\Delta_Q\iota \\
&= (\mu_P \otimes 1_P - 1_P \otimes \mu_P)\Delta_P,
\end{aligned}
$$

since $\theta\mu_Q = \mu_P\theta$. It follows, since $(-1)^m \mu_Q\psi_{f\theta} + f\theta\psi_Q \sim 0$, that by the definitions of ψ_f and of ψ_P above,

$$
\begin{aligned}
(-1)^m \mu_P\psi_f + f\psi_P &= (-1)^m \mu_Q\psi_{f\theta}\iota + f\theta\psi_Q\iota \\
&= ((-1)^m \mu_Q\psi_{f\theta} + f\theta\psi_Q)\iota \quad \sim \quad 0,
\end{aligned}
$$

that is, ψ_f satisfies (6.3.5). Therefore ψ_f is a homotopy lifting of f with respect to Δ_P, and we may similarly define a homotopy lifting of g.

Finally, formula (6.3.1) applied to f, g on P yields

$$
\begin{aligned}
[f, g] &= f\theta\psi_{g\theta}\iota - (-1)^{(m-1)(n-1)}g\theta\psi_{f\theta}\iota \\
&= [f\theta, g\theta]\iota,
\end{aligned}
$$

so the chain map ι takes $[f\theta, g\theta]$ to $[f, g]$. Thus the bracket does not depend on choice of resolution. $\qquad\square$

As a consequence of Theorem 6.3.10, the bracket given by formula (6.3.1) agrees with the Gerstenhaber bracket of Definition 1.4.1 on Hochschild cohomology.

Theorem 6.3.11. *Let P be a projective resolution of A as an A^e-module. Let $f \in \mathrm{Hom}_{A^e}(P_m, A)$ and $g \in \mathrm{Hom}_{A^e}(P_n, A)$ be cocycles on P, and let ψ_f and ψ_g be homotopy liftings of f and g, as in Definition 6.3.6. The bracket given at the chain level by*

$$
[f, g] = f\psi_g - (-1)^{(m-1)(n-1)}g\psi_f
$$

induces the Gerstenhaber bracket on Hochschild cohomology $\mathrm{HH}^*(A)$.

Proof. In Example 6.3.8, we saw that taking P to be the bar resolution recovers the Gerstenhaber bracket of Definition 1.4.1 from formula (6.3.1). By Theorem 6.3.10, it is independent of choices. $\square$

Remark 6.3.12. In practice, often $(\mu_P \otimes 1_P)\Delta_P = 1_P = (1_P \otimes \mu_P)\Delta_P$, in which case Definition 6.3.6 of homotopy lifting can be simplified by taking $\psi = 0$. See Example 6.3.8, the next Section 6.4, and [220, Remark 1].

Exercise 6.3.13. Let ψ_f, ψ_g be homotopy liftings of f, g as in Definition 6.3.6.

(a) Check directly that $[f, g]$ as defined in (6.3.1) is a cocycle.

(b) Check directly that if either f or g is a coboundary, then so is $[f, g]$ as defined in (6.3.1). (See (6.3.7) for a homotopy lifting of a coboundary and verify first that it is indeed a homotopy lifting.)

Exercise 6.3.14. Verify that in the context of Example 6.3.8, the formula for ψ_f indeed yields the classical Gerstenhaber bracket as defined on the bar resolution in Definition 1.4.1.

6.4. Differential graded coalgebras

Some of the results of the previous sections lead to effective computational techniques for the Lie structure on Hochschild cohomology. In particular, we explain in this section some settings in which the theory of homotopy liftings can be simplified for computational purposes. Formula (6.4.2) below gives homotopy liftings for all cocycles f in terms of a diagonal map Δ_P and an additional function ϕ_P under some conditions that we discuss next.

Let $\phi_P : P \otimes_A P \to P[1]$ be a homotopy for $\mu_P \otimes 1_P - 1_P \otimes \mu_P$. That is,

$$(6.4.1) \qquad \mathbf{d}(\phi_P) = \mu_P \otimes 1_P - 1_P \otimes \mu_P.$$

To see that such a homotopy exists, consider the quasi-isomorphism μ_P from $\mathrm{Hom}_{A^e}(P \otimes_A P, P)$ to $\mathrm{Hom}_{A^e}(P \otimes_A P, A)$, which takes $\mu_P \otimes 1_P - 1_P \otimes \mu_P$ to 0. Since μ_P is a chain map, $\mathbf{d}(\mu_P \otimes 1_P - 1_P \otimes \mu_P) = 0$. It follows that in $\mathrm{Hom}_{A^e}(P \otimes_A P, A)$, the map $\mu_P \otimes 1_P - 1_P \otimes \mu_P$ is a cocycle that becomes 0 (which is a coboundary) after applying the quasi-isomorphism μ_P, so it must be a coboundary in $\mathrm{Hom}_{A^e}(P \otimes_A P, A)$.

Let

$$\Delta_P^{(2)} = (\Delta_P \otimes 1_P)\Delta_P.$$

Note that in general this may not be the same as $(1_P \otimes \Delta_P)\Delta_P$. Let $f \in \mathrm{Hom}_{A^e}(P_m, A)$ with $fd = 0$, and let $\psi_f : P \to P[1-m]$ be defined by

$$(6.4.2) \qquad \psi_f = \phi_P(1_P \otimes f \otimes 1_P)\Delta_P^{(2)}.$$

Here the map $1_P \otimes f \otimes 1_P$ is considered to be a map from $P \otimes_A P \otimes_A P$ to $P \otimes_A P$ (upon applying the canonical isomorphism $P \otimes_A A \otimes_A P \cong P \otimes_A P$). We will see next that under some conditions, ψ_f is a homotopy lifting of f with respect to Δ_P as in Definition 6.3.6. In this case, homotopy liftings ψ_f for all cocycles f may be found in terms of these two maps Δ_P and ϕ_P via formula (6.4.2).

Example 6.4.3. Consider the bar resolution $B = B(A)$ of A. We may identify $B_i \otimes_A B_j$ with $A \otimes A^{\otimes i} \otimes A \otimes A^{\otimes j} \otimes A$ under the isomorphism

$$(A \otimes A^{\otimes i} \otimes A) \otimes_A (A \otimes A^{\otimes j} \otimes A) \xrightarrow{\sim} A \otimes A^{\otimes i} \otimes A \otimes A^{\otimes j} \otimes A$$

$$(a_0 \otimes \cdots \otimes a_{i+1}) \otimes_A (a_0' \otimes \cdots \otimes a_{j+1}') \mapsto$$

$$a_0 \otimes (a_1 \otimes \cdots \otimes a_i) \otimes (a_{i+1} a_0') \otimes (a_1' \otimes \cdots \otimes a_j') \otimes a_{j+1}'$$

for all $a_0, \ldots, a_{i+1}, a_0', \ldots, a_{j+1}' \in A$. Define $\phi_B : B \otimes_A B \to B[1]$ by

$$\phi_B(a_0 \otimes (a_1 \otimes \cdots \otimes a_i) \otimes (a_{i+1}) \otimes (a_{i+2} \otimes \cdots \otimes a_{i+j+1}) \otimes a_{i+j+2})$$

$$= (-1)^i a_0 \otimes \cdots \otimes a_{i+j+2}$$

for all $a_0, \ldots, a_{i+j+2} \in A$, that is, up to sign, we have just removed parentheses. Then ψ_f defined as in (6.4.2) agrees with that given in Example 6.3.8.

In the rest of this section we prove results under some hypotheses that will be satisfied for example by Koszul algebras as defined in Section 3.4. A result of Buchweitz, Green, Snashall, and Solberg [**39**] for Koszul algebras guarantees that the standard embedding $\iota : P \to B$ of the Koszul resolution P into the bar resolution B of A (see (3.4.4)) is preserved by the diagonal map in the sense that the diagonal map Δ_B of the bar resolution given by formula (2.3.3) takes $\iota(P)$ to $\iota(P) \otimes_A \iota(P)$. Thus we may define a diagonal map Δ_P on P via this embedding. It follows that Δ_P is coassociative, that is, $(\Delta_P \otimes 1_P)\Delta_P = (1_P \otimes \Delta_P)\Delta_P$ and $(\mu_P \otimes 1_P)\Delta_P = 1_P = (1_P \otimes \mu_P)\Delta_P$, and we say that P is a *counital differential graded coalgebra*. In this case, in Definition 6.3.6 of homotopy lifting of an m-cocycle f on P, we may take $\psi = 0$, and so condition (6.3.5) becomes $\mu_P \psi_f \sim 0$. We may in fact assume under these conditions that $\psi_f|_{P_{m-1}} = 0$. This simplifies the work of finding homotopy liftings (and it simplifies many of the proofs of the previous section under these additional hypotheses). In fact formula (6.4.2) always defines a homotopy lifting in the case that Δ_P, μ_P give P a coalgebra structure, as we see next.

Lemma 6.4.4. *Let $P \xrightarrow{\mu_P} A$ be a projective resolution of the A^e-module A that is a counital differential graded coalgebra. Let $f \in \operatorname{Hom}_{A^e}(P_m, A)$ be a cocycle, and let $\psi_f : P \to P[1-m]$ be defined by formula (6.4.2). Then ψ_f is a homotopy lifting of f with respect to Δ_P.*

Proof. Let $\phi_P : P \otimes_A P \to P[1]$ be a homotopy for $\mu_P \otimes 1_P - 1_P \otimes \mu_P$, that is, equation (6.4.1) holds. Set $\Delta_P^{(2)} = (\Delta_P \otimes 1_P)\Delta_P = (1_P \otimes \Delta_P)\Delta_P$. Since $\Delta_P^{(2)}$ and $1_P \otimes f \otimes 1_P$ are chain maps,

$$
\begin{aligned}
\mathbf{d}(\psi_f) &= \mathbf{d}(\phi_P)(1_P \otimes f \otimes 1_P)\Delta_P^{(2)} \\
&= (\mu_P \otimes 1_P - 1_P \otimes \mu_P)(1_P \otimes f \otimes 1_P)\Delta_P^{(2)} \\
&= (\mu_P \otimes f \otimes 1_P - 1_P \otimes f \otimes \mu_P)\Delta_P^{(2)} \\
&= ((f \otimes 1_P)(\mu_P \otimes 1_P \otimes 1_P) - (1_P \otimes f)(1_P \otimes 1_P \otimes \mu_P))\Delta_P^{(2)} \\
&= (f \otimes 1_P - 1_P \otimes f)\Delta_P.
\end{aligned}
$$

Note that $\psi_f|_{P_{m-1}} = 0$ by its definition, and as explained above, we may take $\psi = 0$ in Definition 6.3.6. Therefore ψ_f is a homotopy lifting of f. $\square$

The following theorem is a reworked version of [**165**, Theorem 3.2.5], which has stronger hypotheses, and of [**165**, Lemma 3.4.1], which has somewhat different hypotheses.

Theorem 6.4.5. *Let $P \xrightarrow{\mu_P} A$ be a projective resolution of the A^e-module A that is a counital differential graded coalgebra. Let $f \in \operatorname{Hom}_{A^e}(P_m, A)$ and $g \in \operatorname{Hom}_{A^e}(P_n, A)$ be cocycles. Define ψ_f by formula (6.4.2), and similarly ψ_g. Then*

$$
[f, g] = f\psi_g - (-1)^{(m-1)(n-1)}g\psi_f
$$

represents the Gerstenhaber bracket of f and g on Hochschild cohomology.

Proof. This follows immediately from Lemma 6.4.4 and Theorem 6.3.11. $\square$

Remark 6.4.6. If the hypotheses of the theorem do not hold, a homotopy ϕ_P for $\mu_P \otimes 1_P - 1_P \otimes \mu_P$ may still be used to define the Gerstenhaber bracket in a similar way, with the addition of some error terms. See [**220**, Corollary 5 and Remark 1] in which the Gerstenhaber bracket is given generally as

$$
\begin{aligned}
(6.4.7) \quad [f, g] &= -f\phi_P(g \otimes 1_P \otimes 1_P - 1_P \otimes g \otimes 1_P + 1_P \otimes 1_P \otimes g)\Delta_P^{(2)} \\
&\quad + (-1)^{(m-1)(n-1)}g\phi_P(f \otimes 1_P \otimes 1_P - 1_P \otimes f \otimes 1_P + 1_P \otimes 1_P \otimes f)\Delta_P^{(2)}.
\end{aligned}
$$

We caution that the function

$$-\phi_P(f \otimes 1_P \otimes 1_P - 1_P \otimes f \otimes 1_P + 1_P \otimes 1_P \otimes f)\Delta_P^{(2)}$$

is not necessarily a homotopy lifting of f; the formula (6.4.7) instead results from a more complicated homotopy lifting as explained in the proof of [**220**, Corollary 5].

In the remainder of this section, we apply Theorem 6.4.5 to an example, a polynomial ring in two indeterminates. The case of n indeterminates is similar, if more notationally unwieldy, and is handled in [**165**, Section 4], showing that formula (6.3.1) indeed yields the familiar Gerstenhaber bracket on Hochschild cohomology of a polynomial ring. In other settings the first computation of Gerstenhaber brackets, or of a related Batalin-Vilkovisky structure, used these techniques (see, e.g., [**101, 102, 166, 220**]).

Example 6.4.8. Let $A = k[x, y]$, and let P be its Koszul resolution (3.1.4), that is, $P = A \otimes \bigwedge^{\bullet} V \otimes A$, where $V = \mathrm{Span}_k\{x, y\}$. Identify $P_i \otimes_A P_j$ with $A \otimes \bigwedge^i V \otimes A \otimes \bigwedge^j V \otimes A$ for each i, j, and identify $\bigwedge^0 V$ with k and $\bigwedge^1 V$ with V. Thus for example, $P_0 \otimes_A P_1 \cong A \otimes k \otimes A \otimes V \otimes A \cong A \otimes A \otimes V \otimes A$, and we use such expressions in our definitions of maps below. We first find a homotopy $\phi_P : P \otimes_A P \to P[1]$ for $\mu_P \otimes 1_P - 1_P \otimes \mu_P$. In degree 2, the map ϕ_P is necessarily 0 since $P_3 = 0$. We define ϕ_P in degrees 0 and 1 on free basis elements:

$$\phi_P(1 \otimes x^i y^j \otimes 1) = \sum_{l=1}^{j} x^i y^{j-l} \otimes y \otimes y^{l-1} + \sum_{l=1}^{i} x^{i-l} \otimes x \otimes x^{l-1} y^j,$$

$$\phi_P(1 \otimes x^i y^j \otimes x \otimes 1) = -\sum_{l=1}^{j} x^i y^{j-l} \otimes x \wedge y \otimes y^{l-1},$$

$$\phi_P(1 \otimes x^i y^j \otimes y \otimes 1) = 0,$$

$$\phi_P(1 \otimes x \otimes x^i y^j \otimes 1) = 0,$$

$$\phi_P(1 \otimes y \otimes x^i y^j \otimes 1) = \sum_{l=1}^{i} x^{i-l} \otimes x \wedge y \otimes x^{l-1} y^j.$$

We use this function ϕ_P, formula (6.4.2) for ψ_f, and the formula of Theorem 6.4.5 to compute some brackets in degree 1. The diagonal map Δ_P is defined by the standard embedding (3.4.12) of P into the bar resolution, followed by the standard diagonal map (2.3.3) on the bar resolution. Consider the cocycles in degree 1 denoted by $x^i y^j x^*$ and $x^i y^j y^*$, where $\{x^*, y^*\}$ is the dual basis to $\{x, y\}$ (e.g., $x^i y^j x^*$ takes x to $x^i y^j$ and y to 0, where x, y are identified with their images in $\bigwedge^1 V$). First we find some values of $\psi_{x^i y^j x^*}$

and $\psi_{x^i y^j y^*}$ via formula (6.4.2):

$$\begin{aligned}
\psi_{x^i y^j x^*}(1 \otimes x \otimes 1) &= \phi_P(1 \otimes x^i y^j \otimes 1), \\
\psi_{x^i y^j x^*}(1 \otimes y \otimes 1) &= 0, \\
\psi_{x^i y^j y^*}(1 \otimes x \otimes 1) &= 0, \\
\psi_{x^i y^j y^*}(1 \otimes y \otimes 1) &= \phi_P(1 \otimes x^i y^j \otimes 1).
\end{aligned}$$

It follows that, for example,

$$\begin{aligned}
&[x^i y^j x^*, \ x^m y^n x^*](1 \otimes x \otimes 1) \\
&= x^i y^j x^* \psi_{x^m y^n x^*}(1 \otimes x \otimes 1) - x^m y^n x^* \psi_{x^i y^j x^*}(1 \otimes x \otimes 1) \\
&= x^i y^j x^* \phi_P(1 \otimes x^m y^n \otimes 1) - x^m y^n x^* \phi_P(1 \otimes x^i y^j \otimes 1) \\
&= \sum_{l=1}^{m} x^i y^j x^{m-l} x^{l-1} y^n - \sum_{l=1}^{i} x^m y^n x^{i-l} x^{l-1} y^j \\
&= m x^i y^j x^{m-1} y^n - i x^m y^n x^{i-1} y^j \\
&= x^i y^j \frac{\partial}{\partial x}(x^m y^n) - x^m y^n \frac{\partial}{\partial x}(x^i y^j).
\end{aligned}$$

Another calculation shows that the value of this bracket function on $1 \otimes y \otimes 1$ is zero. Therefore, for all $p, q \in A$, we have

$$[px^*, \ qx^*] = (p \frac{\partial}{\partial x}(q) - q \frac{\partial}{\partial x}(p))x^*.$$

Similarly we find that

$$\begin{aligned}
[px^*, \ qy^*] &= p \frac{\partial}{\partial x}(q)y^* - q \frac{\partial}{\partial y}(p)x^*, \\
[py^*, \ qy^*] &= (p \frac{\partial}{\partial y}(q) - q \frac{\partial}{\partial y}(p))y^*.
\end{aligned}$$

We may calculate other brackets using the same techniques (cf. Example 6.2.7 and Exercise 6.2.12).

Exercise 6.4.9. Verify that ψ_f as defined in Example 6.4.3 agrees with that given in Example 6.3.8.

Exercise 6.4.10. Let $P \xrightarrow{\mu_P} A$ be a projective resolution of the A^e-module A that is a counital differential graded coalgebra. Show that one choice of chain map $g.$ corresponding to a cocycle $g \in \operatorname{Hom}_{A^e}(P_n, A)$ is given by $g_m = (1 \otimes g)(\Delta_P)_m$ for all m. (This draws a parallel between the cup product as given by formula (2.2.1) and the Gerstenhaber bracket as given by formula (6.3.1) using (6.4.2).)

Exercise 6.4.11. Verify the formulas for $[px^*, qy^*]$ and $[py^*, qy^*]$ in Example 6.4.8. Find $[px^*, qx^* \wedge y^*]$ and compare with Example 6.2.7.

6.5. Extensions

In this section, we consider Schwede's exact sequence interpretation of the Lie structure on Hochschild cohomology [194]. Hermann [110] generalized Schwede's results to some exact monoidal categories, and gave a description of the bracket with degree 0 elements in [108], completing Schwede's interpretation. We refer to these papers for most of the technical details and proofs, and present just a skimming here.

Let $n \geq 1$, and let $\mathcal{E}xt^n_{A^e}(A, A)$ denote the category whose objects are n-extensions of A by A as an A^e-module, and morphisms are maps of n-extensions. (See Section A.3 for a discussion of maps and equivalence classes of n-extensions.) View $\operatorname{HH}^n(A) = \operatorname{Ext}^n_{A^e}(A, A)$ as equivalence classes of objects in $\mathcal{E}xt^n_{A^e}(A, A)$. Consider an m-extension and an n-extension of A by A,

$$\mathbf{f}: \qquad 0 \longrightarrow A \xrightarrow{i_M} M_{m-1} \longrightarrow \cdots \longrightarrow M_0 \xrightarrow{\mu_M} A \longrightarrow 0,$$

$$\mathbf{g}: \qquad 0 \longrightarrow A \xrightarrow{i_N} N_{n-1} \longrightarrow \cdots \longrightarrow N_0 \xrightarrow{\mu_N} A \longrightarrow 0.$$

We will not need notation for the unlabeled maps. We assume that all M_j, N_j are projective as left A-modules, and as right A-modules, where needed. See [194] for a discussion about such an assumption.

Let P be a projective resolution of A as an A^e-module. Let f and g be an m-cocycle and an n-cocycle on P, corresponding to the generalized extensions $\mathbf{f}$ and $\mathbf{g}$, respectively. So $f \in \operatorname{Hom}_{A^e}(P_m, A)$ may be defined via the following commuting diagram, which exists by the Comparison Theorem (Theorem A.2.7); see also Section A.3. The map 1 from A to A is the identity map. We denote by $\hat{f}. : P. \to M.$ the chain map indicated below, so that $f = \hat{f}_m$.

$$\cdots \longrightarrow P_m \xrightarrow{d_m} P_{m-1} \xrightarrow{d_{m-1}} \cdots \xrightarrow{d_1} P_0 \xrightarrow{\mu_P} A \longrightarrow 0$$
$$\Big\downarrow{f} \qquad\quad \Big\downarrow{\hat{f}_{m-1}} \qquad\qquad\quad \Big\downarrow{\hat{f}_0} \qquad \Big\downarrow{1}$$
$$0 \longrightarrow A \xrightarrow{i_M} M_{m-1} \longrightarrow \cdots \longrightarrow M_0 \xrightarrow{\mu_M} A \longrightarrow 0$$

As in Section 2.4, we write $\mathbf{g} \smile \mathbf{f}$ for the following Yoneda splice. In [**194**], it is denoted $\mathbf{g}\#\mathbf{f}$.

$\mathbf{g} \smile \mathbf{f}$:

$$0 \longrightarrow A \xrightarrow{i_M} M_{m-1} \longrightarrow \cdots \longrightarrow M_0 \xrightarrow{i_N \mu_M} N_{n-1} \longrightarrow$$

$$\cdots \longrightarrow N_0 \xrightarrow{\mu_N} A \longrightarrow 0$$

Note that an element of $\mathrm{Hom}_{A^e}(P_{m+n}, A)$ corresponding to this $(m + n)$-extension is $(-1)^{mn} f \smile g$ as discussed in Section 2.4. We caution that our definition of cup product is opposite that in [**194**], leading to some sign differences.

We write $\mathbf{f} \otimes_A \mathbf{g}$ for the $(m + n)$-extension corresponding to the total complex of the tensor product of the two truncated sequences as discussed in Section 2.4. In degree $m + n - 1$, for example, we have the module $(M_m \otimes_A N_{n-1}) \oplus (M_{m-1} \otimes_A N_n) \cong N_{n-1} \oplus M_{m-1}$, and the extension is

$\mathbf{f} \otimes_A \mathbf{g}$:

$$0 \longrightarrow A \longrightarrow M_{m-1} \oplus N_{n-1} \longrightarrow \cdots$$

$$\xrightarrow{\quad} \begin{array}{c}(M_1 \otimes_A N_0) \oplus \\ (M_0 \otimes_A N_1)\end{array} \longrightarrow M_0 \otimes_A N_0 \longrightarrow A \longrightarrow 0$$

The cup product of f and g corresponds to any of the $(m+n)$-extensions $\mathbf{g} \smile \mathbf{f}$, $\mathbf{f} \otimes_A \mathbf{g}$, $(-1)^{mn}\mathbf{f} \smile \mathbf{g}$, $(-1)^{mn}\mathbf{g} \otimes_A \mathbf{f}$. (For an additive inverse $-\mathbf{f} \smile \mathbf{g}$ to the extension $\mathbf{f} \smile \mathbf{g}$, we take here the extension whose modules agree with those of $\mathbf{f} \smile \mathbf{g}$, the map μ_M is replaced by $-\mu_M$, and all other maps agree with those of $\mathbf{f} \smile \mathbf{g}$.) These extensions are all equivalent and there are maps as indicated in the following diagram:

(6.5.1)

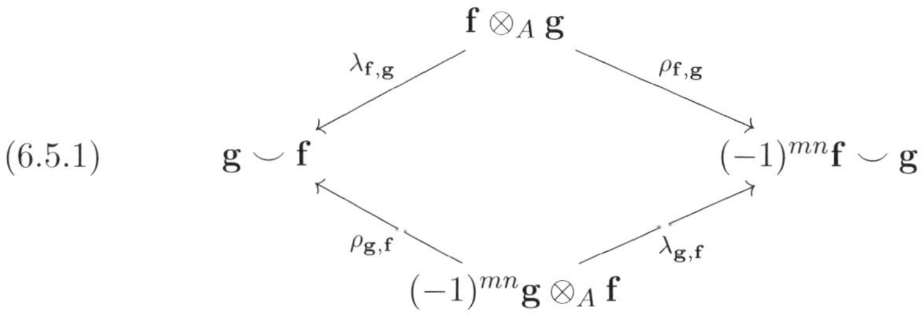

Such maps may be described as follows. Consider the augmented double complex:

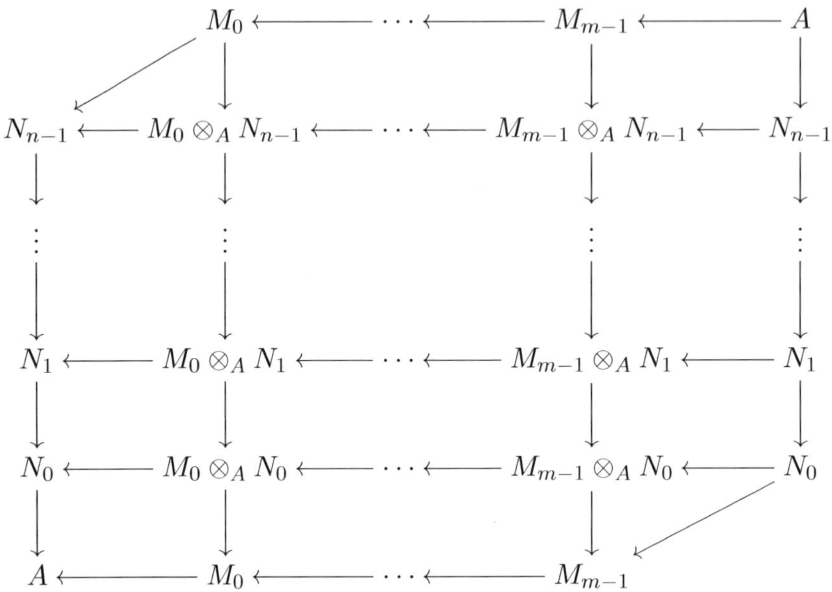

All but the leftmost column and bottom row constitute the truncated double complex corresponding to $\mathbf{f} \otimes_A \mathbf{g}$, and the outermost rows and columns are $\mathbf{g} \smile \mathbf{f}$ (left column and top row) and $(-1)^{mn}\mathbf{f} \smile \mathbf{g}$ (right column and bottom row). The maps $\lambda_{\mathbf{f},\mathbf{g}} : \mathbf{f} \otimes_A \mathbf{g} \to \mathbf{g} \smile \mathbf{f}$ and $\rho_{\mathbf{f},\mathbf{g}} : \mathbf{f} \otimes_A \mathbf{g} \to (-1)^{mn}\mathbf{f} \smile \mathbf{g}$ are given as follows. See also Section 2.4 where these maps are used to show that two definitions of associative product on $\mathrm{HH}^*(A)$ agree.

For $n \leq i \leq m+n$, $\lambda_{\mathbf{f},\mathbf{g}}$ projects $(\mathbf{f} \otimes_A \mathbf{g})_i$ onto $M_{i-n} \otimes_A N_n \cong M_{i-n}$. For $0 \leq i \leq n-1$, $\lambda_{\mathbf{f},\mathbf{g}}$ first projects $(\mathbf{f} \otimes_A \mathbf{g})_i$ onto $M_0 \otimes_A N_i$, then maps to $A \otimes_A N_i \cong N_i$ via $\mu_N \otimes 1$. For $m \leq i \leq m+n$, $\rho_{\mathbf{f},\mathbf{g}}$ projects $(\mathbf{f} \otimes_A \mathbf{g})_i$ onto $M_m \otimes_A N_{i-m} \cong N_{i-m}$ followed by multiplication by $(-1)^{m(m+n-i)}$. For $0 \leq i \leq m-1$, $\rho_{\mathbf{f},\mathbf{g}}$ projects $(\mathbf{f} \otimes_A \mathbf{g})_i$ onto $M_i \otimes_A N_0$ followed by multiplication by $(-1)^{mn}$, then maps to $M_i \otimes_A A \cong M_i$ via $1 \otimes \mu_N$. Similarly define maps $\rho_{\mathbf{g},\mathbf{f}} : (-1)^{mn}\mathbf{g} \otimes_A \mathbf{f} \to \mathbf{g} \smile \mathbf{f}$ and $\lambda_{\mathbf{g},\mathbf{f}} : (-1)^{mn}\mathbf{g} \otimes_A \mathbf{f} \to (-1)^{mn}\mathbf{f} \smile \mathbf{g}$.

Diagram (6.5.1) represents a loop in the classifying space of the extension category $\mathcal{E}xt_{A^e}^{m+n}(A, A)$. This information can be realized combinatorially without reference to the topology, as explained in [**194**]. Basically, paths in the space are zigzags of maps, that is, you can travel either way along a map (from the domain to the codomain or vice versa) and compose such trips (recalling the equivalence relation generated by maps of generalized extensions).

Loops in the classifying space of $\mathcal{E}xt_{A^e}^{m+n}(A, A)$ are in one-to-one correspondence with $\mathrm{Ext}_{A^e}^{m+n-1}(A, A)$ [184]. Under this correspondence, the loop (6.5.1) corresponds to the Gerstenhaber bracket $[f, g]$ [194]. We refer to Retakh [184] and Schwede [194] for most details and proofs. Here we give some of the algebraic ideas underlying Schwede's result. Specifically, for the projective resolution P of A as an A^e-module, we look closely at some maps $P \to \mathbf{g} \smile \mathbf{f}$ arising from the maps comprising the loop (6.5.1).

Replace the loop (6.5.1) with another equivalent loop as follows. Starting with $(-1)^{mn} f \smile g \in \mathrm{Hom}_{A^e}(P_{m+n}, A)$, define an $(m + n)$-extension of A by A by a pushout diagram as in Section A.3. The $(m + n)$-extension $K = K((-1)^{mn} f \smile g)$ is given by

$$0 \longrightarrow A \longrightarrow K_{m+n-1} \longrightarrow P_{m+n-2} \longrightarrow \cdots \longrightarrow P_0 \longrightarrow A \longrightarrow 0,$$

where

$$K_{m+n-1} = (P_{m+n-1} \oplus A)/\{(-d_{m+n}(x),\ (-1)^{mn}(f \smile g)(x)) \mid x \in P_{m+n}\}.$$

The proof in [194] has $P = B$, the bar resolution, since the goal there is to show that the loop (6.5.1) corresponds to the historical definition of Gerstenhaber bracket on the bar resolution. A tedious comparison of maps shows that there is a map ε for which the rightmost quadrilateral in the following diagram commutes:

(6.5.2)

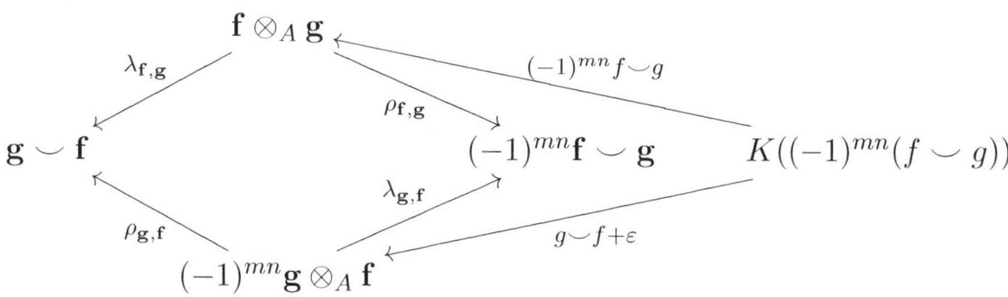

We define the map $(-1)^{mn}(f \smile g)$ (and similarly $g \smile f$), as in [194], to be the map induced by all $(\hat{f}_i \otimes \hat{g}_j)\Delta_B$, where Δ_B is the diagonal map (2.3.3) on the bar resolution B.

Delete the component $(-1)^{mn}\mathbf{f} \smile \mathbf{g}$ from diagram (6.5.2), thus replacing loop (6.5.1) with loop (6.5.3) below. For this purpose, in Schwede's general theory [194], commutativity of the rightmost quadrilateral in diagram (6.5.2) is needed, and so the error correction term ε in diagram (6.5.2) is crucial.

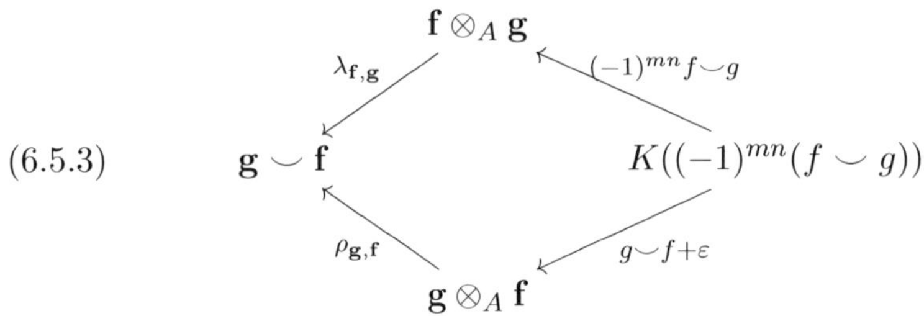

The following theorem is due to Schwede [**194**].

Theorem 6.5.4. *Let* **f** *and* **g** *be an m- and an n-extension, respectively, of A by A as an A^e-module. Let $f \in \mathrm{Hom}_{A^e}(B_m, A)$ and $g \in \mathrm{Hom}_{A^e}(B_n, A)$ be corresponding cocycles, where $B = B(A)$ is the bar resolution (1.1.4) of A. There is a chain homotopy $s : B \to \mathbf{g} \smile \mathbf{f}$, factoring through $K((-1)^{mn}(f \smile g))$, between $\lambda_{\mathbf{f},\mathbf{g}}((-1)^{mn}(f \smile g))$ and $\rho_{\mathbf{g},\mathbf{f}}(g \smile f + \varepsilon)$, for which the Gerstenhaber bracket is represented at the chain level by*

$$[f, g] = (-1)^{n-1} s_{m+n-1}.$$

Exercise 6.5.5. Verify that $\lambda_{\mathbf{f},\mathbf{g}}$, $\rho_{\mathbf{f},\mathbf{g}}$, $\lambda_{\mathbf{g},\mathbf{f}}$, $\rho_{\mathbf{g},\mathbf{f}}$ of diagram (6.5.1) are all indeed maps of $(m + n)$-extensions.

Exercise 6.5.6. Use the definitions of the maps $f \smile g$, $g \smile f$, $\rho_{\mathbf{f},\mathbf{g}}$, and $\lambda_{\mathbf{g},\mathbf{f}}$ in diagram (6.5.2) to find a map ε for which the rightmost quadrilateral commutes (cf. [**194**, p. 171]).

Chapter 7

Infinity Algebras

There are several appearances in Hochschild cohomology of higher order operations, the original idea of which is due to Stasheff [**206**]. Some of these operations extend those on underlying chain complexes, such as the cup product operation, giving rise to infinity algebras. In this chapter, we look at a few settings where such infinity algebras arise in relation to Hochschild cohomology. We define A_∞-algebras and infinitesimal n-deformations, deriving a correspondence between infinitesimal n-deformations and Hochschild n-cocycles. We define minimal models and formality of A_∞-algebras, presenting a characterization of Koszul algebras in this setting. We define the A_∞-center of an A_∞-algebra and look closely at the case of Hochschild cohomology of an augmented algebra. We define L_∞-algebras and formality of associative algebras, making a connection with deformations of algebras and Deligne's Conjecture. There are many further applications of infinity structures that we will not present here.

Indexing and sign conventions vary somewhat in the literature; we make some of the more standard choices.

7.1. A_∞-algebras

In this section we define A_∞-algebras (also called strongly homotopy associative algebras) and their morphisms, and give some examples relevant to Hochschild cohomology.

Definition 7.1.1. An A_∞-*algebra* is a graded vector space $A = \bigoplus_{i\in\mathbb{Z}} A_i$ together with graded linear maps

$$m_n : A^{\otimes n} \to A$$

of degree $|m_n| = 2 - n$ for all $n \geq 1$ such that

$$(7.1.2) \qquad \sum_{r+s+t=n} (-1)^{r+st} m_{r+1+t}(1^{\otimes r} \otimes m_s \otimes 1^{\otimes t}) = 0.$$

The equations (7.1.2) are called *Stasheff identities*. Sometimes we denote an A_∞-algebra by $(A, m_\bullet)$ to emphasize the notation chosen for these higher multiplication maps. If $m_1 = 0$, then A is called *minimal*.

We consider the implications of equation (7.1.2) for small values of n: if $n = 1$, we must take $s = 1$ and $r = t = 0$, and so the equation is

$$m_1^2 = 0,$$

that is, m_1 is a differential on A. We will thus sometimes write $d = m_1$. If $n = 2$, we may take $s = 2$ and $r = t = 0$, or $s = 1$ and $r + t = 1$, to obtain

$$m_1 m_2 - m_2(m_1 \otimes 1) - m_2(1 \otimes m_1) = 0.$$

To express this equation on elements of A, we may write $m_1(a) = d(a)$ and $m_2(a \otimes b) = a \cdot b$, and the equation becomes

$$d(a \cdot b) = d(a) \cdot b + (-1)^{|a|} a \cdot d(b)$$

for all homogeneous $a, b \in A$, due to the standard sign convention (2.3.1). That is, $m_1 = d$ is a graded derivation with respect to m_2. If $n = 3$, equation (7.1.2) becomes

$$m_1 m_3 + m_2(m_2 \otimes 1) - m_2(1 \otimes m_2)$$
$$+ m_3(m_1 \otimes 1^{\otimes 2}) + m_3(1 \otimes m_1 \otimes 1) + m_3(1^{\otimes 2} \otimes m_1) = 0,$$

which may be rewritten

$$\delta(m_3) = m_2(1 \otimes m_2) - m_2(m_2 \otimes 1),$$

where δ is the differential induced by $m_1 = d$ on the complex $\mathrm{Hom}_k(A^{\otimes 3}, A)$. That is, m_2 is associative up to a coboundary in this Hom complex. It follows that the cohomology of (A, m_1) is a graded associative algebra with multiplication induced by m_2.

We give some examples.

Example 7.1.3. If A is any differential graded algebra, we may take m_1 to be its differential, m_2 its multiplication, and $m_n = 0$ for $n \geq 3$, to define an A_∞-algebra structure on A. In this way, the space $C^*(B, B)$ of Hochschild cochains on an associative algebra B becomes an A_∞-algebra (using (1.3.4)). An associative algebra itself may be viewed as a differential graded algebra with zero differential, and thus as an A_∞-algebra in this way. If an A_∞-algebra A is concentrated in degree 0, that is, $A_i = 0$ for all $i \neq 0$, then the maps m_n are necessarily zero maps for all $n \neq 2$ since $|m_n| = 2 - n$, so A is simply an associative algebra.

Next are some examples with $m_n \neq 0$ for arbitrary values of n, the first due to Penkava and Schwarz [**171**].

Example 7.1.4. Let B be an associative algebra, let $A = B[x]/(x^2)$, and let n be a positive integer. We take A to be graded with $|b| = 0$ for all $b \in B$ and $|x| = 2 - n$. Let $g : B^{\otimes n} \to B$ be a Hochschild n-cocycle. We will put an A_∞-algebra structure on A, depending on g. Define a linear function $xg : A^{\otimes n} \to A$ by

$$(xg)(a_1 \otimes \cdots \otimes a_n) = \begin{cases} xg(a_1 \otimes \cdots \otimes a_n), & \text{if } a_1, \ldots, a_n \in B, \\ 0, & \text{if } x \text{ is a factor of } a_1 \cdots a_n. \end{cases}$$

Let $\pi : A^{\otimes 2} \to A$ denote multiplication on A. If $n = 2$, let $m_2 = \pi + xg$ and $m_i = 0$ for all $i \neq 2$. If $n \neq 2$, let $m_2 = \pi$, $m_n = xg$, and $m_i = 0$ for all $i \notin \{2, n\}$. Calculations show that A is an A_∞-algebra. We will derive a connection to algebraic deformation theory via Definition 7.1.7 below; see Theorem 7.1.8.

Example 7.1.5. Let n be a positive integer, $n > 2$, let $B = k[x]/(x^n)$, and let $A = \mathrm{Ext}_B^*(k, k)$. As shown in Example 2.5.10, $A \cong k[y, z]/(y^2)$ with $|y| = 1$ and $|z| = 2$. We will put an A_∞-algebra structure on A. We take m_2 to be multiplication on A, m_i to be the zero map if $i \notin \{2, n\}$, and

$$m_n(y^{i_1} z^{j_1} \otimes \cdots \otimes y^{i_n} z^{j_n}) = \begin{cases} z^{j_1 + \cdots + j_n + 1}, & \text{if } i_1 = \cdots = i_n = 1, \\ 0, & \text{otherwise,} \end{cases}$$

for all nonnegative integers $j_1, \ldots, j_n$ and all $i_1, \ldots, i_n \in \{0, 1\}$. Calculations show that A is an A_∞-algebra for each $n > 2$. This example may also be constructed via the general method outlined in the proof of Theorem 7.2.2 below. In Section 7.3 we will discuss a distinction between this A_∞-structure on $\mathrm{Ext}_B^*(k, k)$ and the structure of $\mathrm{Ext}_{B'}^*(k, k)$ when $B' = k[x]/(x^2)$, which is by contrast a Koszul algebra.

We say that an A_∞-algebra A is *generated* by a subset S if A coincides with its smallest subspace containing S that is closed under all m_n.

Example 7.1.6. In Example 7.1.5, $m_n(y \otimes \cdots \otimes y) = z$. Since y, z generate A as an associative algebra, we see that as an A_∞-algebra, A is generated by y alone.

The next definition generalizes that of an infinitesimal deformation in Definition 5.2.1.

Definition 7.1.7. Let n be a positive integer. Let B be an associative algebra, and let $A = B[x]/(x^2)$, where $|b| = 0$ for all $b \in B$ and $|x| = 2 - n$. An *infinitesimal n-deformation* of B is a $k[x]/(x^2)$-multilinear A_∞-algebra structure on A that lifts the multiplication of B. That is, under composition with the vector space quotient map from A to $A/(x) \cong B$, the map $m_2|_B$ agrees with multiplication on B and $m_i|_B$ becomes 0 for all $i > 2$.

A calculation shows that an infinitesimal 2-deformation may be identified with an infinitesimal deformation as in Definition 5.2.1. Writing

$$m_2(b_1 \otimes b_2) = m_2'(b_1 \otimes b_2) + m_2''(b_1 \otimes b_2)x$$

for all $b_1, b_2 \in B$, it follows from the definitions that m_2' is the original multiplication on B and m_2'' is a Hochschild 2-cocycle. Similarly, an infinitesimal n-deformation corresponds to a Hochschild n-cocycle. Since $|x| = 2 - n$ and $|m_i| = 2 - i$, the only possible nonzero operations m_i are m_2 and m_n. Additionally, m_n takes elements of $B^{\otimes n}$ to Bx if $n > 2$. Calculations show that the resulting coefficient function of x must be an n-cocycle. Thus this observation is essentially a converse to Example 7.1.4, and is a proof of the following theorem.

Theorem 7.1.8. *Let B be an associative algebra, and let $n \geq 2$. The Hochschild n-cocycles on B are in one-to-one correspondence with the infinitesimal n-deformations of B.*

Generalizing the case $n = 2$, a calculation shows that cohomologous Hochschild n-cocycles correspond to isomorphic infinitesimal n-deformations; the appropriate notion of isomorphism is given by Definition 7.1.9 below. Specifically, let $(A_g, m_\bullet)$ be the infinitesimal n-deformation of B given in Example 7.1.4. Suppose $g' = g + hd$ for some $(n-1)$-cochain h and $(A_{g'}, m_\bullet')$ is the infinitesimal n-deformation of B corresponding to g'. Set $f_1 : A_g \to A_{g'}$ to be the identity map, set $f_i = 0$ if $i \notin \{1, n-1\}$, and $f_{n-1} = -xh$. Recalling that $m_1 = 0$ and $m_1' = 0$, we see that the only conditions (7.1.10) below with nonzero terms are the second and $(n-1)$st such equations. The second such equation automatically holds since f_1 is the identity map and $m_2 = m_2'$. The $(n-1)$st equation holds since $g' = g + hd$.

By way of Theorem 7.1.8, for each $n > 2$, we relate the Hochschild cohomology space $\mathrm{HH}^n(B)$ of an associative algebra B with infinitesimal n-deformations of B as an A_∞-algebra, in the same way that the Hochschild cohomology space $\mathrm{HH}^2(B)$ corresponds to infinitesimal deformations of B as an associative algebra.

We now return to the general setting of A_∞-algebras.

Definition 7.1.9. Let $(A, m_\bullet^A)$, $(B, m_\bullet^B)$ be A_∞-algebras. A *morphism* of A_∞-algebras $f_\bullet : (A, m_\bullet^A) \to (B, m_\bullet^B)$ consists of graded linear maps

$$f_n : A^{\otimes n} \to B$$

of degree $|f_n| = 1 - n$ for all $n \geq 1$ such that

$$(7.1.10)$$
$$\sum_{r+s+t=n} (-1)^{r+st} f_{r+1+t}(1^{\otimes r} \otimes m_s^A \otimes 1^{\otimes t}) = \sum_{i_1+\cdots+i_r=n} (-1)^u m_r^B(f_{i_1} \otimes \cdots \otimes f_{i_r}),$$

where $u = (r-1)(i_1-1) + (r-2)(i_2-1) + \cdots + 2(i_{r-2}-1) + (i_{r-1}-1)$. The *identity morphism* $f_{\bullet} : A \to A$ is defined by $f_1 = 1_A$ and $f_n = 0$ for $n \neq 1$. The composition of two morphisms $g : A \to B$ and $f : B \to C$ is given by

$$(fg)_n = \sum_{i_1+\cdots+i_r=n} (-1)^u f_r(g_{i_1} \otimes \cdots \otimes g_{i_r})$$

for all n, with $u = u(i_1, \ldots, i_r)$ as above. An A_∞-morphism $f_{\bullet}$ is a *quasi-isomorphism* if f_1 is a quasi-isomorphism, that is, f_1 induces an isomorphism on cohomology, $\mathrm{H}^*(A) \overset{\sim}{\longrightarrow} \mathrm{H}^*(B)$.

We interpret the definition of A_∞-morphism for small values of n. If $n = 1$, equation (7.1.10) is

$$f_1 m_1^A = m_1^B f_1,$$

in other words, f_1 is a cochain map. If $n = 2$, it is

$$f_1 m_2^A = m_2^B(f_1 \otimes f_1) + m_1^B f_2 + f_2(m_1^A \otimes 1 + 1 \otimes m_1^A),$$

which may be rewritten

$$f_1 m_2^A = m_2^B(f_1 \otimes f_1) + \delta(f_2).$$

That is, up to the coboundary $\delta(f_2)$, the map f_1 is an algebra homomorphism with respect to multiplication m_2.

Exercise 7.1.11. Find an expression for the Stasheff identity (7.1.2) when $n = 4$.

Exercise 7.1.12. Verify that $A = B[x]/(x^2)$ of Example 7.1.4 is indeed an A_∞-algebra.

Exercise 7.1.13. Verify that $A = \mathrm{Ext}_B^*(k, k)$ of Example 7.1.5 is indeed an A_∞-algebra.

Exercise 7.1.14. Prove Theorem 7.1.8 following the outline indicated in the text preceding its statement.

7.2. Minimal models

Recall that an A_∞-algebra is called minimal if $m_1 = 0$. For some applications, an A_∞-algebra may be replaced by a minimal A_∞-algebra to which it is quasi-isomorphic (see Definition 7.1.9). We outline this technique here.

Definition 7.2.1. Let A be an A_∞-algebra. A *minimal model* for A is a minimal A_∞-algebra B together with a quasi-isomorphism of A_∞-algebras $f_\bullet : B \to A$.

Let A be an A_∞-algebra, and $\mathrm{H}^*(A)$ its cohomology. The following theorem of Kadeishvili [**123**] states that the cohomology $\mathrm{H}^*(A)$ has the structure of a minimal A_∞-algebra. The theorem thus implies existence of a minimal model.

Theorem 7.2.2. *The cohomology* $\mathrm{H}^*(A)$ *of an* A_∞*-algebra* A *may be given the structure of an* A_∞*-algebra under which it is a minimal model for* A. *This structure is unique up to isomorphism of* A_∞*-algebras.*

Proof. We give a proof only in the special case that $(A, m_\bullet)$ is a differential graded algebra, that is, $m_n = 0$ for $n > 2$. For the general case, see [**123**].

We will define maps $m'_n : \mathrm{H}^*(A)^{\otimes n} \to \mathrm{H}^*(A)$ for each n under which $(\mathrm{H}^*(A), m'_\bullet)$ becomes an A_∞-algebra. At the same time we will define maps $f_n : \mathrm{H}^*(A)^{\otimes n} \to A$ that will constitute a quasi-isomorphism $f_\bullet : (\mathrm{H}^*(A), m'_\bullet) \to (A, m_\bullet)$. Set $m'_1 = 0$ and let $f_1 : \mathrm{H}^*(A) \to A$ be any k-linear section of the surjection $p : Z^*(A) \to \mathrm{H}^*(A)$ from the space of cocycles $Z^*(A)$ to the cohomology $\mathrm{H}^*(A)$ of A. That is, f_1 takes values in $Z^*(A)$ and $pf_1 = 1_{\mathrm{H}^*(A)}$. Let m'_2 be multiplication on $\mathrm{H}^*(A)$ as induced by multiplication m_2 on A. Then by definition, for each $\alpha, \beta \in \mathrm{H}^*(A)$, the elements $m_2(f_1(\alpha) \otimes f_1(\beta))$ and $f_1(m'_2(\alpha \otimes \beta))$ are cohomologous in A. Put another way, letting $\Phi_2 = m_2(f_1 \otimes f_1)$, we see that $f_1 m'_2 - \Phi_2$ is a coboundary, that is, there is some k-linear map $f_2 : \mathrm{H}^*(A)^{\otimes 2} \to A$ for which

$$f_1 m'_2 - \Phi_2 = m_1 f_2.$$

Since $m'_1 = 0$, we may rewrite this as

$$\Phi_2 = f_1 m'_2 - \delta(f_2),$$

the required condition (7.1.10), with $n = 2$, for an A_∞-morphism. Since $m'_1 = 0$ and m'_2 is associative, condition (7.1.2) holds with $n = 3$.

The remainder of the proof proceeds by induction on n. We explain the case $n = 3$ first for clarity. Let

$$\Phi_3 = m_2(f_1 \otimes f_2 - f_2 \otimes f_1) + f_2(1 \otimes m'_2 - m'_2 \otimes 1).$$

A calculation shows that Φ_3 takes values in the space $Z^*(A)$ of cocycles of A. Let $m'_3 : \mathrm{H}^*(A)^{\otimes 3} \to \mathrm{H}^*(A)$ be a k-linear function such that $m'_3(\alpha \otimes \beta \otimes \gamma)$ represents $\Phi_3(\alpha \otimes \beta \otimes \gamma)$ for all $\alpha, \beta, \gamma \in \mathrm{H}^*(A)$. Then by definition of m'_3, the elements $f_1 m'_3(\alpha \otimes \beta \otimes \gamma)$ and $\Phi_3(\alpha \otimes \beta \otimes \gamma)$ are cohomologous. It follows that

$$f_1 m'_3 - \Phi_3 = m_1 f_3$$

for some $f_3 : \mathrm{H}^*(A)^{\otimes 3} \to A$. Thus equation (7.1.10) holds when $n = 3$. Now consider the left side of equation (7.1.2) with $n = 4$ for $m_1' = 0$, m_2', m_3':

$$(7.2.3) \quad m_2'(-m_3' \otimes 1 - 1 \otimes m_3') + m_3'(m_2' \otimes 1 \otimes 1 - 1 \otimes m_2' \otimes 1 + 1 \otimes 1 \otimes m_2').$$

Compose with f_1 and apply (7.1.10) repeatedly to obtain a coboundary in A. Since f_1 is a section of the quotient map from $Z^*(A)$ to $\mathrm{H}^*(A)$, this implies that the expression (7.2.3) is equal to 0.

More generally, let $n > 3$ and suppose we have defined m_i', f_i for all $i < n$. Let

$$\Phi_n = \sum_{i_1+i_2=n} (-1)^{i_1-1} m_2(f_{i_1} \otimes f_{i_2}) - \sum_{\substack{r+s+t=n \\ s>1,\ r+t>0}} (-1)^{r+st} f_{r+1+t}(1^{\otimes r} \otimes m_s' \otimes 1^{\otimes t}),$$

that is, the difference of the right and left sides of equation (7.1.10), excluding the terms $f_1 m_n'$ and $m_1 f_n$ (since $m_i = 0$ for $i > 2$). A calculation shows that Φ_n takes values in the space $Z^*(A)$ of cocycles. Let $m_n' : \mathrm{H}^*(A)^{\otimes n} \to \mathrm{H}^*(A)$ be a k-linear function such that $m_n'(\alpha_1 \otimes \cdots \otimes \alpha_n)$ represents $\Phi_n(\alpha_1 \otimes \cdots \otimes \alpha_n)$ for all $\alpha_1, \ldots, \alpha_n \in \mathrm{H}^*(A)$. Then $f_1 m_n'(\alpha_1 \otimes \cdots \otimes \alpha_n)$ and $\Phi_n(\alpha_1 \otimes \cdots \otimes \alpha_n)$ are cohomologous, so

$$f_1 m_n' - \Phi_n = m_1 f_n$$

for some $f_n : \mathrm{H}^*(A)^{\otimes n} \to A$. Thus equation (7.1.10) holds for n.

By construction, $f_\bullet$ is an A_∞-morphism. Condition (7.1.2) automatically holds for $(\mathrm{H}^*(A), m_\bullet')$ when $n = 1$ or $n = 2$ since $m_1' = 0$, and as we saw before, it holds when $n = 3$ since m_2' is an associative multiplication on $\mathrm{H}^*(A)$. More generally, apply f_1 to the left side of equation (7.1.2) for $m_\bullet'$. The result is seen to be a coboundary in A by repeated use of equation (7.1.10) to eliminate terms involving $m_\bullet'$ from the expression, as explained for $n = 4$ above. Since f_1 is a quasi-isomorphism and $m_1' = 0$, the left side of equation (7.1.2) for $m_\bullet'$ is indeed 0 for each n.

The uniqueness statement can be proven by invoking a k-linear projection from A to $\mathrm{H}^*(A)$ whose composition with f_1 is the identity on $\mathrm{H}^*(A)$. This may be extended to an A_∞-morphism from A to $\mathrm{H}^*(A)$; in fact any A_∞-quasi-isomorphism has a homotopy inverse [**119**]. Given two copies of $\mathrm{H}^*(A)$, with possibly different higher multiplications $m_\bullet'$ and A_∞-morphisms $f_\bullet$, by mapping each to A and then to the other, we obtain maps between the two copies whose compositions in both directions must be identity maps. $\quad\square$

We illustrate Theorem 7.2.2 with a specific example.

Example 7.2.4. Let $B = k[x]/(x^n)$, $n > 2$, and let $P_\bullet$ be the standard periodic free resolution of k as a B-module as in Example 2.5.10. Let $A = \mathrm{Hom}_B(P_\bullet, P_\bullet)$, so that $\mathrm{H}^*(A) \cong \mathrm{Ext}_B^*(k, k) \cong k[y, z]/(y^2)$. Applying the algorithm suggested by the proof of Theorem 7.2.2 leads to the A_∞-algebra structure on $\mathrm{H}^*(A)$ that was given in Example 7.1.5.

Remark 7.2.5. In Theorem 7.2.2, letting B be an associative algebra, we may take A to be the differential graded algebra $\bigoplus_{i\geq 0} \mathrm{Hom}_{B^e}(B^{\otimes(i+2)}, B)$, that is, $C^*(B, B)$, as in Example 7.1.3. Then $\mathrm{H}^*(A)$ is the Hochschild cohomology $\mathrm{HH}^*(B)$, and Theorem 7.2.2 implies that we may realize this Hochschild cohomology as a minimal model. The proof of the theorem indicates how to define the needed higher operations. See also [**119**] for a proof using homological perturbation and further results.

Exercise 7.2.6. In the proof of Theorem 7.2.2, verify that Φ_3 takes values in $Z^*(A)$. More generally, verify that Φ_n takes values in $Z^*(A)$.

Exercise 7.2.7. In the proof of Theorem 7.2.2, verify that applying f_1 to the left side of equation (7.1.2) for $m'_\bullet$ results in a coboundary on A. First check the case $n = 4$, then general n.

Exercise 7.2.8. Verify the details in Example 7.2.4.

7.3. Formality and Koszul algebras

In this section, we present a special case of results of Keller [**126**] on Koszul algebras and formality as defined next.

Definition 7.3.1. An A_∞-algebra A is *formal* if its minimal model has the property that $m_n = 0$ for all $n \geq 3$.

We will see next that a large class of formal A_∞-algebras is given by cohomology of Koszul algebras.

Let $B = T(V)/(R)$ for a finite-dimensional vector space V with homogeneous relations R (where R is a subset of $\bigoplus_{n\geq 2} T_n(V)$). Consider k to be a B-module on which each element of V acts as 0, and let $P_\bullet$ be a projective resolution of k as a B-module. Then $\mathrm{Hom}_B(P_\bullet, P_\bullet)$ is a differential graded algebra with homology $\mathrm{Ext}_B^*(k, k)$. View $\mathrm{Hom}_B(P_\bullet, P_\bullet)$ as an A_∞-algebra with higher multiplication maps 0. Theorem 7.2.2 implies that $\mathrm{Ext}_B^*(k, k)$ has the structure of an A_∞-algebra for which it is the minimal model for $\mathrm{Hom}_B(P_\bullet, P_\bullet)$. In general, $\mathrm{Ext}_B^*(k, k)$ will have higher multiplication maps. When we mention the A_∞-algebra $\mathrm{Ext}_B^*(k, k)$, it is this A_∞-structure that is intended. A use of comparison maps between resolutions shows that, up to isomorphism, this A_∞-structure will not depend on the choice of resolution $P_\bullet$.

We will need the following lemma about a generating set for this A_∞-algebra.

Lemma 7.3.2. *The A_∞-algebra* $\mathrm{Ext}_B^*(k, k)$ *is generated in degree* 1.

A detailed proof was given by Conner [**54**]. See also [**126**, §2.2] for the statement and some techniques needed in a proof.

Example 7.3.3. Let $B = k[x]/(x^n)$ as in Example 7.1.6. As we saw there, $\mathrm{Ext}_B^*(k, k)$ is generated by a single element y as an A_∞-algebra. The degree of y is 1, in accordance with Lemma 7.3.2.

Recall from Theorem 3.4.6 that the Ext algebra $\mathrm{Ext}_B^*(k, k)$ of a Koszul algebra B is generated in degree 1 as an associative algebra, and this is essentially a defining characteristic of Koszul algebras. We thus obtain the following formality result.

Theorem 7.3.4. *Let B be a finitely generated graded connected algebra. The A_∞-algebra $\mathrm{Ext}_B^*(k, k)$ is formal if and only if B is a Koszul algebra.*

Proof. First assume that B is Koszul. The Koszul resolution $\widetilde{K}_\bullet = \widetilde{K}_\bullet(B)$ defined by (3.4.5) gives rise to the complex $\mathrm{Hom}_B(\widetilde{K}_\bullet, \widetilde{K}_\bullet)$, which has grading both by homological degree and that induced by the grading on the algebra B. There is a subcomplex of $\mathrm{Hom}_B(\widetilde{K}_\bullet, \widetilde{K}_\bullet)$ consisting of all elements whose homological degree is the opposite of grading degree. By Theorem 3.4.6 and Lemma 3.4.8, since B is Koszul, $\mathrm{Ext}_B^i(k, k) = \mathrm{Ext}_B^{i,-i}(k, k)$ and $\mathrm{Ext}_B^*(k, k)$ embeds into $\mathrm{Hom}_B(\widetilde{K}_\bullet, \widetilde{K}_\bullet)$ as an associative algebra. Therefore $\mathrm{Ext}_B^*(k, k)$ is formal.

Now assume that $\mathrm{Ext}_B^*(k, k)$ is formal. By Lemma 7.3.2, it is generated in degree 1 as an A_∞-algebra. Since $\mathrm{Ext}_B^*(k, k)$ is formal, we conclude that it is generated in degree 1 as an associative algebra. By Theorem 3.4.6, B is Koszul. $\qquad\square$

We illustrate Theorem 7.3.4 with an example.

Example 7.3.5. See Example 7.1.5 for a distinction made by the theorem. Let $B = k[x]/(x^n)$. If $n = 2$, then B is Koszul and $\mathrm{Ext}_B^*(k, k)$ is formal. If $n > 2$, then B is not Koszul and $\mathrm{Ext}_B^*(k, k)$ is not formal. Accordingly we had found a nonzero higher multiplication m_n in this latter case.

Exercise 7.3.6. Let $B = k[x]/(x^n)$. Let $P_\bullet$ be the free resolution of k as a B-module given in Example 2.5.10. Verify that the A_∞-algebra structure on $\mathrm{Ext}_B^*(k, k) = \mathrm{H}^*(\mathrm{Hom}_B(P_\bullet, P_\bullet))$ provided by the proof of Theorem 7.2.2 agrees with Example 7.1.5.

7.4. A_∞-center

There is a notion of center of an A_∞-algebra that plays an important role in Hochschild cohomology, as we will see in Theorem 7.4.4. There is more than one reasonable way to define the center of an A_∞-algebra. We follow

Briggs and Gélinas [**34**] for a definition, in the special case of a minimal A_∞-algebra, that is invariant under quasi-isomorphism.

We first introduce the graded symmetric and exterior algebras that will be used for the rest of this chapter. The notation $S(V)$ and $\bigwedge(V)$ below agree with earlier uses of the notation in this book in the case that V is concentrated in degree 0. We assume for most of this section that the characteristic of the field k is not 2, and we will point out the few instances when we will take it to be 2.

Let V be a graded vector space over k (graded by $\mathbb{N}$ or by $\mathbb{Z}$). The *graded symmetric algebra* is

$$S(V) = T(V)/(u \otimes v - (-1)^{|u||v|}v \otimes u \mid u, v \text{ homogeneous elements of } V).$$

The *graded exterior algebra* is

$$\textstyle\bigwedge(V) = T(V)/(u \otimes v + (-1)^{|u||v|}v \otimes u \mid u, v \text{ homogeneous elements of } V).$$

It follows from the definitions that $S(V)$ is universal with respect to graded symmetric maps and $\bigwedge(V)$ is universal with respect to graded antisymmetric maps (see Exercise 7.4.6).

Remark 7.4.1. If instead the characteristic of k is 2, the above definitions of $S(V)$ and $\bigwedge(V)$ may be modified to obtain algebras that satisfy these universal properties. In the former case we must additionally mod out by all $v \otimes v$ for which $|v|$ is odd, and in the latter by all $v \otimes v$ for which $|v|$ is even. There is some resulting redundancy in these larger sets of relations.

In the rest of this chapter, all our symmetric and exterior algebras will be graded in the above sense.

Let S_n denote the symmetric group on n symbols. For each $\sigma \in S_n$ and homogeneous $v_1, \ldots, v_n \in V$, define the scalar $\chi(\sigma; v_1, \ldots, v_n)$ by the following equation involving elements of the graded exterior algebra $\bigwedge(V)$:

$$(7.4.2) \qquad\qquad v_{\sigma(1)} \cdots v_{\sigma(n)} = \chi(\sigma; v_1, \ldots, v_n)v_1 \cdots v_n.$$

We sometimes write $\chi(\sigma)$ when it is clear which vectors $v_1, \ldots, v_n$ are involved. If V is concentrated in degree 0, then $\chi(\sigma)$ is simply $\mathrm{sgn}(\sigma)$.

For an A_∞-algebra A, define *higher commutators* $[-;-]_{p,q} : A^{\otimes p} \otimes A^{\otimes q} \to A$ by

$$[a_1, \ldots, a_p; a_{p+1}, \ldots, a_{p+q}]_{p,q}$$

$$= \sum_{\sigma \in S_{p,q}} \chi(\sigma; a_1, \ldots, a_{p+q})m_{p+q}\big(a_{\sigma^{-1}(1)} \otimes \cdots \otimes a_{\sigma^{-1}(p+q)}\big)$$

for all homogeneous $a_1, \ldots, a_n \in A$, where $S_{p,q}$ is the set of all (p, q)-shuffles in the symmetric group S_{p+q} as in Definition 1.5.3. Note that $[-;-]_{1,1}$ is the usual commutator for m_2, that is,

$$[a_1; a_2]_{1,1} = m_2(a_1 \otimes a_2) - (-1)^{|a_1||a_2|} m_2(a_2 \otimes a_1).$$

Definition 7.4.3. Let A be a minimal A_∞-algebra, and let a be a homogeneous element of A. Then a is *central* in A if for all $n \geq 1$, there are k-linear maps $p_i : A^{\otimes i} \to A$ of degrees $|p_i| = |a| - i$ $(i \geq 1)$ for which

$$[a; -]_{1,n} = \sum_{r+s+t=n} (-1)^{r(|a|+s)+t(|a|+1)} m_{r+1+t}(1^{\otimes r} \otimes p_s \otimes 1^{\otimes t})$$
$$- (-1)^{|a|}(-1)^{rs+t} p_{r+1+t}(1^{\otimes r} \otimes m_s \otimes 1^{\otimes t}).$$

Put more concisely, $\mathrm{ad}(a) = \partial(p)$ for suitable notions of adjoint map ad and differential ∂ (see [34] for this and more details). The A_∞-*center* of A, denoted $Z_\infty(A)$, is the vector space spanned by all homogeneous central elements of A.

The maps $p_\bullet$ are called *homotopy derivations*. The related notion of *strong homotopy derivation* [124] is such a map $p_\bullet$ for which the right side of the equation in the definition above is 0, that is, $\partial(p) = 0$. Note that the higher commutator on the left side of the equation has the effect of inserting the element a between factors in all possible ways. For example,

$$[a; a_1, a_2]_{1,2} = m_3(a \otimes a_1 \otimes a_2)$$
$$- (-1)^{|a||a_1|} m_3(a_1 \otimes a \otimes a_2) + (-1)^{|a|(|a_1|+|a_2|)} m_3(a_1 \otimes a_2 \otimes a).$$

Recalling that A is assumed to be minimal, it follows from Definition 7.4.3 of a central element a in an A_∞-algebra A that $[a; -]_{1,1} = 0$, and thus a is in fact in the graded center of A (see Definition 2.5.8). In other words, $Z_\infty(A) \subseteq Z_{\mathrm{gr}}(A)$.

The following theorem expands on Corollary 2.5.9 in the case $M = k$, and is due to Briggs and Gélinas [34], as a consequence of more general results. For a proof, see [34]. Recall that an augmented algebra B is one for which there is an algebra homomorphism to k, and k is considered to be a module via this map. The map from Hochschild cohomology $\mathrm{HH}^*(B)$ to $\mathrm{Ext}_B^*(k, k)$ to which the theorem refers is the map induced by the chain level map ϕ_k defined by equation (2.5.3).

Theorem 7.4.4. *Let B be an augmented algebra over the field k. The image of the Hochschild cohomology ring $\mathrm{HH}^*(B)$ in the Ext algebra $\mathrm{Ext}_B^*(k, k)$ is precisely the A_∞-center $Z_\infty(\mathrm{Ext}_B^*(k, k))$.*

The theorem generalizes Theorem 3.4.14 from Koszul algebras to augmented algebras as follows. If B is a Koszul algebra, then the image of Hochschild cohomology $\mathrm{HH}^*(B)$ in the Ext algebra $\mathrm{Ext}_B^*(k, k)$ is precisely the graded center $Z_{\mathrm{gr}}(\mathrm{Ext}_B^*(k, k))$ as stated in Theorem 3.4.14. As we saw in the last section, the Ext algebra of a Koszul algebra B is formal, and thus the graded center of $\mathrm{Ext}_B^*(k, k)$ coincides with its A_∞-center.

Example 7.4.5. Let $B = k[x]/(x^n)$, $n > 2$. Then $\mathrm{Ext}_B^*(k, k) \cong k[y, z]/(y^2)$ as in Example 2.5.10 and

$$Z_\infty(\mathrm{Ext}_B^*(k, k)) = \begin{cases} k[z], & \text{if char}(k) \nmid n, \\ k[y, z]/(y^2), & \text{if char}(k) \mid n. \end{cases}$$

(Recall that $\mathrm{char}(k) \neq 2$ in this section. See [**34**] for details.)

There are many more applications of A_∞-algebras, as well as the dual notion of A_∞-coalgebras, that we do not consider here. See, for example, [**111**].

Exercise 7.4.6. Verify the universality of the graded symmetric algebra $S(V)$ and the graded exterior algebra $\bigwedge(V)$ as defined in this section in case $\mathrm{char}(k) \neq 2$:

(a) Given a graded vector space V and a k-algebra B, a *graded symmetric* map $f : V \to B$ is a k-linear map such that $f(v)f(w) = (-1)^{|v||w|} f(w)f(v)$ for all homogeneous $v, w \in V$. Show that for all k-algebras B and graded symmetric maps $f : V \to B$, there is a unique k-algebra homomorphism $F : S(V) \to B$ such that $F|_V = f$.

(b) Given a graded vector space V and a k-algebra B, a *graded anti-symmetric* map $f : V \to B$ is a k-linear map such that $f(v)f(w) = -(-1)^{|v||w|} f(w)f(v)$ for all homogeneous $v, w \in V$. Show that for all k-algebras B and graded anti-symmetric maps $f : V \to B$, there is a unique k-algebra homomorphism $F : \bigwedge(V) \to B$ such that $F|_V = f$.

Exercise 7.4.7. Let k be a field of characteristic 2. Define $S(V)$ and $\bigwedge(V)$ by universal properties such as in Exercise 7.4.6(a) and (b). Deduce that $S(V)$ and $\bigwedge(V)$ are isomorphic to particular quotients of $T(V)$ as outlined in Remark 7.4.1.

Exercise 7.4.8. Verify that $Z_\infty(A) \subseteq Z_{\mathrm{gr}}(A)$ by comparing the definitions of these two notions of center when A is an A_∞-algebra.

Exercise 7.4.9. Find an expression for $[a_1, a_2; a]_{2,1}$ as a sum over $(2, 1)$-shuffles.

7.5. L_∞-algebras

Analogous to A_∞-algebras are L_∞-algebras in which the operations are higher order Lie brackets. This notion first appeared in the paper [**191**] by Schlessinger and Stasheff.

Definition 7.5.1. An L_∞-*algebra* is a graded vector space L together with graded linear maps

$$\ell_n : \bigwedge{}^n L \to L$$

of degree $|\ell_n| = 2 - n$ for all n such that

$$\sum_{i=1}^{n} \sum_{\sigma \in S_{i,n-i}} (-1)^{i(n-i)} \chi(\sigma) \ell_{n-i+1}(\ell_i(v_{\sigma(1)}, \ldots, v_{\sigma(i)}), v_{\sigma(i+1)}, \ldots, v_{\sigma(n)}) = 0$$

for all homogeneous $v_1, \ldots, v_n \in L$, where $\chi(\sigma) = \chi(\sigma; v_1, \ldots, v_n)$ is defined by (7.4.2) and $S_{i,n-i}$ is the set of all $(i, n-i)$-shuffles in the symmetric group S_n as in Definition 1.5.3. For simplicity of notation, we have separated elements of L by commas when their product in $\bigwedge(L)$ is the argument of a function. Sometimes we denote the L_∞-algebra by $(L, \ell_\bullet)$ to emphasize the notation chosen for the higher operations. An L_∞-algebra is *minimal* if $\ell_1 = 0$.

Remark 7.5.2. In the literature, the elements of $S_{i,n-i}$ are sometimes called *i-unshuffles* in this context. Also, a graded symmetric algebra sometimes appears in definitions of L_∞-structures, in place of the graded exterior algebra we have here, due to different choices of grading and indexing.

We interpret these equations in low degrees. Write $d(v) = \ell_1(v)$ and $[u, v] = \ell_2(u, v)$, $[u, v, w] = \ell_3(u, v, w)$, etc., for elements u, v, w of L. If $n = 1$, the equation in Definition 7.5.1 is simply $d^2(v) = 0$ for all $v \in L$, and so d is a differential on L. If $n = 2$, the condition is

$$d([u, v]) = [d(u), v] + (-1)^{|u|}[u, d(v)]$$

for all homogeneous $u, v \in L$, that is, d is a graded derivation with respect to $\ell_2 = [\ ,\ ]$. If $n = 3$, the condition may be written

$$(-1)^{|u||w|}[[u, v], w] + (-1)^{|u||v|}[[v, w], u] + (-1)^{|v||w|}[[w, u], v]$$

$$= (-1)^{|u||w|}(d\ell_3 + \ell_3 d)(u, v, w)$$

for all homogeneous $u, v, w \in L$, that is, up to homotopy, the graded Jacobi identity holds.

Example 7.5.3. If L is a differential graded Lie algebra, we may take ℓ_1 to be its differential, ℓ_2 to be its Lie bracket, and $\ell_n = 0$ for all $n \geq 3$, for an L_∞-structure on L. In particular, a graded Lie algebra may be viewed as a differential graded Lie algebra with zero differential, and thus as an L_∞-algebra in this way. If an L_∞-algebra is concentrated in degree 0, that is, $L_i = 0$ for all $i \neq 0$, the maps ℓ_n are necessarily zero maps for all $n \neq 2$ since $|\ell_n| = 2 - n$, so L is simply a Lie algebra.

A large class of examples is provided by A_∞-algebras together with graded commutators, as the following theorem of Lada and Markl [138] shows. This generalizes a relationship between associative algebras and Lie algebras. The theorem may be proven by direct computation, or by invoking a connection between L_∞-structures and coderivations as in [138].

Theorem 7.5.4. *Let $(A, m_\bullet)$ be an A_∞-algebra. Let*

$$\ell_n(a_1, \ldots, a_n) = \sum_{\sigma \in S_n} \chi(\sigma) m_n(a_{\sigma(1)}, \ldots, a_{\sigma(n)})$$

for all homogeneous $a_1, \ldots, a_n \in A$. Then $(A, \ell_\bullet)$ is an L_∞-algebra.

Morphisms of L_∞-algebras involve the following generalization of (p, q)-shuffles. There are more conceptual alternative descriptions, as well as equivalent formulas in characteristic 0 that instead involve sums over all permutations and division by factorials. See, for example, [5] or [124].

Let $i_1, \ldots, i_t$ be positive integers. A permutation σ of $S_{i_1 + \cdots + i_j}$ is an $(i_1, \ldots, i_t)$-*shuffle* if

$$\sigma(1) < \cdots < \sigma(i_1),$$
$$\sigma(i_1 + 1) < \cdots < \sigma(i_1 + i_2), \ldots,$$
$$\text{and} \quad \sigma(i_1 + \cdots + i_{t-1} + 1) < \cdots < \sigma(i_1 + \cdots + i_t).$$

Definition 7.5.5. Let $(L, \ell_\bullet)$, $(L', \ell'_\bullet)$ be L_∞-algebras. A *morphism* of L_∞-algebras $f_\bullet : L \to L'$ consists of graded linear maps

$$f_n : \bigwedge^n L \to L'$$

of degree $|f_n| = 1 - n$ for all $n \geq 1$ such that for all homogeneous $v_1, \ldots, v_n \in L$,

$$\sum_{i=1}^{n} \sum_{\sigma \in S_{i, n-i}} (-1)^{i(n-i)} \chi(\sigma) f_{n-i+1}(\ell_i \otimes 1^{\otimes(n-i)})(v_{\sigma(1)}, \ldots, v_{\sigma(n)})$$

$$= \sum_{\substack{1 \leq r \leq n \\ i_1 + \cdots + i_r = n}} \sum_{\tau} (-1)^u \chi(\tau) \ell'_t(f_{i_1} \otimes \cdots \otimes f_{i_r})(v_{\tau(1)}, \ldots, v_{\tau(n)}),$$

where τ runs over all $(i_1, \ldots, i_t)$-shuffles for which

$$\tau(i_1 + \cdots + i_{l-1} + 1) < \tau(i_1 + \cdots + i_l + 1)$$

if $i_l = i_{l+1}$, and $u = (r-1)(i_1-1) + (r-2)(i_2-1) + \cdots + 2(i_{r-2}-1) + (i_{r-1}-1)$. An L_∞-morphism $f_\bullet$ is a *quasi-isomorphism* if f_1 is a quasi-isomorphism.

We interpret the definition for some small values of n. If $n = 1$, the equation is

$$f_1 \ell_1 = \ell'_1 f_1,$$

that is, f_1 is a cochain map. If $n = 2$, we obtain

$$- f_2(d(v_1), v_2) - (-1)^{|v_1||v_2|} f_2(d(v_2), v_1) + f_1([v_1, v_2])$$
$$= d'(f_2(v_1, v_2)) + [f_1(v_1), f_1(v_2)]$$

for all homogeneous $v_1, v_2 \in L$. Rewriting, the equation is

$$f_1([v_1, v_2])$$
$$= [f_1(v_1), f_1(v_2)] + d'(f_2(v_1, v_2)) + f_2(d(v_1), v_2) + (-1)^{|v_1||v_2|} f_2(d(v_2), v_1),$$

that is, f_1 preserves the bracket up to the coboundary $\partial(f_2)$.

Just as for A_∞-algebras, there is a notion of minimal model (see Definition 7.2.1) and a version of Theorem 7.2.2 on existence (see [**120, 193**]). For more general results, see for example [**113**, Lemma 4.2.1].

Exercise 7.5.6. Verify that the conditions on ℓ_n given in Definition 7.5.1 for $n = 1, 2, 3$ are as claimed in this section (that is, ℓ_1 is a differential that is a graded derivation with respect to ℓ_2, and ℓ_2 satisfies the graded Jacobi identity up to homotopy).

Exercise 7.5.7. Understand the proof of Theorem 7.5.4 either by direct computation or by looking up the proof technique used in [**138**].

Exercise 7.5.8. Verify that the conditions on f_n implied by Definition 7.5.5 for $n = 1, 2$ are as claimed in this section (that is, f_1 is a cochain map that preserves the bracket up to addition of a coboundary).

7.6. Formality and algebraic deformations

Just as for A_∞-algebras, there is a notion of formality for L_∞-algebras, and in this section we discuss it specifically in the case of Hochschild cohomology. We assume that the characteristic of the field k is 0 so that exponential maps are defined. The main ideas we present here are due to Kontsevich [**133**]. Let A be an associative algebra and consider $C^*(A, A) = \bigoplus_{i \geq 0} \mathrm{Hom}_k(A^{\otimes i}, A)$, a differential graded Lie algebra as described in Section 1.4.

Definition 7.6.1. An *HKR map* is a graded k-linear injective map

$$\phi : \mathrm{HH}^*(A) \to C^*(A, A),$$

of degree $|\phi| = 0$, with image contained in the space of cocycles, that is a section of the quotient map from the space of cocycles to $\mathrm{HH}^*(A)$.

By Lemma 1.4.3 and the discussion following it, we may view both $\mathrm{HH}^*(A)$ and $C^*(A, A)$ as differential graded Lie algebras, the former having differential 0. An HKR map is in general not a morphism of differential graded Lie algebras. However, viewing $\mathrm{HH}^*(A)$ and $C^*(A, A)$ as L_∞-algebras (with higher brackets 0), an HKR map can sometimes be extended to a quasi-isomorphism of L_∞-algebras, as we will see below. In this case we say that A is formal.

Definition 7.6.2. The associative algebra A is *formal* if there is a quasi-isomorphism of L_∞-algebras $\Phi_{\bullet} : (\mathrm{HH}^*(A), \ell_{\bullet}) \to (C^*(A, A), \ell'_{\bullet})$ for which Φ_1 is an HKR map. Such a map Φ is called a *formality map*.

This definition is analogous to Definition 7.3.1 when we replace the A_∞-structure with the L_∞-structure, taking $\mathrm{HH}^*(A)$ to be the minimal model of $C^*(A, A)$. Which version of formality is intended should be clear from the context.

Note that the grading on $\mathrm{HH}^*(A)$ and on $C^*(A, A)$ in Definition 7.6.2 is shifted by 1 since we are dealing with the Lie structure. So for example, the degree of a Hochschild 2-cocycle is now 1, an important distinction to make in the proof of the next theorem. The hypothesis on the infinitesimal deformation α in the theorem recalls, in the commutative setting, a connection to Poisson brackets (see, e.g., equation (5.3.3)).

Theorem 7.6.3. *Let A be a formal associative algebra over a field of characteristic 0, and let α be an infinitesimal deformation of A for which the first obstruction (5.1.8) vanishes as a cochain. Then α lifts to a formal deformation of A.*

Proof. Let α be a Hochschild 2-cocycle on A for which the first obstruction vanishes as a cochain, that is, $[\alpha, \alpha] = 0$. Consider the following element in $S(\mathrm{HH}^*(A))[[t]]$:

$$\exp(t\alpha) = 1 + t\alpha + \frac{1}{2!}t^2\alpha^2 + \frac{1}{3!}t^3\alpha^3 + \cdots,$$

where we consider the ith term to be in $S^i(\mathrm{HH}^*(A))[[t]]$, starting with $i = 0$. Let Φ be a formality map for A. Consider the image of $\exp(t\alpha)$ under Φ extended to formal power series in t. Explicitly, we write this as

$$\Phi(\exp(t\alpha)) = 1 + t\Phi_1(\alpha) + \frac{1}{2!}t^2\Phi_2(\alpha^2) + \frac{1}{3!}t^3\Phi_3(\alpha^3) + \cdots,$$

where $\Phi_2(\alpha^2)$ may also be written $\Phi_2(\alpha, \alpha)$, and $\Phi_3(\alpha^3)$ as $\Phi_3(\alpha, \alpha, \alpha)$, and so on. Due to the degree requirement on L_∞-morphisms, the elements $\Phi_i(\alpha^i)$ each have degree 1, that is, they are Hochschild 2-cochains. Since Φ is an L_∞-morphism and $[\alpha, \alpha] = 0$, it follows from the definitions that $\Phi(\exp(t\alpha))$ satisfies the Maurer-Cartan equation (5.3.1). Write $\mu_i = \frac{1}{i!}\Phi_i(\alpha^i)$ and $\mu = \mu_0 + \mu_*$, where $\mu_* = t\mu_1 + t^2\mu_2 + \cdots$. Then $(A[[t]], \mu)$ is a formal deformation of (A, μ_0). $\qquad\square$

The following is a special case of general results of Kontsevich [133] about Poisson manifolds. See also [58, 129]. We take k to be $\mathbb{C}$ or $\mathbb{R}$ here.

Theorem 7.6.4. *Let V be a finite-dimensional vector space. Its (ungraded) symmetric algebra $S(V)$ is formal.*

For an outline of the proof, see [**32**, §5.2], where the theorem is then generalized to universal enveloping algebras of some Lie algebras. The proof is combinatorial and geometric, involving sums over some planar graphs. For more details and geometric context, see the excellent survey [**189**].

There are further infinity structures arising on Hochschild cohomology $HH^*(A)$ and the Hochschild complex $C^*(A, A)$. In particular, extending the A_∞- and L_∞-structures we obtain a G_∞-algebra [**104**], that is an infinity analog of a Gerstenhaber algebra (see Definition 1.4.8). Deligne conjectured that the Hochschild complex $C^*(A, A)$ is an algebra over an operad of little disks. An algebraic version of Deligne's Conjecture states that the Hochschild complex $C^*(A, A)$ is a G_∞-algebra. This is now a theorem and there are various proofs in the literature, for example [**134**, **157**, **214**, **221**, **222**]. Combining the algebra and coalgebra structures, we obtain a B_∞-algebra [**127**], that is an infinity analog of a bialgebra. The Hochschild complex $C^*(A, A)$ is a B_∞-algebra. This structure has implications for invariance of Hochschild cohomology and points to an isomorphism between the Hochschild cohomology ring of a Koszul algebra and its dual [**127**].

Exercise 7.6.5. Let $A = k[x]$. Find an HKR map $\phi : HH^*(A) \to C^*(A, A)$. (Recall that $HH^*(A)$ was found in Example 1.1.18 and that Hochschild 1-cocycles correspond to derivations as discussed in Section 1.2.)

Exercise 7.6.6. Look up a proof of a version or generalization of Theorem 7.6.4, for example in [**32**, **58**, **133**, **189**].

Chapter 8

Support Varieties for Finite-Dimensional Algebras

In this chapter we present an application of Hochschild cohomology to the representation theory of finite-dimensional algebras, that is, algebras that are finite-dimensional as vector spaces over the field k. For many finite-dimensional algebras A, though not all, Hochschild cohomology $\mathrm{HH}^*(A)$ is finitely generated as an associative algebra under cup product. Since $\mathrm{HH}^*(A)$ is also graded commutative, its maximal ideals may be equipped with the geometric structure of an affine variety. This variety has subvarieties associated to A-modules through actions of Hochschild cohomology $\mathrm{HH}^*(A)$, and the geometry of these subvarieties has implications in representation theory. These "varieties for modules" based on Hochschild cohomology were introduced by Snashall and Solberg [201] to mimic support varieties for finite groups (based on group cohomology). The theory is particularly well-behaved for self-injective algebras, as further developed by Erdmann, Holloway, Snashall, Solberg, and Taillefer [65]. See Solberg's excellent survey [203]. A related survey on support varieties for finite-dimensional Hopf algebras, including finite group algebras, is [224].

A good support variety theory is an important tool particularly for algebras of wild representation type, that is, those whose indecomposable modules cannot be classified in a meaningful way (see, e.g., [21]). There are connections between the structure of Hochschild cohomology and the representation type of an algebra that we do not explore here. See, for example, [31, 36, 78].

We begin this chapter by briefly introducing the needed geometric notions and finite generation conditions for support variety theory. We take A to be a finite-dimensional algebra and k an algebraically closed field, although support variety theory has been developed for other algebras and fields as well. In some of the references, A is assumed to be indecomposable. We do not make this assumption since we aim at more general applications, and there are some minor differences. We refer the reader to [**93, 158**] for the general theory of noncommutative noetherian rings, although we will only need this theory in the graded commutative setting where it is essentially the same as for commutative rings.

8.1. Affine varieties

In this section we give a very brief introduction to the geometry that we will use in this chapter. For more details, see any text on algebraic geometry or commutative algebra, or see [**22**, Section 5.4] for a longer introduction to what is needed specifically for support variety theory. Let k be an algebraically closed field.

Let H be a finitely generated commutative algebra over k. Equivalently, $H \cong k[x_1, \ldots, x_n]/I$ for some ideal I of a polynomial ring $k[x_1, \ldots, x_n]$. Let $\mathrm{Max}(H)$ denote the set of maximal ideals of H, so $\mathrm{Max}(H)$ is in one-to-one correspondence with the set of maximal ideals of $k[x_1, \ldots, x_n]$ containing I. In particular, since k is algebraically closed,

$$\mathrm{Max}(k[x_1, \ldots, x_n]) = \{(x_1 - a_1, \ldots, x_n - a_n) \mid a_1, \ldots, a_n \in k\},$$

so that the set of maximal ideals of the polynomial ring $k[x_1, \ldots, x_n]$ is in one-to-one correspondence with k^n.

The set $\mathrm{Max}(H)$ becomes a topological space under the *Zariski topology*: closed sets are the sets

$$V(J) = \{J' \in \mathrm{Max}(H) \mid J' \supset J\},$$

determined by ideals J of H. Sometimes we write $V_H(J)$ in place of $V(J)$ to emphasize dependence on H. These sets satisfy the relations

$$(8.1.1) \quad V(J_1 J_2) = V(J_1) \cup V(J_2) \quad \text{and} \quad V\left(\sum_\alpha J_\alpha\right) = \bigcap_\alpha V(J_\alpha),$$

where α ranges over an indexing set and J_1, J_2, J_α are ideals of H. We call $\mathrm{Max}(H)$ with this topology the *maximal ideal spectrum* of H, also called an *affine variety*. In particular, $\mathrm{Max}(k[x_1, \ldots, x_n])$ is the *affine space* k^n. Prime ideals and projective varieties are also of interest in representation theory, but we will not need them here.

The following lemma will be used in a definition of dimension of an affine variety.

Lemma 8.1.2 (Noether Normalization Lemma). *Let H be a finitely generated commutative algebra over k. There are elements $y_1, \ldots, y_n \in H$ generating a subalgebra of H that is isomorphic to the polynomial ring $k[y_1, \ldots, y_n]$ and over which H is finitely generated as a module.*

For a proof, see [**156**, §33]. By its definition, the integer n in the lemma is unique.

Definition 8.1.3. Let H be a finitely generated commutative algebra over k, and $\mathrm{Max}(H)$ its maximal ideal spectrum. The *dimension* of $\mathrm{Max}(H)$ is the integer n of the Noether Normalization Lemma (Lemma 8.1.2).

It can be shown that if H is finitely generated, then the dimension of $\mathrm{Max}(H)$ defined above is the same as the Krull dimension of H, defined next. (See [**156**].)

Definition 8.1.4. For any commutative ring H, its *Krull dimension* is the largest nonnegative integer n for which there exist prime ideals

$$I_0 \supset I_1 \supset \cdots \supset I_n$$

of H such that $I_j \neq I_{j+1}$ for $0 \leq j \leq n-1$.

There is another notion of size that we will need.

Definition 8.1.5. Let $V = \bigoplus_{i \geq 0} V_i$ be a graded vector space. The *rate of growth* $\gamma(V)$ is the smallest nonnegative integer c such that there is a real number b and a positive integer m for which $\dim_k V_n \leq bn^{c-1}$ for all $n \geq m$.

The Krull dimension of a finitely generated graded commutative algebra is precisely its rate of growth. To see this, note that a polynomial ring in n indeterminates has rate of growth n. Now apply Lemma 8.1.2 and the above comments.

Exercise 8.1.6. Let $H = k[x, y]$, $J_1 = (x)$, and $J_2 = (y)$. Find $V(J_1 J_2)$ and $V(J_1 + J_2)$ and verify that the equalities (8.1.1) hold.

Exercise 8.1.7. Let $H = k[x, y]/(xy)$. What is the Krull dimension of H? Find elements of H satisfying the conclusion of the Noether Normalization Lemma.

Exercise 8.1.8. Show directly that the Krull dimension of $k[x]$ is 1 by applying the definition.

Exercise 8.1.9. Justify the statement that $k[x_1, \ldots, x_n]$ has rate of growth n by examining an expression for $\dim_k(k[x_1, \ldots, x_n]_m)$ for each m.

8.2. Finiteness properties

Let $\mathfrak{r} = \mathrm{rad}(A)$, the Jacobson radical of the finite-dimensional algebra A, that is, the intersection of all maximal left ideals. For details on the Jacobson radical and related representation theory of finite-dimensional algebras over algebraically closed fields, see, e.g., [**6**, Sections 1 and 2] or [**11**].

We will use the action of $\mathrm{HH}^*(A)$ on $\mathrm{Ext}^*_A(M, M)$ for an A-module M, as described in Section 2.5, beginning with the special case $M = A/\mathfrak{r}$. We will assume in Sections 8.3 through 8.5 that A satisfies the following finiteness condition:

(fg) $\mathrm{HH}^*(A)$ is a noetherian ring and $\mathrm{Ext}^*_A(A/\mathfrak{r}, A/\mathfrak{r})$ is a finitely generated $\mathrm{HH}^*(A)$-module.

Since $\mathrm{HH}^*(A)$ is graded commutative, if $\mathrm{HH}^*(A)$ is noetherian, then it is finitely generated as an algebra by homogeneous elements. Conversely, if $\mathrm{HH}^*(A)$ is finitely generated as an algebra, then it is noetherian, since free graded commutative rings are noetherian and quotients of noetherian rings are noetherian. We will refine these statements in Theorem 8.2.3 below.

There are many finite-dimensional algebras A that satisfy condition (fg), as well as some that do not, as we will see.

Example 8.2.1. Let $A = k[x_1, \ldots, x_m]/(x_1^{n_1}, \ldots, x_m^{n_m})$. We saw in Example 3.1.5 that $\mathrm{HH}^*(A)$ is finitely generated. (See also Example 3.2.6 for some related noncommutative examples.) The Jacobson radical $\mathfrak{r}$ of A is generated by $x_1, \ldots, x_m$, and so $A/\mathfrak{r} \cong k$. If $m = 1$, as a consequence of our work in Example 2.5.10, the Ext algebra $\mathrm{Ext}^*_A(k, k)$ is finitely generated as an $\mathrm{HH}^*(A)$-module. The same is seen to be true if $m > 1$ by combining the techniques of Examples 2.5.10 and 3.1.5.

Many other algebras satisfy condition (fg), such as finite group algebras [**75**, **92**, **217**] and Hecke algebras [**24**, **142**]. Others satisfy related finiteness conditions, such as monomial algebras [**97**], self-injective algebras of finite representation type [**98**], algebras of finite global dimension [**105**], and some special biserial algebras [**202**]. A 2009 survey of finite-dimensional algebras known at that time to satisfy condition (fg) is [**200**]. Necessary and sufficient conditions for radical cube algebras to satisfy (fg) are given in [**70**]. A categorical context for such finiteness conditions and other references are in [**178**]. Some types of Hopf algebras satisfy condition (fg); Hopf algebras generally are discussed in Chapter 9.

We next give a class of algebras that do not satisfy condition (fg). In characteristic 2, the example below is due to Xu [**225**], who presented it as the first counterexample to a related conjecture of Snashall and Solberg. Here we describe a generalization of Xu's example to arbitrary characteristic, due to Snashall [**200**]. It was generalized further by Gawell and Xantcha [**81**]. See Section 3.6 for algebras defined by quivers and relations as in the example.

Example 8.2.2. Let $A = kQ/I$, where Q is the quiver

with two vertices as indicated, arrows a, b, c, and $I = (a^2,\ b^2,\ ab - ba,\ ac)$. Using techniques similar to those of Section 3.4, it can be shown that A is a Koszul algebra (in a more general sense) and that $\operatorname{Ext}^*_A(A/\mathfrak{r}, A/\mathfrak{r}) \cong kQ^{\mathrm{op}}/I'$, where Q^{op} is the quiver Q with arrows reversed, labeled α, β, γ, and $I' = (\alpha\beta + \beta\alpha,\ \gamma\beta)$ [**200**]. Calculations show that the graded center of $\operatorname{Ext}^*_A(A/\mathfrak{r}, A/\mathfrak{r})$ is $k \oplus k[\alpha, \beta]\beta$ if $\operatorname{char}(k) = 2$, and $k \oplus k[\alpha^2, \beta^2]\beta^2$ if $\operatorname{char}(k) \neq 2$, where β has degree 1, $\alpha\beta$ has degree 2, β^2 has degree 2, and $\alpha^2\beta^2$ has degree 4. By [**39**, Theorem 4.1], the image of $\operatorname{HH}^*(A)$ under the map $\phi_{A/\mathfrak{r}}$ defined by (2.5.3) is precisely this graded center (cf. Theorem 3.4.14). This graded center is not finitely generated as an algebra, since for each i, the element $\alpha^i\beta$ in characteristic 2 (respectively, $\alpha^{2i}\beta^2$ in characteristic not 2) is not in any subalgebra generated by elements of lower homological degree. Therefore $\operatorname{HH}^*(A)$ is not finitely generated as an algebra, and consequently does not satisfy condition (fg).

For the purpose of defining affine varieties, one can essentially ignore nilpotent elements, and indeed Snashall and Solberg [**201**] had originally conjectured that the quotient of $\operatorname{HH}^*(A)$ by its ideal generated by all homogeneous nilpotent elements is noetherian. Example 8.2.2 is a counterexample. Hermann [**109**] asked if a weaker condition might be satisfied by more finite-dimensional algebras: replace the condition that $\operatorname{HH}^*(A)$ be noetherian with the condition that the quotient by its Gerstenhaber ideal generated by all homogeneous nilpotent elements be noetherian. (By this ideal, we mean the ideal generated via the binary operations of cup product and Gerstenhaber bracket.) We will not consider these weaker conditions here.

We will prove the following theorem from [**65**, Proposition 1.4] and [**203**, Proposition 5.7]. We will then use it to define support varieties in the next section. The flexibility in choosing an algebra H satisfying the conditions of the theorem below will be helpful. We will use the action of $\operatorname{HH}^*(A)$ on $\operatorname{Ext}^*_A(M, N)$ for any two A-modules M, N that is described in Section 2.5.

Theorem 8.2.3. *The finite-dimensional algebra A satisfies condition* (fg) *if and only if there exists a graded subalgebra H of* $\mathrm{HH}^*(A)$ *such that*

(fg1) *H is finitely generated commutative and $H^0 = \mathrm{HH}^0(A)$, and*

(fg2) *$\mathrm{Ext}_A^*(A/\mathfrak{r}, A/\mathfrak{r})$ is a finitely generated H-module.*

There is a trade-off between these two conditions (fg1) and (fg2): condition (fg1) says that H is small enough for some geometric applications, and condition (fg2) says that H is large enough for others. The two taken together say that H is just right for the theory of support varieties that we will define in Section 8.3.

In order to prove the theorem, we need the following lemmas. If M, N are A-modules, then $\mathrm{Hom}_k(M, N)$ is an A-bimodule with action given by

$$(afb)(m) = a(f(bm))$$

for all $f \in \mathrm{Hom}_k(M, N)$, $a, b \in A$, and $m \in M$. The action of $\mathrm{HH}^*(A)$ on $\mathrm{HH}^*(A, B)$ for any A-bimodule B is given by Yoneda product as described in Section 2.5.

Lemma 8.2.4. *For all finite-dimensional A-modules M, N, there is a graded vector space isomorphism*

$$\mathrm{Ext}_A^*(M, N) \cong \mathrm{HH}^*(A, \mathrm{Hom}_k(M, N)).$$

Moreover, the actions of the Hochschild cohomology ring $\mathrm{HH}^*(A)$ correspond under this isomorphism.

Proof. Let $P_\bullet \to A$ be a projective resolution of A as an A^e-module. Then $P_\bullet \otimes_A M$ is a projective resolution of the A-module M. For each i, define a function $\phi_i : \mathrm{Hom}_A(P_i \otimes_A M, N) \to \mathrm{Hom}_{A^e}(P_i, \mathrm{Hom}_k(M, N))$ by $\phi_i(f)(x)(m) = f(x \otimes m)$ for all $f \in \mathrm{Hom}_A(P_i \otimes_A M, N)$, $x \in P_i$, and $m \in M$. A calculation shows that $\phi_i(f)$ is an A^e-module homomorphism and $\phi_\bullet$ is a cochain map. The inverse maps are $\psi_i : \mathrm{Hom}_{A^e}(P_i, \mathrm{Hom}_k(M, N)) \to \mathrm{Hom}_A(P_i \otimes_A M, N)$ given by $\psi_i(f)(x \otimes m) = f(x)(m)$. A calculation now shows that the actions of $\mathrm{HH}^*(A)$ correspond. $\square$

The next lemma is [**65**, Proposition 2.4].

Lemma 8.2.5. *Let H be a finitely generated commutative subalgebra of $\mathrm{HH}^*(A)$. The following are equivalent:*

(i) *$\mathrm{Ext}_A^*(A/\mathfrak{r}, A/\mathfrak{r})$ is a finitely generated H-module.*

(ii) *$\mathrm{Ext}_A^*(M, N)$ is a finitely generated H-module for all finite-dimensional A-modules M, N.*

(iii) *$\mathrm{HH}^*(A, B)$ is a finitely generated H-module for all finite-dimensional A-bimodules B.*

Proof. By Lemma 8.2.4, $\mathrm{Ext}_A^*(M, N) \cong \mathrm{HH}^*(A, \mathrm{Hom}_k(M, N))$ and the actions of $\mathrm{HH}^*(A)$ on these two spaces correspond, so (iii) implies (ii). By setting $M = N = A/\mathfrak{r}$, we see that (ii) implies (i). It remains to show that (i) implies (iii). By Lemma 8.2.4 with $M = N = A/\mathfrak{r}$, there is an isomorphism

$$\mathrm{Ext}_A^*(A/\mathfrak{r}, A/\mathfrak{r}) \cong \mathrm{HH}^*(A, \mathrm{Hom}_k(A/\mathfrak{r}, A/\mathfrak{r})).$$

Each simple A-module S is a direct summand of $A/\mathfrak{r}$, and simple A^e-modules are all of the form $\mathrm{Hom}_k(S, T)$ for simple A-modules S, T. Letting B be any finite-dimensional A^e-module, it has a composition series with simple factors B_i. By (i) and the above observations, $\mathrm{HH}^*(A, B_i)$ is a finitely generated H-module for each i. By induction on the length of the composition series of B and the first long exact sequence for Ext (Theorem A.4.4), the H-module $\mathrm{HH}^*(A, B)$ is finitely generated. Specifically, for the induction step, suppose $0 \to U \to V \to W \to 0$ is a short exact sequence of A^e-modules and $\mathrm{HH}^*(A, U)$, $\mathrm{HH}^*(A, W)$ are both finitely generated H-modules. Choose a finite set of generators for $\mathrm{HH}^*(A, U)$, and consider their images in $\mathrm{HH}^*(A, V)$ under the map induced by $U \to V$. Since H is noetherian and the image of $\mathrm{HH}^*(A, V)$ in $\mathrm{HH}^*(A, W)$, under the map induced by $V \to W$, is an H-submodule, the image of $\mathrm{HH}^*(A, V)$ in $\mathrm{HH}^*(A, W)$ is finitely generated. Choose a finite set of generators of this image, and choose an inverse image of each one in $\mathrm{HH}^*(A, V)$. Then the finite set of all these generators taken together generates $\mathrm{HH}^*(A, V)$. $\qquad\square$

Proof of Theorem 8.2.3. Assume A satisfies condition (fg). If $\mathrm{char}(k) = 2$, let $H = \mathrm{HH}^*(A)$, and if $\mathrm{char}(k) \neq 2$, let $H = \mathrm{HH}^{\mathrm{ev}}(A)$, the subalgebra of $\mathrm{HH}^*(A)$ generated by all homogeneous elements of even degree. Then $H^0 = \mathrm{HH}^0(A)$ and H is commutative since $\mathrm{HH}^*(A)$ is graded commutative. In addition, H is finitely generated: take a finite set of homogeneous generators of $\mathrm{HH}^*(A)$, and replace those of odd degree by all products of pairs of odd degree generators. The resulting finite set generates H. So (fg1) holds. Condition (fg2) also holds since H is a subalgebra of $\mathrm{HH}^*(A)$ and we have assumed that $\mathrm{Ext}_A^*(A/\mathfrak{r}, A/\mathfrak{r})$ is finitely generated over the noetherian ring $\mathrm{HH}^*(A)$.

Conversely, assume there is a graded subalgebra H of $\mathrm{HH}^*(A)$ that satisfies (fg1) and (fg2). Note that (fg2) is precisely condition (i) of Lemma 8.2.5. By Lemma 8.2.5(iii) with $B = A$, $\mathrm{HH}^*(A)$ is a finitely generated H-module, and so it is finitely generated as an algebra (take the algebra generators of H together with the H-module generators of $\mathrm{HH}^*(A)$). Since $\mathrm{Ext}_A^*(A/\mathfrak{r}, A/\mathfrak{r})$ is finitely generated as an H-module, it will be finitely generated as an $\mathrm{HH}^*(A)$-module, and so condition (fg) holds. $\qquad\square$

We will be interested in the maximal ideal spectrum of H and of $\mathrm{HH}^*(A)$. Due to (graded) commutativity of these algebras, in either case the nilpotent elements constitute an ideal that is contained in all maximal ideals. Thus for the purpose of considering the maximal ideal spectrum, we may as well work with the quotient by this ideal. The following theorem, due to Snashall and Solberg [**201**, Proposition 4.6], gives some information about nilpotent elements of $\mathrm{HH}^*(A)$. We will need the notion of the *radical* of an ideal I of a (graded) commutative ring R, that is,

$$\sqrt{I} = \{x \in R \mid x^n \in I \text{ for some positive integer } n\}.$$

By Corollary 2.5.9 with $M = A/\mathfrak{r}$, the map $\phi_{A/\mathfrak{r}}$ defined in (2.5.3) is a ring homomorphism from $\mathrm{HH}^*(A)$ to $\mathrm{Ext}^*_A(A/\mathfrak{r}, A/\mathfrak{r})$ with image in the graded center. We now examine this ring homomorphism further.

Theorem 8.2.6. *Let $\mathcal{N}$ be the ideal of $\mathrm{HH}^*(A)$ generated by all homogeneous nilpotent elements. Then*

$$\mathcal{N} = \sqrt{\mathrm{Ker}(\phi_{A/\mathfrak{r}})}.$$

Proof. A nilpotent element is in the radical of every ideal by definition, so the containment $\mathcal{N} \subseteq \sqrt{\mathrm{Ker}(\phi_{A/\mathfrak{r}})}$ is automatic. It remains to prove that $\mathrm{Ker}(\phi_{A/\mathfrak{r}}) \subseteq \mathcal{N}$, and the containment $\sqrt{\mathrm{Ker}(\phi_{A/\mathfrak{r}})} \subseteq \mathcal{N}$ will follow since $\mathcal{N}$ contains all nilpotent elements. Let η be a homogeneous element in $\mathrm{Ker}(\phi_{A/\mathfrak{r}})$, and suppose η is represented by a function $f : P_m \to A$, where $P_\bullet$ is a projective resolution of A as an A^e-module. Since $\phi_{A/\mathfrak{r}}(\eta) = 0$, at the chain level, $\phi_{A/\mathfrak{r}}(f)$ has image in $\mathfrak{r}$. For each m, the power η^m is represented by f^m, whose image is thus in $\mathfrak{r}^m$. Since $\mathfrak{r}$ is a nilpotent ideal, it follows that η is nilpotent. $\qquad\square$

Exercise 8.2.7. Let $A = k[x]/(x^n)$. Use the results of Example 2.5.10 to give the structure of $\mathrm{HH}^*(A)$, of $\mathrm{Ext}^*_A(A/\mathfrak{r}, A/\mathfrak{r})$, and the action of the former on the latter, thus verifying directly that A satisfies condition (fg).

Exercise 8.2.8. Let $A = k[x]/(x^n)$. Find a graded subalgebra H of $\mathrm{HH}^*(A)$ that satisfies conditions (fg1) and (fg2) of Theorem 8.2.3.

Exercise 8.2.9. Verify the last statement in the proof of Lemma 8.2.4, that the actions correspond.

Exercise 8.2.10. Let $A = k[x]/(x^n)$. What is the ideal $\mathcal{N}$ of $\mathrm{HH}^*(A)$ generated by all homogeneous nilpotent elements?

8.3. Support varieties

We are now ready to define support varieties. Assume the finite-dimensional algebra A satisfies condition (fg) of the previous section. Applying Theorem 8.2.3, we fix a subalgebra H of $\mathrm{HH}^*(A)$ for which conditions (fg1) and (fg2) hold. We do not assume that A is indecomposable, as is done in some of the references. The main difference is that in applications, it may be necessary to keep track of extra points in support varieties of nonindecomposable modules, corresponding to idempotents in A that are not in the annihilators of the modules. Specifically, as in isomorphism (1.2.5), the Hochschild cohomology ring of A decomposes as a direct sum of Hochschild cohomology rings of the algebras $e_j A$, where $\{e_1, \ldots, e_i\}$ is a set of orthogonal central idempotents of A for which $1 = e_1 + \cdots + e_i$. Assuming that each e_j is primitive, that is, it is not the sum of two nonzero orthogonal central idempotents, the algebras $e_j A$ are themselves indecomposable.

For finite-dimensional A-modules M, N, let $I_H(M, N)$ be the annihilator in H of $\mathrm{Ext}_A^*(M, N)$, that is,

$$I_H(M, N) = \{\alpha \in H \mid \alpha \cdot \beta = 0 \text{ for all } \beta \in \mathrm{Ext}_A^*(M, N)\},$$

where the (left) action of the subalgebra H of $\mathrm{HH}^*(A)$ on $\mathrm{Ext}_A^*(M, N)$ is defined in Section 2.5. By its definition, $I_H(M, N)$ is an ideal of H. Let $I_H(M) = I_H(M, M)$. There is an analogous definition of support variety of a right module, and we will sometimes use it as well.

Definition 8.3.1. Let M, N be finite-dimensional A-modules. The *support variety* of the pair M, N is

$$V_H(M, N) = V_H(I_H(M, N)) \cong \mathrm{Max}(H/I_H(M, N)),$$

the maximal ideal spectrum of the quotient ring $H/I_H(M, N)$. The *support variety* of M is $V_H(M) = V_H(M, M)$.

Note that by its definition, a support variety may be canonically identified with a subvariety of $\mathrm{Max}(H)$, the maximal ideal spectrum of H.

Example 8.3.2. Let $A = k[x]/(x^n)$. Let $M = k$, the trivial module (on which x acts as 0). As we saw in Example 2.5.10, $\mathrm{Ext}_A^*(k, k)$ is isomorphic to $k[y]$ in case $n = 2$ and is isomorphic to $k[y, z]/(y^2)$ otherwise. We also found the image of $\mathrm{HH}^*(A)$ in $\mathrm{Ext}_A^*(k, k)$ under the map ϕ_k defined in (2.5.3). Let $H = \mathrm{HH}^*(A)$ in characteristic 2 and $H = \mathrm{HH}^{\mathrm{ev}}(A)$ otherwise, where $\mathrm{HH}^{\mathrm{ev}}(A)$ is the subalgebra of $\mathrm{HH}^*(A)$ generated by homogeneous elements of even degree. In both cases, $H/I_H(k)$ has Krull dimension 1 and so $V_H(k)$ is a line.

The following lemma gives a relationship between the support variety of a pair of modules and the support varieties of the modules.

Lemma 8.3.3. *For all finite-dimensional A-modules M and N,*

$$V_H(M, N) \subseteq V_H(M) \cap V_H(N).$$

Proof. This is true by the definitions. The (left) action of H on the space $\operatorname{Ext}_A^*(M, N)$ factors through that on $\operatorname{Ext}_A^*(N, N)$. By Theorem 2.5.5, this is the same, up to a sign, as a right action factoring through that on $\operatorname{Ext}_A^*(M, M)$, which in turn is the same, up to a sign, as a left action on $\operatorname{Ext}_A^*(M, M)$. $\qquad\square$

Let $\operatorname{Irr} A$ denote a set of representatives of isomorphism classes of simple A-modules. The following lemma gives a relationship between the support variety of a module and those of the simple A-modules, and a relationship among support varieties for modules in a short exact sequence. Recall that $\mathfrak{r} = \operatorname{rad}(A)$ denotes the Jacobson radical of A.

Proposition 8.3.4. *Let M, M_1, M_2, M_3 be finite-dimensional A-modules. Then:*

(i) $V_H(M) = \bigcup_{S \in \operatorname{Irr}(A)} V_H(M, S) = \bigcup_{S \in \operatorname{Irr}(A)} V_H(S, M)$.

(ii) *If $0 \to M_1 \to M_2 \to M_3 \to 0$ is a short exact sequence, then*

$$V_H(M_i) \subseteq V_H(M_j) \cup V_H(M_l)$$

whenever $\{i, j, l\} = \{1, 2, 3\}$.

(iii) $V_H(M) = V_H(M, A/\mathfrak{r}) = V_H(A/\mathfrak{r}, M)$.

Proof. (i) If $0 \to U' \to U \to U'' \to 0$ is a short exact sequence of A-modules, then $V_H(M, U) \subseteq V_H(M, U') \cup V_H(M, U'')$ since the annihilator in H of $\operatorname{Ext}_A^*(M, U)$ contains the product of the annihilators of $\operatorname{Ext}_A^*(M, U')$ and $\operatorname{Ext}_A^*(M, U'')$ in light of the first long exact sequence for Ext (Theorem A.4.4). Therefore $V_H(M, N) \subseteq \bigcup_S V_H(M, S)$, the union over all simple modules in a composition series for M, and similarly the containment $V_H(M, N) \subseteq \bigcup_S V_H(S, N)$ holds. Setting $M = N$ and $V_H(M) = V_H(M, M)$, Lemma 8.3.3 establishes the required reverse containments.

(ii) The above argument using the first long exact sequence for Ext (Theorem A.4.4) shows that $V_H(M_i) \subseteq V_H(M_i, M_j) \cup V_H(M_i, M_l)$, and this is contained in $V_H(M_j) \cup V_H(M_l)$ by Lemma 8.3.3.

(iii) By the above arguments, $V_H(M) \subseteq V_H(M, A/\mathfrak{r})$ and $V_H(M) \subseteq V_H(A/\mathfrak{r}, M)$. On the other hand, by Lemma 8.3.3, the reverse inclusions also are true. $\qquad\square$

We will need the notions of projective cover and minimal projective resolution. For a finite-dimensional algebra A with Jacobson radical $\mathfrak{r} = \operatorname{rad}(A)$, we may take the following definitions, equivalent to those in Section A.2. For an A-module M, write $\mathrm{rad}(M) = \mathrm{rad}(A)M$, the radical of M. A projective cover of M is a projective A-module P for which $P/\mathrm{rad}(P) \cong M/\mathrm{rad}(M)$. A minimal projective resolution $P_\bullet$ of M is one for which P_0 is a projective cover of M and P_n is a projective cover of K_n for all $n \geq 1$, where K_n is the nth syzygy module (notation as in Section A.2). Note this implies that K_n embeds into $\mathrm{rad}(P_{n-1})$ for all $n \geq 1$. We will also need the following property of projective modules for finite-dimensional algebras. A finite-dimensional projective A-module P is a direct sum of copies of the projective covers of simple modules. (See, e.g., [**11**, Corollary 4.5].)

Next we define the complexity of an A-module M and derive a connection to the dimension of its support variety.

Definition 8.3.5. Let $\cdots \to P_1 \to P_0 \to M \to 0$ be a minimal projective resolution of a finite-dimensional A-module M, and view $\bigoplus_{i \geq 0} P_i$ as a graded vector space. The *complexity* of M is $\mathrm{cx}_A(M) = \gamma(P_\bullet) = \gamma(\bigoplus_{i \geq 0} P_i)$, the rate of growth of the resolution (see Definition 8.1.5).

Recall that the dimension of $V_H(M)$ is given by Definition 8.1.3, equivalently by Definition 8.1.4 as the Krull dimension of $H/I_H(M)$, equivalently by Definition 8.1.5 as the rate of growth of $H/I_H(M)$. The following theorem makes a connection with the rate of growth of a minimal projective resolution.

Theorem 8.3.6. *Let M be a finite-dimensional A-module. Then*

$$\dim V_H(M) = \mathrm{cx}_A(M).$$

Proof. The proof is essentially the same as that of [**22**, Proposition 5.7.2], which is the case that A is a group algebra of a finite group. By Lemma 8.2.5(ii), $\mathrm{Ext}_A^*(M, M)$ is finitely generated as a module over $H/I_H(M)$. It follows that

$$\dim V_H(M) = \gamma(H/I_H(M)) = \gamma(\mathrm{Ext}_A^*(M, M)).$$

We will show that $\gamma(\mathrm{Ext}_A^*(M, M)) = \gamma(P_\bullet)$, where $P_\bullet$ is a minimal projective resolution of the A-module M. This is by definition $\mathrm{cx}_A(M)$.

Let K_n denote the nth syzygy module of $P_\bullet$. For any simple A-module S, the multiplicity of its projective cover $P(S)$ as a direct summand of P_n is

$$\dim_k(\mathrm{Hom}_A(P_n, S)) = \dim_k(\mathrm{Hom}_A(K_n, S)) = \dim_k(\mathrm{Ext}_A^n(M, S))$$

since $S \cong P(S)/\mathrm{rad}(P(S))$, the radical $\mathfrak{r}$ acts trivially on S, and K_n embeds into $\mathrm{rad}(P_{n-1})$ for all $n \geq 1$. So

$$\dim_k P_n = \sum_{S \in \mathrm{Irr}(A)} \dim_k P(S) \cdot \dim_k(\mathrm{Ext}_A^n(M, S)).$$

It follows that

$$\gamma(P_\bullet) \leq \max\{\gamma(\mathrm{Ext}^*_A(M, S)) \mid S \in \mathrm{Irr}(A)\}.$$

Now for each simple A-module S, the action of H on $\mathrm{Ext}^*_A(M, S)$ factors through $\mathrm{Ext}^*_A(M, M)$, and both are finitely generated as H-modules. Therefore

$$\gamma(\mathrm{Ext}^*_A(M, S)) \leq \gamma(\mathrm{Ext}^*_A(M, M))$$

for each S. It follows that $\gamma(P_\bullet) \leq \gamma(\mathrm{Ext}^*_A(M, M))$. On the other hand, by definition of Ext,

$$\dim_k \mathrm{Ext}^n_A(M, M) \leq \dim_k \mathrm{Hom}_k(P_n, M) = \dim_k(P_n) \dim_k(M)$$

for each n, and so $\gamma(\mathrm{Ext}^*_A(M, M)) \leq \gamma(P_\bullet)$. It follows that $\gamma(\mathrm{Ext}^*_A(M, M)) = \gamma(P_\bullet)$, and this is by definition the complexity of M. $\square$

Remark 8.3.7. For completeness, we mention a duality result, although we will not use it. For each A-module M, let $D(M) = \mathrm{Hom}_k(M, k)$, an A^{op}-module with action given by $(a \cdot f)(m) = f(a \cdot m)$ for all $a \in A$, $f \in D(M)$, and $m \in M$. It can be shown that $V_H(M) = V_H(D(M))$ for all finite-dimensional A-modules M (see [**201**, Proposition 3.5]).

Exercise 8.3.8. Verify some details of Example 8.3.2: find $I_H(k)$ explicitly and show that $H/I_H(k)$ has Krull dimension 1.

Exercise 8.3.9. Let $A = k[x]/(x^n)$, and let k be the A-module on which x acts as 0. Let H be as in Example 8.3.2. Verify directly that $\dim V_H(k) = \mathrm{cx}_A(k)$ by finding the rate of growth of a minimal projective resolution of k and comparing with Exercise 8.3.8.

8.4. Self-injective algebras and realization

For the rest of this chapter, assume A is a finite-dimensional self-injective algebra satisfying condition (fg) from Section 8.2. A *self-injective* algebra A is one for which the left A-module A, under multiplication, is an injective A-module. Some examples of self-injective algebras are finite group algebras and finite-dimensional Hopf algebras.

Fix a subalgebra H of $\mathrm{HH}^*(A)$ satisfying conditions (fg1) and (fg2) of Theorem 8.2.3. Under these conditions, in Theorem 8.4.4 below we state a tensor product property that has some important consequences. First we characterize the support varieties of projective modules.

Lemma 8.4.1. *Let M be a finite-dimensional A-module. Then M is projective if and only if $\dim V_H(M) = 0$.*

Proof. If M is projective, then $0 \to M \to M \to 0$ is a projective resolution of M, and so it must be that $\mathrm{cx}_A(M) = 0$. By Theorem 8.3.6, $\dim V_H(M) = \mathrm{cx}_A(M) = 0$. Conversely, assume $\dim V_H(M) = 0$, that is, $\mathrm{cx}_A(M) = 0$, so that if $P_\bullet$ is a minimal projective resolution of M, then $P_n = 0$ for some n. That is,

$$0 \to P_{n-1} \to \cdots \to P_0 \to M \to 0$$

is a projective resolution of M. Since A is self-injective, each P_i is also injective, so the sequence splits. Thus M is a direct summand of P_0, and so is projective. $\qquad\square$

The Heller operator $\Omega = \Omega_A$ of the following lemma is defined in Section A.2. We assume, without loss of generality, that it refers specifically to the syzygy module of a projective cover in this setting of self-injective algebras.

Lemma 8.4.2. *Let M be a finite-dimensional indecomposable A-module. Then $V_H(M) = V_H(\Omega_A M)$.*

Proof. Let P be a projective cover of the A-module M, so that $P/\mathrm{rad}(P) \cong M/\mathrm{rad}(M)$. Since M is indecomposable, there is a unique primitive central idempotent e_j of A for which $e_j M = M$ and $e_j P = P$. By the Wedderburn Theorem, $V_H(P)$ consists of precisely one point corresponding to e_j that is also in $V_H(M)$ and in $V_H(\Omega_A M)$. Now apply Proposition 8.3.4(ii) to the sequence $0 \to \Omega_A M \to P \to M \to 0$. $\qquad\square$

Next we give a realization result, namely that any closed homogeneous subvariety of $\mathrm{Max}(H)$ is the support variety of some A-module. This makes use of some bimodules attached to elements of H, analogous to Carlson's modules L_ζ in the group algebra case (see, e.g., [**22**, §5.9]). We follow the treatment of these modules in [**65**] for self-injective algebras.

Let $P_\bullet$ be a minimal resolution of A as an A^e-module. Let $\eta \in \mathrm{HH}^n(A)$, $n \geq 1$, so that η is represented by an element $\hat{\eta}$ of the space $\mathrm{Hom}_{A^e}(\Omega^n_{A^e} A, A)$. Define the A^e-module M_η to be the pushout (see Section A.1) of the inclusion $\Omega^n_{A^e} A \hookrightarrow P_{n-1}$ and $\hat{\eta} : \Omega^n_{A^e} A \to A$. By definition of pushout, there is a commuting diagram with both rows exact:

$$
\begin{array}{ccccccccc}
0 & \longrightarrow & \Omega^n_{A^e} A & \longrightarrow & P_{n-1} & \longrightarrow & \Omega^{n-1}_{A^e} A & \longrightarrow & 0 \\
& & \downarrow{\hat{\eta}} & & \downarrow & & \downarrow{1} & & \\
0 & \longrightarrow & A & \longrightarrow & M_\eta & \longrightarrow & \Omega^{n-1}_{A^e} A & \longrightarrow & 0
\end{array}
$$

Note that M_η is projective as a right A-module and as a left A-module by its definition, since both A and $\Omega^{n-1}_{A^e} A$ have this property. We will frequently use the bottom row of the diagram, the short exact sequence

$$(8.4.3) \qquad 0 \to A \to M_\eta \to \Omega_{A^e}^{n-1} A \to 0.$$

In the following theorem, by $V_H(\eta)$, we mean the variety of the ideal (η) generated by η, that is, all maximal ideals containing η.

Theorem 8.4.4. *Let M be an A-module and $\eta \in \mathrm{HH}^n(A)$, $n \geq 1$. Then*

$$V_H(M_\eta \otimes_A M) = V_H(\eta) \cap V_H(M).$$

Proof. For the proof, we may assume that M is indecomposable so that it is a module for Ae_j for some primitive central idempotent e_j of A, and that $\eta \in \mathrm{HH}^*(e_j A)$. (See isomorphism (1.2.5) for the decomposition of $\mathrm{HH}^*(A)$ into a direct sum of such components.)

By the definitions, a maximal ideal $\mathfrak{m}$ of H is in $V_H(M, N)$ if and only if $I_H(M, N) \subseteq \mathfrak{m}$ if and only if $\mathrm{Ext}_A^*(M, N)_\mathfrak{m} \neq 0$, where $\mathrm{Ext}_A^*(M, N)_\mathfrak{m}$ is the localization at $\mathfrak{m}$ of $\mathrm{Ext}_A^*(M, N)$.

By Proposition 8.3.4(i), $V_H(M_\eta \otimes_A M) = \bigcup_{S \in \mathrm{Irr}(A)} V_H(M_\eta \otimes_A M, S)$. We will first show that for each simple A-module S,

$$V_H(\eta) \cap V_H(M, S) \subseteq V_H(M_\eta \otimes_A M, S),$$

from which it will follow that $V_H(\eta) \cap V_H(M) \subseteq V_H(M_\eta \otimes_A M)$.

Let $\mathfrak{m}$ be a maximal ideal in $V_H(\eta) \cap V_H(M, S)$, that is, $\mathfrak{m}$ contains (η) and $I_H(M, S)$. We want to show that $I_H(M_\eta \otimes_A M, S) \subseteq \mathfrak{m}$. Suppose this is not true. Then after localization, $\mathrm{Ext}_A^*(M_\eta \otimes_A M, S)_\mathfrak{m} = 0$. Apply $- \otimes_A M$ to the short exact sequence (8.4.3). Since each A^e-module in the sequence is free as a right A-module, we obtain a short exact sequence of A-modules,

$$(8.4.5) \qquad 0 \to M \to M_\eta \otimes_A M \to \Omega_{A^e}^{n-1} A \otimes_A M \to 0.$$

For each simple A-module S, apply $\mathrm{Ext}_A^*(-, S)$ and consider the corresponding second long exact sequence for Ext (Theorem A.4.5). By dimension shifting (Theorem A.3.3) and the observation that $\Omega_{A^e}^{n-1} A \otimes_A M$ and $\Omega_A^{n-1} M$ agree up to projective direct summands, it is:

$$(8.4.6) \quad \cdots \to \mathrm{Ext}_A^{i+n-1}(M, S) \xrightarrow{\phi} \mathrm{Ext}_A^i(M_\eta \otimes_A M, S) \longrightarrow \mathrm{Ext}_A^i(M, S) \xrightarrow{\tilde{\eta}}$$
$$\mathrm{Ext}_A^{i+n}(M, S) \to \cdots,$$

where $\tilde{\eta}$ is the action of η on $\mathrm{Ext}_A^*(M, S)$. Let $z \in \mathrm{Ext}_A^{i+n-1}(M, S)$, and consider $\phi(z) \in \mathrm{Ext}_A^i(M_\eta \otimes_A M, S)$. Since $\mathrm{Ext}_A^*(M_\eta \otimes_A M, S)_\mathfrak{m} = 0$ by assumption, there is a homogeneous element $a \notin \mathfrak{m}$ such that $\phi(az) = a\phi(z) = 0$. In light of the above long exact sequence, $az = \tilde{\eta}(y)$ for some $y \in \mathrm{Ext}_A^{|a|+i}(M, S)$. Upon localizing, we have $z = a^{-1}\tilde{\eta}(y)$, so

$$\mathrm{Ext}_A^*(M, S)_\mathfrak{m} = \tilde{\eta}(\mathrm{Ext}_A^*(M, S)_\mathfrak{m}).$$

Since $\eta \in \mathfrak{m}$ and $\mathrm{Ext}_A^*(M, S)$ is finitely generated over H, it now follows by Nakayama's Lemma that $\mathrm{Ext}_A^*(M, S)_{\mathfrak{m}} = 0$. This contradicts the assumption that $I_H(M, S) \subseteq \mathfrak{m}$. So $I_H(M_\eta \otimes_A M, S) \subseteq \mathfrak{m}$, and therefore $V_H(\eta) \cap V_H(M, S) \subseteq V_H(M_\eta \otimes_A M, S)$, as claimed. Thus $V_H(\eta) \cap V_H(M) \subseteq V_H(M_\eta \otimes_A M)$.

To prove the reverse inclusion $V_H(M_\eta \otimes_A M) \subseteq V_H(\eta) \cap V_H(M)$, we will first show that $V_H(M_\eta \otimes_A M) \subseteq V_H(\eta)$. By Proposition 8.3.4(i), it suffices to show that $V_H(S, M_\eta \otimes_A M) \subseteq V_H(\eta)$ for each simple A-module S. Equivalently, it suffices to show that if $\mathrm{Ext}_A^*(S, M_\eta \otimes_A M)_{\mathfrak{m}} \neq 0$ for a maximal ideal $\mathfrak{m}$ of H, then $\eta \in \mathfrak{m}$. Suppose on the contrary that $\eta \notin \mathfrak{m}$. Then multiplication by η induces an isomorphism on $\mathrm{Ext}_A^*(S, M_\eta \otimes_A M)_{\mathfrak{m}}$ since it is invertible in $H_{\mathfrak{m}}$. As localization is exact, existence of the short exact sequence (8.4.3) implies that $\mathrm{Ext}_A^*(S, M_\eta \otimes_A M)_{\mathfrak{m}}$ is the kernel of the isomorphism $\eta : \mathrm{Ext}_A^*(S, M)_{\mathfrak{m}} \to \mathrm{Ext}_A^{*+n}(S, M)_{\mathfrak{m}}$. So $\mathrm{Ext}_A^*(S, M_\eta \otimes_A M)_{\mathfrak{m}} = 0$, as desired.

Finally we will show that $V_H(M_\eta \otimes_A M) \subseteq V_H(M)$. This is true by Proposition 8.3.4(ii) and Lemma 8.4.2 applied to the sequence (8.4.5) since $\Omega_{A^e}^n A \otimes_A M$ is the same as $\Omega_A^n M$ up to projective direct summands. $\square$

We obtain the aforementioned realization result as a corollary.

Corollary 8.4.7. *For any homogeneous ideal I of H, there exists an A-module M such that $V_H(M) = V_H(I)$.*

Proof. Let I be a homogeneous ideal of H. Since H is noetherian, I is finitely generated by homogeneous elements, say $I = (\eta_1, \ldots, \eta_r)$. Let

$$M = M_{\eta_1} \otimes_A \cdots \otimes_A M_{\eta_r} \otimes_A A/\mathfrak{r}.$$

Then $V_H(M) = V_H(I)$ by Theorem 8.4.4. $\square$

Exercise 8.4.8. Let $A = k[x]/(x^n)$. Find $\Omega_A k$, where k is an A-module on which x acts as 0. Letting H be as in Example 8.3.2, verify directly that $V_H(k) = V_H(\Omega_A k)$.

Exercise 8.4.9. In the long exact sequence (8.4.6), verify that $\tilde{\eta}$ is indeed the action of η on $\mathrm{Ext}_A^*(M, S)$ as claimed.

8.5. Self-injective algebras and indecomposable modules

We continue under the assumption that A is a finite-dimensional self-injective algebra satisfying condition (fg) from Section 8.2. In accordance with Theorem 8.2.3, fix a subalgebra H of $\mathrm{HH}^*(A)$ for which conditions (fg1) and (fg2) hold. A consequence of Theorem 8.5.6 below is that the support variety of an indecomposable module is connected. (More precisely, the corresponding projective variety is connected.) We will introduce periodic modules and see that the indecomposable periodic modules are those whose support varieties have dimension 1. These results appeared in [65], based on techniques from support variety theory for finite groups in [44–46, 62].

We will need some lemmas. For each A-module M, let

$$M^{\#} = \mathrm{Hom}_A(M, A),$$

an A^{op}-module under the action $(af)(m) = f(am)$ for all $a \in A^{\mathrm{op}}$, $m \in M$, and $f \in \mathrm{Hom}_A(M, A)$. We may view $M^{\#}$ alternatively as a right A-module via $(fa)(m) = f(am)$, and consider the action of H on $\mathrm{Ext}_A^*(M^{\#}, M^{\#})$. Note that $(M^{\#})^{\#} \cong M$.

Lemma 8.5.1. *Let M be a finite-dimensional A-module. Then*

$$V_H(M) = V_H(M^{\#}).$$

Proof. The proof uses properties of the duality $(\ )^{\#}$ and adjoint functors, similarly to the proof of Lemma 8.2.4; see [201, Proposition 3.6] for details. We give only an outline here. Let $P_{\bullet}$ be a minimal projective resolution of A as an A^e-module. Let $\eta \in I_H(M)$, so that $\eta \otimes 1_M : P_n \otimes_A M \to M$ factors through $d_n \otimes 1_M : P_n \otimes_A M \to P_{n-1} \otimes_A M$. Consider $1_{M^{\#}} \otimes_A \eta : M^{\#} \otimes_A P_n \to M^{\#}$. We wish to show that $1_{M^{\#}} \otimes_A \eta$ factors through $1_{M^{\#}} \otimes d_n$. This is a consequence of a sequence of isomorphisms for all i:

$$
\begin{aligned}
\mathrm{Hom}_{A^{\mathrm{op}}}(M^{\#} \otimes_A P_i, M^{\#}) &\cong \mathrm{Hom}_A(M, \mathrm{Hom}_{A^{\mathrm{op}}}(M^{\#} \otimes_A P_i, A)) \\
&\cong \mathrm{Hom}_A(M, \mathrm{Hom}_A(P_i, M)) \\
&\cong \mathrm{Hom}_A(P_i \otimes_A M, M).
\end{aligned}
$$

$\square$

For the next lemma, recall the definition of the modules M_η via sequence (8.4.3) as part of a pushout diagram.

Lemma 8.5.2. *Let η be a homogeneous element of H of positive degree n, and let M be a finite-dimensional A-module. Then $\eta \in I_H(M)$ if and only if*

$$M_\eta \otimes_A M \cong M \oplus \Omega_A^{n-1}(M) \oplus Q$$

for some projective A-module Q.

Proof. We may assume M is nonprojective, as the lemma is automatically true in case M is projective. Let $\eta \in H^n$, represented by an element of $\mathrm{Hom}_{A^e}(\Omega_{A^e}^n A, A) \cong \mathrm{Ext}_{A^e}^1(\Omega_{A^e}^{n-1} A, A)$ (see Theorem A.3.3). Note that $\eta \in I_H(M)$ if and only if the sequence (8.4.5) splits, that is, if and only if

$$M_\eta \otimes_A M \cong M \oplus (\Omega_{A^e}^{n-1} A \otimes_A M).$$

Since $\Omega_{A^e}^{n-1} A \otimes_A M$ and $\Omega_A^{n-1} M$ agree up to projective direct summands, the statement follows. $\square$

As a consequence of the lemma, we next show a connection between the intersection of support varieties of two modules and their generalized extensions.

Lemma 8.5.3. *Let M, N be finite-dimensional A-modules. If*

$$\dim(V_H(M) \cap V_H(N)) = 0,$$

then $\operatorname{Ext}_A^i(M, N) = 0$ *for all $i > 0$.*

Proof. Since H is noetherian, $I_H(M)$ is finitely generated by homogeneous elements. Suppose $I_H(M) = (\eta_1, \ldots, \eta_r)$. By Lemma 8.5.2 and induction on r, up to projective direct summands, the A-module M is a direct summand of

$$M_{\eta_1} \otimes_A \cdots \otimes_A M_{\eta_r} \otimes_A M.$$

So $\operatorname{Ext}_A^i(M, N)$ is a direct summand of

$$\operatorname{Ext}_A^i(M_{\eta_1} \otimes_A \cdots \otimes_A M_{\eta_r} \otimes_A M, N),$$

and this Ext space is isomorphic to

$$\operatorname{Ext}_A^i(M, (M_{\eta_1} \otimes_A \cdots \otimes_A M_{\eta_r})^\# \otimes_A N)$$

by an argument similar to the proof of Lemma 8.2.4. Also,

$$\operatorname{Ext}_A^i(A/\mathfrak{r}, (M_{\eta_1} \otimes_A \cdots \otimes_A M_{\eta_r})^\# \otimes_A N)$$
$$\cong \operatorname{Ext}_A^i(M_{\eta_1} \otimes_A \cdots \otimes_A M_{\eta_r} \otimes_A A/\mathfrak{r}, N).$$

So $V_H((M_{\eta_1} \otimes_A \cdots \otimes_A M_{\eta_r})^\# \otimes_A N)$ is contained in the intersection of the varieties $V_H(N)$ and $V_H((M_{\eta_1} \otimes_A \cdots \otimes_A M_{\eta_r}) \otimes_A A/\mathfrak{r})$, by Lemma 8.3.3 and Proposition 8.3.4(iii). The latter is contained in $V_H(M)$ by Theorem 8.4.4, since $I_H(M) = (\eta_1, \ldots, \eta_r)$, so by hypothesis, the variety of the A-module $(M_{\eta_1} \otimes_A \cdots \otimes_A M_{\eta_r})^\# \otimes_A N$ has dimension 0. It follows that the module is projective by Lemma 8.4.1, and therefore it is injective, and so

$$\operatorname{Ext}_A^i(M, (M_{\eta_1} \otimes_A \cdots \otimes_A M_{\eta_r})^\# \otimes_A N) = 0$$

for all $i > 0$. The same is then true of the direct summand $\operatorname{Ext}_A^i(M, N)$. $\square$

We need one more lemma before stating the main theorem of this section. The lemma states existence of a short exact sequence corresponding to two homogeneous elements of H.

Lemma 8.5.4. *Let $\eta_1 \in H^m$ and $\eta_2 \in H^n$. There is a projective A^e-module P for which there is a short exact sequence*

$$0 \to \Omega_{A^e}^n(M_{\eta_1}) \to M_{\eta_2\eta_1} \oplus P \to M_{\eta_2} \to 0.$$

Proof. Starting with the short exact sequence (8.4.3) with n replaced by m and η by η_1, we claim that there are projective A^e-modules Q, Q' for which there is an exact sequence

$$(8.5.5) \qquad 0 \to \Omega_{A^e}(M_{\eta_1}) \oplus Q' \to \Omega_{A^e}^m(A) \oplus Q \to A \to 0,$$

where the map $\Omega_{A^e}^m(A) \to A$ is $\hat{\eta}_1$. To see this, first recall the definition (8.4.3) of M_{η_1} via a pushout diagram. In accordance with the expression (A.1.2) of a pushout module, since $\Omega_{A^e}^m(A)$ maps injectively to P_{m-1}, there is a short exact sequence

$$0 \to \Omega_{A^e}^m(A) \to A \oplus P_{m-1} \to M_{\eta_1} \to 0.$$

Let P be a projective cover of M_{η_1}, with kernel $\Omega_{A^e}(M_{\eta_1})$, expressed as a short exact sequence $0 \to \Omega_{A^e}(M_{\eta_1}) \to P \to M_{\eta_1} \to 0$. This sequence maps to the previous short exact sequence, where M_{η_1} is mapped to itself by the identity map, since P is projective:

$$
\begin{array}{ccccccccc}
0 & \longrightarrow & \Omega_{A^e}(M_{\eta_1}) & \longrightarrow & P & \longrightarrow & M_{\eta_1} & \longrightarrow & 0 \\
& & \downarrow & & \downarrow & & \downarrow{\scriptstyle 1} & & \\
0 & \longrightarrow & \Omega_{A^e}^m(A) & \longrightarrow & A \oplus P_{m-1} & \longrightarrow & M_{\eta_1} & \longrightarrow & 0
\end{array}
$$

It may be checked that the leftmost square is a pullback diagram, and so there is a short exact sequence

$$0 \to \Omega_{A^e}(M_{\eta_1}) \to \Omega_{A^e}^m(A) \oplus P \to A \oplus P_{m-1} \to 0.$$

Split off the direct summand P_{m-1} of $A \oplus P_{m-1}$. Any projective direct summand of $\Omega_{A^e}^m(A)$ that is lost in the process can be added back in, by adding it instead to $\Omega_{A^e}(M_{\eta_1})$, in order to preserve $\Omega_{A^e}^m(A)$, resulting in a sequence of the form (8.5.5), as claimed.

Now $\widehat{\eta_2\eta_1}$ is the composition $\hat{\eta}_2(\hat{\eta}_1)_n$, where $(\hat{\eta}_1)_n$ may be viewed as a map from $\Omega_{A^e}^{m+n}A$ to $\Omega_{A^e}^m A$. Therefore, there is a commutative diagram with exact rows:

$$
\begin{array}{ccccccccc}
0 & \longrightarrow & A & \longrightarrow & M_{\eta_2\eta_1} & \longrightarrow & \Omega_{A^e}^{m+n-1}A & \longrightarrow & 0 \\
& & \downarrow{\scriptstyle 1} & & \downarrow & & \downarrow & & \\
0 & \longrightarrow & A & \longrightarrow & M_{\eta_2} & \longrightarrow & \Omega_{A^e}^{n-1}A & \longrightarrow & 0
\end{array}
$$

Apply the Horseshoe Lemma (Lemma A.4.3) to the sequence (8.5.5) to see that there is a projective module P' such that the map

$$\Omega_{A^e}^{m+n-1}A \oplus P' \to \Omega_{A^e}^{n-1}A$$

is surjective, with kernel $\Omega_{A^e}^n(M_{\eta_1})$. Now in the above diagram, replace $M_{\eta_2\eta_1}$ with $M_{\eta_2\eta_1} \oplus P'$ and $\Omega_{A^e}^{m+n-1}A$ with $\Omega_{A^e}^{m+n-1}A \oplus P'$. We have just identified the kernel of the resulting rightmost vertical map as $\Omega_{A^e}^n(M_{\eta_1})$.

By the Snake Lemma (Lemma A.4.1), the kernel of the middle vertical map is also $\Omega^n_{A^e}(M_{\eta_1})$, completing the proof. $\qquad\square$

We are now ready to state the main theorem of this section.

Theorem 8.5.6. *Let A be a finite-dimensional self-injective algebra satisfying condition* (fg), *and let H be a subalgebra of* $\mathrm{HH}^*(A)$ *satisfying conditions* (fg1) *and* (fg2) *of Theorem 8.2.3. Let M be a finite-dimensional A-module for which $V_H(M) = V_1 \cup V_2$ for some homogeneous varieties V_1 and V_2 with $\dim(V_1 \cap V_2) = 0$. Then there are A-modules M_1 and M_2 with $V_H(M_1) = V_1$, $V_H(M_2) = V_2$, and $M \cong M_1 \oplus M_2$.*

A consequence of the theorem is that the projective variety of an indecomposable module is connected, where the projective variety of H is the space of lines through the origin in $\mathrm{Max}(H)$ and that of a module M is lines through the origin in $\mathrm{Max}(H/I_H(M))$. This makes sense as $I_H(M)$ is a homogeneous ideal by its definition.

Proof. Let $m_1 = \dim V_1$ and $m_2 = \dim V_2$. We will prove the statement by induction on $m_1 + m_2$. If $m_1 = 0$ or $m_2 = 0$, the result is clear, so assume $m_1 > 0$ and $m_2 > 0$. Then there are homogeneous elements η_1 and η_2 of H such that $V_1 \subseteq V_H(\eta_1)$, $V_2 \subseteq V_H(\eta_2)$, and

$$\dim(V_2 \cap V_H(\eta_1)) = m_2 - 1, \quad \dim(V_1 \cap V_H(\eta_2)) = m_1 - 1.$$

Now $V_H(\eta_2\eta_1) = V_H(\eta_1) \cup V_H(\eta_2) \supseteq V_1 \cup V_2 = V_H(M)$, so $(\eta_2\eta_1)^s \in I_H(M)$ for some s. We may assume $\eta_2\eta_1 \in I_H(M)$ by replacing each η_i with η_i^s if $s > 1$. By Lemma 8.5.2,

$$M_{\eta_2\eta_1} \otimes_A M \cong M \oplus \Omega^{n-1}_A M \oplus Q$$

for some projective A-module Q. By Lemma 8.5.4, there is a short exact sequence

$$0 \to \Omega^n_{A^e} M_{\eta_1} \to M_{\eta_2\eta_1} \oplus P \to M_{\eta_2} \to 0$$

for some projective A^e-module P. Apply $- \otimes_A M$ to this sequence to obtain

$$0 \to \Omega^n_{A^e} M_{\eta_1} \otimes_A M \to (M_{\eta_2\eta_1} \otimes_A M) \oplus (P \otimes_A M) \to M_{\eta_2} \otimes_A M \to 0.$$

Replacing $M_{\eta_2\eta_1} \otimes_A M$ via the above isomorphism, we have a short exact sequence

$$
\begin{aligned}
(8.5.7) \qquad 0 \to \Omega^n_{A^e} M_{\eta_1} \otimes_A M &\to M \oplus \Omega^{n-1}_A M \oplus Q \oplus (P \otimes_A M) \\
&\to M_{\eta_2} \otimes_A M \to 0.
\end{aligned}
$$

By Theorem 8.4.4,

$$
\begin{aligned}
V_H(M_{\eta_2} \otimes_A M) &= V_H(\eta_2) \cap V_H(M) \\
&= V_H(\eta_2) \cap (V_1 \cup V_2) \\
&= (V_H(\eta_2) \cap V_1) \cup V_2,
\end{aligned}
$$

as $V_2 \subseteq V_H(\eta_2)$. The intersection of the varieties $V_H(\eta_2) \cap V_1$ and V_2 is contained in $V_1 \cap V_2$, and thus has dimension 0. Further, the sum of the dimensions of these two varieties is $m_1 - 1 + m_2$. By the induction hypothesis, $\Omega^n_{A^e} M_{\eta_1} \otimes_A M \cong N_1 \oplus N_2$ for A-modules N_1, N_2 with $V_H(N_1) = V_H(\eta_2) \cap V_1$ and $V_H(N_2) = V_2$. We also find that by Theorem 8.4.4 similarly,

$$V_H(\Omega^n_{A^e} M_{\eta_1} \otimes_A M) = V_H(\eta_1) \cap V_H(M) = V_1 \cup (V_H(\eta_1) \cap V_2),$$

and so $\Omega^n_{A^e} M_{\eta_1} \otimes_A M \cong N'_1 \oplus N'_2$ for A-modules N'_1, N'_2 with $V_H(N'_1) = V_1$ and $V_H(N'_2) = V_H(\eta_1) \cap V_2$. Thus the sequence (8.5.7) may be rewritten

$$(8.5.8) \qquad\qquad 0 \to N_1 \oplus N_2 \to X \to N'_1 \oplus N'_2 \to 0$$

for $X = M \oplus \Omega^{n-1}_A M \oplus Q \oplus (P \otimes_A M)$.

Now $V_H(N'_1) \cap V_H(N_2) = V_1 \cap V_2$, which has dimension 0, and so by Lemma 8.5.3, $\mathrm{Ext}^i_A(N'_1, N_2) = 0$ for all $i > 0$. Similarly, $V_H(N_1) \cap V_H(N'_2) \subseteq V_1 \cap V_2$, and so has dimension 0, so $\mathrm{Ext}^i_A(N'_2, N_1) = 0$ for all $i > 0$. Then

$$\mathrm{Ext}^1_A(N'_1 \oplus N'_2, N_1 \oplus N_2) \cong \mathrm{Ext}^1_A(N'_1, N_1) \oplus \mathrm{Ext}^1_A(N'_2, N_2)$$

and so (8.5.8) is a direct sum of two sequences $0 \to N_1 \to N''_1 \to N'_1 \to 0$ and $0 \to N_2 \to N''_2 \to N'_2 \to 0$. Moreover, $V_H(N''_1) \subseteq V_1$ and $V_H(N''_2) \subseteq V_2$. Rewriting X as a direct sum:

$$M \oplus \Omega^{n-1}_A M \oplus Q \oplus (P \otimes_A M) \cong N''_1 \oplus N''_2.$$

Now $P \otimes_A M$ is a projective A-module, so

$$M \oplus \Omega^{n-1}_A M \oplus Q'' \cong N''_1 \oplus N''_2$$

for some projective A-module Q''. By the Krull-Schmidt Theorem, $M \cong M_1 \oplus M_2$ for some M_1 and M_2 with $V_H(M_1) \subseteq V_H(N''_1) \subseteq V_1$ and $V_H(M_2) \subseteq V_H(N''_2) \subseteq V_2$. By hypothesis, $V_H(M) = V_1 \cup V_2$, and this forces $V_H(M_1) = V_1$ and $V_H(M_2) = V_2$. $\qquad\square$

As a final topic, we consider periodic indecomposable modules.

Definition 8.5.9. An A-module M is *periodic* if $\Omega^i_A M \cong M$ as A-modules for some i.

Example 8.5.10. The trivial module k for $A = k[x]/(x^n)$ is periodic, as can be seen from our work in Example 2.5.10. If $n = 2$, the period is $i = 1$, and if $n > 2$, the period is $i = 2$.

The following result characterizes indecomposable periodic modules.

Theorem 8.5.11. *Let A be a finite-dimensional self-injective algebra satisfying condition* (fg), *and let H be a subalgebra of* $\mathrm{HH}^*(A)$ *satisfying conditions* (fg1) *and* (fg2) *of Theorem 8.2.3. Let M be a finite-dimensional indecomposable A-module. Then M is periodic if and only if* $\dim V_H(M) = 1$.

Proof. Assume M is periodic, so that $\Omega_A^n M \cong M$ for some n. Let $\zeta \in \mathrm{Ext}_A^n(M, M)$ correspond to an isomorphism $\hat{\zeta} : \Omega_A^n(M) \xrightarrow{\sim} M$. Let $\eta \in \mathrm{Ext}_A^r(M, M)$ be any homogeneous element and $\hat{\eta} : \Omega_A^r(M) \to M$ the corresponding map. Since $\Omega_A^{nr} M \cong M$, the function $\hat{\eta}^n : \Omega_A^{nr} M \to M$ can be identified with an element of $\mathrm{Hom}_A(M, M)$, a local ring since M is indecomposable. Accordingly, $\hat{\eta}^n$ is either an isomorphism or is nilpotent, and moreover if it is an isomorphism, it must be a sum of a scalar multiple of $\hat{\zeta}^r$ and a nilpotent element. It follows that $\mathrm{Ext}_A^*(M, M)$ is a direct sum of $k[\zeta]$ and a nilpotent ideal. Thus $\dim(V_H(M)) = 1$.

Conversely, assume $\dim(V_H(M)) = 1$. Then $H/I_H(M)$ has Krull dimension 1. By hypothesis, for all simple A-modules S, $\mathrm{Ext}_A^*(M, S)$ is finitely generated as a module over $H/I_H(M)$, and so is finitely generated over a polynomial subring $k[\zeta]$ of H, for ζ a homogeneous element of some positive degree n. By the classification of finitely generated modules over a principal ideal domain, action by ζ is an isomorphism

$$\mathrm{Ext}_A^i(M, S) \xrightarrow{\sim} \mathrm{Ext}_A^{i+n}(M, S)$$

for sufficiently large i and all simple A-modules S. Let $P_\bullet$ be a minimal projective resolution of the A-module M. Then $\mathrm{Ext}_A^i(M, S) \cong \mathrm{Hom}_A(P_i, S)$ for all i, and so there is an isomorphism

$$\mathrm{Hom}_A(P_i, S) \xrightarrow{\sim} \mathrm{Hom}_A(P_{i+n}, S)$$

for all i sufficiently large and all simple A-modules S. These homomorphism spaces uniquely determine the projective A-modules P_i, and so $P_i \cong P_{i+n}$ for all i sufficiently large, which further implies that $\Omega_A^i M \cong \Omega_A^{i+n} M$. Choosing just one such value of i, and applying Ω_A^{-i} to this isomorphism, we have $M \cong \Omega_A^n M$. Thus M is periodic. $\square$

There are many further applications of support varieties. For example, all modules in a connected stable component of the Auslander-Reiten quiver have the same variety [**201**, Theorem 3.7]. There is a connection to representation type through complexity of modules [**65**, Section 5]. In some of the best understood settings, some of the structure of module categories can be understood using support variety theory (see, e.g., [**23**]).

Exercise 8.5.12. Let $A = k[x]/(x^n)$. Let $M = k[x]/(x^m)$ for some integer m ($2 \leq m < n$), considered to be an A-module via a quotient map and multiplication. (The case $m = 1$ corresponds to the trivial module k.) Show directly that M is periodic by examining its minimal projective resolution. Compare with Theorem 8.5.11.

Exercise 8.5.13. Suppose $\operatorname{char}(k) = 2$ and $A = k[x,y]/(x^2, y^2)$. Let k be the A-module on which x and y both act as 0. Let $H = \operatorname{HH}^*(A)$.

(a) Show that k is not periodic by showing that $\dim V_H(k) = 2$ and appealing to Theorem 8.5.11.

(b) Show that k is not periodic by examining its minimal projective resolution.

Chapter 9

Hopf Algebras

Hopf algebras comprise an important class of algebras that include group algebras, universal enveloping algebras of Lie algebras, and quantum groups. They act on rings, giving rise to extension rings such as smash and crossed product rings and Hopf Galois extensions. In this chapter, we collect some techniques for understanding homological information about Hopf algebras and related ring extensions. We introduce the Hopf algebra cohomology ring in Section 9.3 and prove that it embeds into the Hochschild cohomology ring in Section 9.4. These rings and these embeddings have many applications, for example: they aid exploration of support varieties of modules, as defined in Chapter 8; they lead to better understanding of cup products in Hochschild cohomology by way of cup products in Hopf algebra cohomology, as discussed in Sections 9.4 and 9.5; they provide tools for constructing spectral sequences, for example a spectral sequence relating the Hochschild cohomology ring of a smash product to cohomology rings of its component parts as we present in Section 9.6.

9.1. Hopf algebras and actions on rings

In this section we give a brief introduction to Hopf algebras and smash product algebras. For details and more general notions such as crossed product algebras and Hopf Galois extensions, see [**162, 182, 192**].

Let A be an algebra over the field k. Let $\pi : A \otimes A \to A$ denote its multiplication map, and let $\eta : k \to A$ denote its unit map (that is, the embedding of the field k into A as scalar multiples of the multiplicative identity). We consider the tensor product algebra $A \otimes A$ with its factorwise multiplication: $(a \otimes b)(c \otimes d) = ac \otimes bd$ for all $a, b, c, d \in A$. In the following definition, we canonically identify the spaces $k \otimes A$ and $A \otimes k$ with A.

Definition 9.1.1. A *Hopf algebra* is an algebra A over k together with algebra homomorphisms $\Delta : A \to A \otimes A$ (called *coproduct* or *comultiplication*) and $\varepsilon : A \to k$ (called *counit* or *augmentation*) and an algebra antihomomorphism (that is, reversing order of multiplication) $S : A \to A$ (called *coinverse* or *antipode*) such that

$$
\begin{aligned}
(\Delta \otimes 1)\Delta &= (1 \otimes \Delta)\Delta, \\
(\varepsilon \otimes 1)\Delta &= 1 = (1 \otimes \varepsilon)\Delta, \\
\pi(S \otimes 1)\Delta &= \eta\varepsilon = \pi(1 \otimes S)\Delta
\end{aligned}
$$

as maps on $A \otimes A$, where 1 denotes the identity map on A. The first equation above is called *coassociativity* and we may view the last as stating that the antipode S is the convolution inverse of the identity map. We say that A is *cocommutative* if $\tau\Delta = \Delta$, where $\tau : A \otimes A \to A \otimes A$ is the map that interchanges tensor factors, that is, $\tau(a \otimes b) = b \otimes a$ for all $a, b \in A$.

It can be shown that the antipode S is a coalgebra antihomomorphism, that is, $\Delta S = (S \otimes S)\tau\Delta$ and $\varepsilon S = \varepsilon$. (See, for example, [**182**, Theorem 7.1.14].)

Some examples of Hopf algebras are the following.

Example 9.1.2 (Group algebras). Let G be a group and $A = kG$, its group algebra. Let $\Delta(g) = g \otimes g$, $\varepsilon(g) = 1$, and $S(g) = g^{-1}$ for all $g \in G$. Then A is a cocommutative Hopf algebra.

Example 9.1.3 (Universal enveloping algebras of Lie algebras). Let $\mathfrak{g}$ be a Lie algebra, and let $A = U(\mathfrak{g})$ be its universal enveloping algebra as defined in Section 5.1. That is,

$$
U(\mathfrak{g}) = T(\mathfrak{g})/(x \otimes y - y \otimes x - [x, y] \mid x, y \in \mathfrak{g}),
$$

where $T(\mathfrak{g})$ denotes the tensor algebra of the underlying vector space of $\mathfrak{g}$. Let $\Delta(x) = x \otimes 1 + 1 \otimes x$, $\varepsilon(x) = 0$, and $S(x) = -x$ for all $x \in \mathfrak{g}$. The maps Δ and ε are extended to be algebra homomorphisms, and S to be an algebra antihomomorphism. Then A is a cocommutative Hopf algebra.

Example 9.1.4 (Quantum enveloping algebras). Let $A = U_q(\mathfrak{g})$, a quantum enveloping algebra. See, for example, [**91**] for the definition in the general case. Here we give just one small example explicitly. Let q be a nonzero scalar, $q^2 \neq 1$. Let $U_q(\mathfrak{sl}_2)$ be the k-algebra generated by $E, F, K^{\pm 1}$ with relations $KK^{-1} = K^{-1}K = 1$, $KE = q^2 EK$, $KF = q^{-2}FK$, and

$$
EF = FE + \frac{K - K^{-1}}{q - q^{-1}}.
$$

Let $\Delta(E) = E \otimes 1 + K \otimes E$, $\Delta(F) = F \otimes K^{-1} + 1 \otimes F$, $\Delta(K) = K \otimes K$, $\varepsilon(E) = 0$, $\varepsilon(F) = 0$, $\varepsilon(K) = 1$, $S(E) = -K^{-1}E$, $S(F) = -FK$, and $S(K) = K^{-1}$. Then A is a noncocommutative Hopf algebra. If q is a primitive nth root of unity, $n > 2$, we may consider the quotient of $U_q(\mathfrak{sl}_2)$ by the ideal generated by E^n, F^n, $K^n - 1$, denoted by $u_q(\mathfrak{sl}_2)$. This quotient is a finite-dimensional (that is, finite-dimensional as a vector space) noncocommutative Hopf algebra, called a small quantum group.

Example 9.1.5 (Quantum elementary abelian groups). Let m and n be positive integers, $n \geq 2$. Let q be a primitive nth root of unity, and let A be the k-algebra generated by $x_1, \ldots, x_m, g_1, \ldots, g_m$ with relations $x_i^n = 0$, $g_i^n = 1$, $x_i x_j = x_j x_i$, $g_i g_j = g_j g_i$, and $g_i x_j = q^{\delta_{i,j}} x_j g_i$ for all i, j. Let $\Delta(x_i) = x_i \otimes 1 + g_i \otimes x_i$, $\Delta(g_i) = g_i \otimes g_i$, $\varepsilon(x_i) = 0$, $\varepsilon(g_i) = 1$, and $S(x_i) = -g_i^{-1} x_i$, $S(g_i) = g_i^{-1}$ for all i. Then A is a finite-dimensional noncocommutative Hopf algebra.

We will use some standard notation for the coproduct, called *Sweedler notation*: write

$$\Delta(a) = \sum_{(a)} a_1 \otimes a_2,$$

or simply $\sum a_1 \otimes a_2$, where the notation a_1, a_2 for tensor factors is symbolic. Some authors dispense with the summation symbol, writing $a_1 \otimes a_2$ to denote this sum. Similarly write $\Delta(1 \otimes \Delta)(a) = \sum a_1 \otimes a_2 \otimes a_3$, and note this is the same as $\Delta(\Delta \otimes 1)(a)$ by coassociativity. The second two equations in Definition 9.1.1 may now be rewritten as

$$\sum \varepsilon(a_1) a_2 = a = \sum a_1 \varepsilon(a_2),$$
$$\sum S(a_1) a_2 = \varepsilon(a) = \sum a_1 S(a_2)$$

for all $a \in A$, where we have identified k with its image in A under η.

If A is a finite-dimensional Hopf algebra, then its vector space dual $A^* = \mathrm{Hom}_k(A, k)$ is also a Hopf algebra under the duals of the defining maps of A. That is, identifying $(A \otimes A)^*$ canonically with $A^* \otimes A^*$, multiplication on A^* is Δ^*, comultiplication is π^*, the unit map is ε^*, the counit map is η^*, and the antipode is S^*. If A is an infinite-dimensional Hopf algebra, there are meaningful dual Hopf algebras such as the finite dual (see [**162**]). If A and B are Hopf algebras, then the tensor product algebra $A \otimes B$ is a Hopf algebra with factorwise coproduct, counit, and antipode, and the opposite algebra A^{op} is a Hopf algebra with the same coproduct, counit, and antipode as A. In this way, $A^e = A \otimes A^{\mathrm{op}}$ is also a Hopf algebra.

An element h of a Hopf algebra A is a *left integral* of A if $a \cdot h = \varepsilon(a)h$ for all $a \in A$. A right integral is defined similarly. It can be shown that the spaces of left and right integrals of a finite-dimensional Hopf algebra are one dimensional and are interchanged by the antipode. Maschke's Theorem for Hopf algebras states that a finite-dimensional Hopf algebra A is semisimple if and only if $\varepsilon(h) \neq 0$ for a nonzero left (respectively, right) integral h.

A finite-dimensional Hopf algebra is a *Frobenius algebra,* and therefore self-injective (see [**162**]). Specifically, a nonzero left integral λ in the dual Hopf algebra A^* gives rise to a nondegenerate associative bilinear form on A given by $(a, b) \mapsto \lambda(ab)$ for all $a, b \in A$. The left A-module A is isomorphic to the left A-module given by the dual of the right A-module A, via the map that takes a to ϕ_a, where $\phi_a(b) = \lambda(ba)$ for all $a, b \in A$.

The quantum elementary abelian groups in Example 9.1.5 are examples of skew group algebras as defined in Section 3.5. We generalize skew group algebras by defining smash products next.

Definition 9.1.6. Let A be a Hopf algebra. An *A-module algebra* is an algebra R over k that is also an A-module for which

$$
\begin{aligned}
a \cdot 1 &= \varepsilon(a)1, \\
a \cdot (rr') &= \sum (a_1 \cdot r)(a_2 \cdot r')
\end{aligned}
$$

for all $a \in A$ and $r, r' \in R$. The *smash product* $R\#A$ of A with R is an algebra that is $R \otimes A$ as a vector space, with multiplication defined by

$$(r \otimes a)(r' \otimes a') = \sum r(a_1 \cdot r') \otimes a_2 a'$$

for all $a, a' \in A$ and $r, r' \in R$.

By definition of multiplication on the smash product algebra, R and A are both subalgebras of $R\#A$. So it should cause no confusion if we sometimes abuse notation and write r for $r \otimes 1$ and a for $1 \otimes a$ as elements of $R\#A$ for $r \in R$ and $a \in A$.

A more general construction than the smash product is a crossed product involving cocycles, and both are examples of yet more general Hopf Galois extensions. We will not use these more general notions in this book.

We rewrite an A-module algebra action of A on R as an internal action within the smash product $R\#A$: for each $a \in A$ and $r \in R$, by the properties of Hopf algebras and definition of multiplication in the smash product,

$$
\begin{aligned}
(a \cdot r) \otimes 1 = \sum ((a_1 \varepsilon(a_2)) \cdot r) \otimes 1 &= \sum a_1 \cdot r \otimes \varepsilon(a_2) \\
&= \sum a_1 \cdot r \otimes a_2 S(a_3) \\
&= \sum (1 \otimes a_1)(r \otimes 1)(1 \otimes S(a_2)).
\end{aligned}
$$

We call this an *adjoint action* of A. Written more compactly, the action of A on R is given by $a \cdot r = \sum a_1 r S(a_2)$ in $R \# A$ for all $a \in A$, $r \in R$.

The quantum elementary abelian groups of Example 9.1.5 are smash products of $R = k[x_1, \ldots, x_m]/(x_1^n, \ldots, x_m^n)$ with $A = kG$, where $G = \langle g_1, \ldots, g_m \rangle$ is a direct product of cyclic groups $\langle g_i \rangle$, each of order n, acting by $g_i \cdot x_j = q^{\delta_{ij}} x_j$. Skew group algebras (Section 3.5) are all smash product algebras; specifically, $R \rtimes G \cong R \# kG$. For an example in which the Hopf algebra is not a group algebra, take $A = U_q(\mathfrak{sl}_2)$ from Example 9.1.4, $R = k\langle x, y \rangle/(xy - qyx)$, and $E \cdot x = 0$, $E \cdot y = x$, $F \cdot x = y$, $F \cdot y = 0$, $K^{\pm 1} \cdot u = q^{\pm 1} u$, $K^{\pm 1} \cdot v = q^{\mp 1} v$. These actions extend to an action of A on R under which R is an A-module algebra, and we may form the smash product algebra $R \# A$.

Exercise 9.1.7. Use Sweedler notation to rewrite the condition that the antipode S is a coalgebra antihomomorphism; in particular, rewrite the equation $\Delta S = (S \otimes S)\tau\Delta$.

Exercise 9.1.8. Let G be a finite group and $A = kG$. What is the structure of the dual Hopf algebra A^*? Describe multiplication, unit, comultiplication, counit, and antipode in terms of the basis dual to G.

Exercise 9.1.9. Let G be a finite group and $A = kG$. Show that $\sum_{g \in G} g$ is a left integral, and that any left integral is a scalar multiple of this one.

Exercise 9.1.10. Verify that R and A are indeed both subalgebras of the smash product algebra $R \# A$.

Exercise 9.1.11. Let G be a group. Show that a kG-module algebra R is simply an algebra R with action of G by algebra automorphisms, and that the smash product $R \# kG$ given by Definition 9.1.6 is precisely the skew group algebra defined in Section 3.5.

Exercise 9.1.12. Let $\mathfrak{g}$ be a Lie algebra. Show that a $U(\mathfrak{g})$-module algebra R is simply an algebra R with action of $\mathfrak{g}$ by derivations. What is the structure of the smash product $R \# U(\mathfrak{g})$?

Exercise 9.1.13. Verify that the quantum plane $R = k\langle x, y \rangle/(xy - qyx)$ is indeed a $U_q(\mathfrak{sl}_2)$-module algebra under the action given at the end of this section.

9.2. Modules for Hopf algebras

Modules for a Hopf algebra A have some important properties: tensor products (over k) of A-modules are A-modules, and Homs (over k) of A-modules are A-modules. The category of A-modules is a tensor category [**71**]. In this section, we define the relevant module structures and relationships among them.

We work primarily with left modules as before, and the term module will mean left module if not otherwise specified. However, we will also need right modules in Section 9.6, and we include some discussion in this section of the properties of right modules that we will need there.

Given a Hopf algebra A, the field k is an A-module via the counit ε, sometimes called the *trivial module*, that is, $a \cdot c = \varepsilon(a)c$ for all $a \in A$, $c \in k$. For any two A-modules V and W, their vector space tensor product $V \otimes W$ is an A-module via the coproduct Δ:

$$(9.2.1) \qquad a \cdot (v \otimes w) = \sum (a_1 \cdot v) \otimes (a_2 \cdot w)$$

for all $a \in A$, $v \in V$, and $w \in W$. Similarly, $\mathrm{Hom}_k(V, W)$ is an A-module via

$$(a \cdot f)(v) = \sum a_1 f(S(a_2)v)$$

for all $a \in A$, $f \in \mathrm{Hom}_k(V, W)$, and $v \in V$. In particular, if V is an A-module, its dual vector space

$$V^* = \mathrm{Hom}_k(V, k)$$

has an A-module structure given by

$$(a \cdot f)(v) = f(S(a)v)$$

for all $a \in A$, $v \in V$, and $f \in V^*$. Note that the tensor product of the trivial module k with V is isomorphic to V, that is, $k \otimes V \cong V$ and $V \otimes k \cong V$ as A-modules due to the properties of the maps Δ, ε.

For any A-module V, let V^A denote the A-submodule of V on which A acts via ε, that is,

$$V^A = \{v \in V \mid a \cdot v = \varepsilon(a)v \text{ for all } a \in A\},$$

called the submodule of *A-invariants* of V. We will use the same notation for A-invariants of a right A-module.

Lemma 9.2.2. *Let V, W be A-modules. There is an isomorphism of vector spaces*

$$\mathrm{Hom}_A(V, W) \cong (\mathrm{Hom}_k(V, W))^A.$$

Proof. This is [**192**, Lemma 4.1]. Let $f \in \mathrm{Hom}_A(V, W)$. Then

$$(a \cdot f)(v) \; = \; \sum a_1 f(S(a_2)v) \; = \; \sum a_1 S(a_2) f(v) \; = \; \varepsilon(a)f(v)$$

for all $a \in A$, $v \in V$. So $f \in (\mathrm{Hom}_k(V,W))^A$. Conversely, let $f \in (\mathrm{Hom}_k(V,W))^A$. Then

$$
\begin{aligned}
f(a \cdot v) \;=\; \sum f(\varepsilon(a_1) a_2 \cdot v) \;&=\; \sum \varepsilon(a_1) f(a_2 \cdot v) \\
&=\; \sum a_1 f(S(a_2) \cdot (a_3 \cdot v)) \\
&=\; \sum a_1 f(\varepsilon(a_2) v) \\
&=\; \sum a_1 \varepsilon(a_2) f(v) \;=\; a \cdot (f(v))
\end{aligned}
$$

for all $a \in A$, $v \in V$. So $f \in \mathrm{Hom}_A(V,W)$. $\qquad\square$

For a Hopf algebra having a bijective antipode, there is an alternative action on Hom, as we describe in the next remark. All finite-dimensional Hopf algebras have bijective antipodes (see [162, Theorem 2.1.3] or [182, Theorem 7.1.4]), and many infinite-dimensional Hopf algebras do as well.

Remark 9.2.3. Under the assumption that the antipode S is bijective with inverse map S^{-1}, we may alternatively define a dual module to V as

$$
{}^*V = \mathrm{Hom}_k(V, k)
$$

with action $(a \cdot f)(v) = f(S^{-1}(a)v)$ for all $a \in A$, $v \in V$, and $f \in {}^*V$. Similarly give $\mathrm{Hom}_k(V,W)$ the alternative A-module structure $(a \cdot f)(v) = \sum a_2 f(S^{-1}(a_1)v)$ for all $a \in A$, $f \in \mathrm{Hom}_k(V,W)$, $v \in V$. Lemma 9.2.2 holds under this alternative action, that is, the subspace of A-homomorphisms in $\mathrm{Hom}_k(V,W)$ is also equal to the A-invariant subspace of $\mathrm{Hom}_k(V,W)$ under this alternative action. To verify this statement, it is helpful to apply S to $\sum a_2 S^{-1}(a_1)$ to obtain $\sum a_1 S(a_2) = \varepsilon(a)$, implying that $\sum a_2 S^{-1}(a_1) = \varepsilon(a)$ for all $a \in A$.

There are right module versions of all of these actions as well.

Remark 9.2.4. If V, W are *right* A-modules, then $\mathrm{Hom}_k(V,W)$ is a right A-module via $(f \cdot a)(v) = \sum f(vS(a_1))a_2$ for all $f \in \mathrm{Hom}_A(V,W)$, $a \in A$, $v \in V$. There is a right module version of Lemma 9.2.2. If S is bijective, then $\mathrm{Hom}_k(V,W)$ is also a right A-module via $(f \cdot a)(v) = \sum f(vS^{-1}(a_2))a_1$ for all $f \in \mathrm{Hom}_k(V,W)$, $a \in A$, $v \in V$, and again there is a corresponding version of Lemma 9.2.2.

We next establish relations among A-modules that are obtained by taking Hom, $\otimes$, and duals.

Lemma 9.2.5. *Let U, V, and W be A-modules. There is a natural isomorphism of A-modules*

$$
\mathrm{Hom}_k(U \otimes V, W) \cong \mathrm{Hom}_k(U, \mathrm{Hom}_k(V, W)),
$$

and a natural isomorphism of vector spaces

$$\mathrm{Hom}_A(U \otimes V, W) \cong \mathrm{Hom}_A(U, \mathrm{Hom}_k(V, W)).$$

Proof. Define a function $\phi : \mathrm{Hom}_k(U \otimes V, W) \to \mathrm{Hom}_k(U, \mathrm{Hom}_k(V, W))$ by

$$(\phi(f)(u))(v) = f(u \otimes v)$$

and a function $\psi : \mathrm{Hom}_k(U, \mathrm{Hom}_k(V, W)) \to \mathrm{Hom}_k(U \otimes V, W)$ by

$$(\psi(g))(u \otimes v) = (g(u))(v).$$

By its definition, ψ is inverse to ϕ. We check that ϕ is an A-module homomorphism. Let $a \in A$ and $f \in \mathrm{Hom}_k(U \otimes V, W)$. Then, as S reverses the order of comultiplication, for all $u \in U$ and $v \in V$,

$$
\begin{aligned}
(\phi(a \cdot f)(u))(v) &= (a \cdot f)(u \otimes v) \\
&= \sum a_1(f(S(a_2) \cdot (u \otimes v))) \\
&= \sum a_1(f(S(a_3)u \otimes S(a_2)v)).
\end{aligned}
$$

On the other hand,

$$
\begin{aligned}
(a \cdot \phi(f))(u)(v) &= \sum (a_1(\phi(f)(S(a_2)u)))(v) \\
&= \sum a_1((\phi(f))(S(a_3)u)(S(a_2)v)) \\
&= \sum a_1(f(S(a_3)u \otimes S(a_2)v)).
\end{aligned}
$$

Therefore $\phi(a \cdot f) = a \cdot \phi(f)$. These isomorphisms are natural by their definitions.

The second statement now follows by Lemma 9.2.2. $\qquad\square$

Remark 9.2.6. A similar proof shows that if U, V, W are *right* A-modules, then there is a (slightly different) right A-module isomorphism

$$\mathrm{Hom}_k(U \otimes V, W) \cong \mathrm{Hom}_k(V, \mathrm{Hom}_k(U, W))$$

under the first action defined in Remark 9.2.4. Taking A-invariants, we obtain an isomorphism

$$\mathrm{Hom}_A(U \otimes V, W) \cong \mathrm{Hom}_A(V, \mathrm{Hom}_k(U, W)).$$

The next lemma gives a relationship between Hom and dual.

Lemma 9.2.7. *Let V, W be A-modules. If V is finite dimensional as a vector space over k, there is a natural A-module isomorphism*

$$\mathrm{Hom}_k(V, W) \cong W \otimes V^*.$$

Proof. Let $\phi : W \otimes V^* \to \mathrm{Hom}_k(V, W)$ and $\psi : \mathrm{Hom}_k(V, W) \to W \otimes V^*$ be defined by $(\phi(w \otimes f))(v) = f(v)w$ and $\psi(f) = \sum_i f(v_i) \otimes v_i^*$, where $\{v_i\}, \{v_i^*\}$ are dual bases for V, V^*, for all $v \in V$, $w \in W$, and $f \in V^*$. Let $a \in A$. Then

$$
\begin{aligned}
\phi(a \cdot (w \otimes f))(v) &= \sum \phi(a_1 w \otimes (a_2 \cdot f))(v) \\
&= \sum ((a_2 \cdot f)(v))(a_1 w) = \sum f(S(a_2)v)a_1 w.
\end{aligned}
$$

On the other hand,

$$
\begin{aligned}
(a \cdot (\phi(w \otimes f)))(v) &= \sum a_1(\phi(w \otimes f)(S(a_2)v)) \\
&= \sum a_1(f(S(a_2)v)w) = \sum f(S(a_2)v)a_1 w.
\end{aligned}
$$

Therefore $\phi(a \cdot (w \otimes f)) = a \cdot (\phi(w \otimes f))$. A calculation shows that ϕ is inverse to ψ. $\qquad\square$

Remark 9.2.8. Under the *alternative A-module* structure described in Remark 9.2.3, there are (slightly different) A-module isomorphisms

$$
\mathrm{Hom}_k(V, W) \cong {}^*V \otimes W
$$

and

$$
\mathrm{Hom}_k(U \otimes V, W) \cong \mathrm{Hom}_k(V, \mathrm{Hom}_k(U, W)).
$$

Now consider only finite-dimensional A-modules. This observation combined with Lemma 9.2.7 implies that V^* is a *left dual* of V in the category of finite-dimensional A-modules and ${}^*V$ is a *right dual* of V in this category (notation and terminology as in [**71**]). It follows that ${}^*(V^*) \cong V$ and $({}^*V)^* \cong V$ for all finite-dimensional A-modules V; these isomorphisms can also be deduced directly from the definitions. We caution that the terms left and right dual are interchanged in comparison to [**16**].

Lemma 9.2.9. *Let P be a projective A-module, and V any A-module. Then $P \otimes V$ is a projective A-module. If the antipode S is bijective, then $V \otimes P$ is a projective A-module. Similar statements hold for right modules, as well as when "projective" is replaced by "flat".*

Proof. We give two proofs of the first two statements. The first proof is essentially that of [**21**, Proposition 3.1.5]. The projective module P is a direct summand of a free module, so it suffices to prove that $A \otimes V$ and $V \otimes A$ are both free as A-modules (the latter under the condition that S is bijective). There is an isomorphism $A \otimes V \xrightarrow{\sim} A \otimes V_{tr}$, where V_{tr} is the underlying vector space of V, but with the trivial A-module structure (via ε). This isomorphism is similar to one in [**162**, Theorem 1.9.4], and is given by $a \otimes v \mapsto \sum a_1 \otimes S(a_2)v$, the inverse function by $a \otimes v \mapsto \sum a_1 \otimes a_2 v$, for all $v \in V$, $a \in A$. Now V_{tr} is a direct sum of copies of the trivial module k, and so $A \otimes V_{tr}$ is a free A-module. Similarly, if S is bijective, there is an isomorphism of left A-modules, $V \otimes A \xrightarrow{\sim} V_{tr} \otimes A$, via the A-module homomorphism $v \otimes a \mapsto \sum S^{-1}(a_1)v \otimes a_2$ whose inverse is $v \otimes a \mapsto \sum a_1 v \otimes a_2$, and so $V \otimes A$ is a free A-module.

The second proof uses properties of functors. As V is projective over the field k and P is projective over A, $\mathrm{Hom}_A(P, \mathrm{Hom}_k(V, -))$ is an exact functor. By Lemma 9.2.5, this is the same as $\mathrm{Hom}_A(P \otimes V, -)$. Therefore $P \otimes V$ is projective. A similar argument applies to $V \otimes P$ if S is bijective, using Remark 9.2.8 and the alternative A-action on Hom given in Remark 9.2.3.

The last statement, for right modules, may be proven similarly, and the statement about flat modules follows since flat modules are direct limits of finitely generated free modules. $\qquad\square$

Exercise 9.2.10. Let A be a Hopf algebra. Verify that $k \otimes V \cong V$ and $V \otimes k \cong V$ for all A-modules V, where k is the trivial A-module.

Exercise 9.2.11. Verify the claim in Remark 9.2.3 that the subspace of A-homomorphisms in $\mathrm{Hom}_k(V, W)$ is its A-invariant subspace under the alternative action defined there.

Exercise 9.2.12. Prove the two isomorphisms stated in Remark 9.2.6.

Exercise 9.2.13. Prove the two isomorphisms stated in Remark 9.2.8.

Exercise 9.2.14. Verify the isomorphism $A \otimes V \cong A \otimes V_{tr}$ in the proof of Lemma 9.2.9.

Exercise 9.2.15. Formulate the last statement of Lemma 9.2.9 more precisely and prove it.

9.3. Hopf algebra cohomology and actions

In this section we define Hopf algebra cohomology and derive many properties of Ext spaces for modules of a Hopf algebra A. We will connect these ideas with Hochschild cohomology in the next section.

Lemma 9.3.1. *Let* U, V, W *be* A-*modules. There is an isomorphism of graded vector spaces,*

$$\mathrm{Ext}_A^*(U \otimes V, W) \cong \mathrm{Ext}_A^*(U, W \otimes V^*).$$

Proof. Let $P_{\bullet}$ be a projective resolution of U. By Lemma 9.2.9, $P_i \otimes V$ is a projective A-module for all i. Since the tensor product is over the field k, the functor $- \otimes V$ is exact, and so $P_{\bullet} \otimes V$ is a projective resolution of the A-module $U \otimes V$. The natural isomorphisms of Lemmas 9.2.5 and 9.2.7 yield a quasi-isomorphism between $\mathrm{Hom}_A(P_{\bullet}, W \otimes V^*)$ and $\mathrm{Hom}_A(P_{\bullet} \otimes V, W)$, and thus the claimed isomorphism on Ext spaces. $\qquad\square$

Remark 9.3.2. Using the alternative A-action on Hom described in Remark 9.2.3, we similarly find that

$$\operatorname{Ext}^*_A(U \otimes V, W) \cong \operatorname{Ext}^*_A(V, {}^*U \otimes W).$$

Now let M, M', N, N' be A-modules. We will next define a *cup product* for each $i, j \geq 0$,

$$(9.3.3) \qquad \smile \; : \operatorname{Ext}^i_A(M, M') \times \operatorname{Ext}^j_A(N, N') \longrightarrow \operatorname{Ext}^{i+j}_A(M \otimes N, M' \otimes N').$$

Let $P_{\bullet}$ be a projective resolution of M, and let $Q_{\bullet}$ be a projective resolution of N. Consider the total complex of the tensor product complex $P_{\bullet} \otimes Q_{\bullet}$, with action of A on each $P_i \otimes Q_j$ given by the coproduct Δ on A as in (9.2.1). Note that the differentials on the total complex of $P_{\bullet} \otimes Q_{\bullet}$ are A-module homomorphisms since the differentials on $P_{\bullet}$ and on $Q_{\bullet}$ are A-module homomorphisms. By Lemma 9.2.9, each module in this tensor product complex is projective. By the Künneth Theorem (Theorem A.5.2), since the tensor product is over the field k and Tor^k_1 is 0, the total complex of $P_{\bullet} \otimes Q_{\bullet}$ is a projective resolution of the A-module $M \otimes N$.

Let $f \in \operatorname{Hom}_A(P_i, M')$ and $g \in \operatorname{Hom}_A(Q_j, N')$ represent elements in the spaces $\operatorname{Ext}^i_A(M, M')$ and $\operatorname{Ext}^j_A(N, N')$, respectively. Then

$$f \otimes g \in \operatorname{Hom}_A(P_i \otimes Q_j, M' \otimes N')$$

and this function may be extended to an element of

$$\operatorname{Hom}_A\left(\bigoplus_{r+s=i+j} (P_r \otimes Q_s), M' \otimes N' \right)$$

by defining it to be the zero map on all components other than $P_i \otimes Q_j$. We follow the sign convention (2.3.1) in using this notation $f \otimes g$. By definition, $d(f \otimes g) = d(f) \otimes g + (-1)^{|f|} f \otimes d(g)$. It follows that if f and g are cocycles, then their tensor product $f \otimes g$ is again a cocycle, and the tensor product of a cocycle with a coboundary is a coboundary. Therefore this tensor product of functions induces a well-defined product on cohomology, namely the cup product also denoted $\smile$.

We will need the following result giving an equivalent definition of this cup product via a *Yoneda composition*, paralleling the equivalent definitions of cup product on Hochschild cohomology in Section 2.4. See also [**21**, Proposition 3.2.1].

Lemma 9.3.4. *If M, M', N, N' are left A-modules and $\alpha \in \operatorname{Ext}^m_A(M, M')$, $\beta \in \operatorname{Ext}^n_A(N, N')$, then the cup product*

$$\alpha \smile \beta \in \operatorname{Ext}^{m+n}_A(M \otimes N, M' \otimes N')$$

of (9.3.3) is equal to the Yoneda composite of

$$\alpha \otimes 1_{N'} \quad \text{and} \quad 1_M \otimes \beta$$

in $\mathrm{Ext}_A^m(M \otimes N', M' \otimes N')$ *and* $\mathrm{Ext}_A^n(M \otimes N, M \otimes N')$, *respectively.*

Proof. Let

$$\mathbf{f}: \qquad 0 \longrightarrow M' \longrightarrow M_{m-1} \longrightarrow \cdots M_0 \longrightarrow M \longrightarrow 0,$$

$$\mathbf{g}: \qquad 0 \longrightarrow N' \longrightarrow N_{n-1} \longrightarrow \cdots N_0 \longrightarrow N \longrightarrow 0$$

be an m- and an n-extension representing α and β, respectively (see Section A.3). By the Künneth Theorem (Theorem A.5.2), taking their tensor product over the field k results in an $(m+n)$-extension of $M \otimes N$ by $M' \otimes N'$:

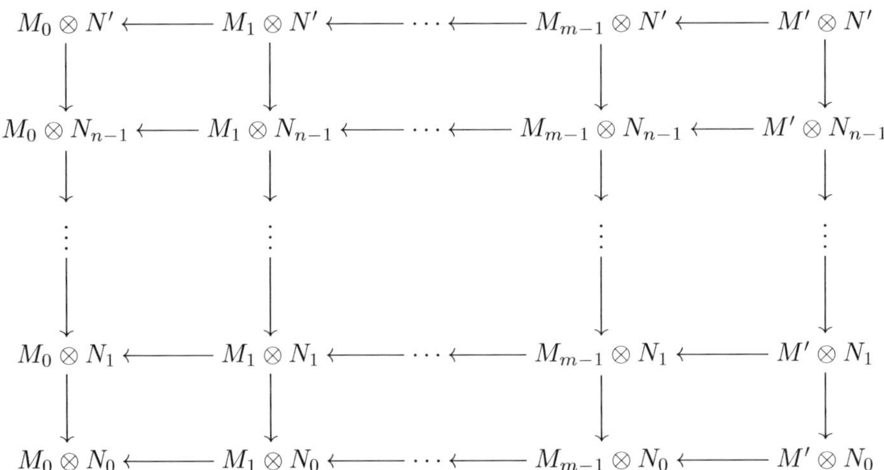

The cup product $\alpha \smile \beta$ is represented by the total complex of this tensor product complex, augmented by $M \otimes N$, denoted $\mathbf{f} \otimes \mathbf{g}$. This can be deduced from examination of the relationship between generalized extensions and elements of Ext as in Section A.3. Just as in Sections 2.4 and 6.5, there is a map from the total complex of this bicomplex to the Yoneda composite of $\mathbf{f} \otimes N'$ with $M \otimes \mathbf{g}$: project the bicomplex onto the leftmost column followed by the map $M_0 \to M$ and onto the top row. Specifically, in degrees $i + j$ with $0 \leq i + j \leq n - 1$, send $M_i \otimes N_j$ to 0 if $i > 0$ and to $M \otimes N_j$ as a projection from $M_0 \otimes N_j$ if $i = 0$. In degrees $i + j$ with $n \leq i + j \leq m + n - 1$, send $M_i \otimes N_j$ to 0 if $j < n$ and to $M_{i+j-n} \otimes N'$ if $j = n$. It can be checked that this is a chain map. $\qquad\square$

Recall the trivial module for A is k with action given by the counit ε. Set $M = M' = N = N' = k$ in Lemma 9.3.4 and identify $k \otimes k$ with k in the following definition.

Definition 9.3.5. The *Hopf algebra cohomology ring* of the Hopf algebra A over the field k is

$$\mathrm{H}^*(A, k) = \mathrm{Ext}_A^*(k, k)$$

under cup product. More generally, we write $\mathrm{H}^*(A, M) = \mathrm{Ext}_A^*(k, M)$ for any A-module M, a graded vector space that is an $\mathrm{H}^*(A, k)$-module under cup product and the identification of $k \otimes M$ with M.

We could equally well have defined Hopf algebra cohomology via right A-modules, and we will use this right module version in Section 9.6. Left and right modules are interchanged by applying the antipode, since it is an algebra antihomomorphism.

By Lemma 9.2.2 and the definitions, the degree 0 Hopf algebra cohomology is

$$\mathrm{H}^0(A, M) \cong \mathrm{Hom}_A(k, M) \cong (\mathrm{Hom}_k(k, M))^A \cong M^A.$$

Remark 9.3.6. Another proof of Lemma 9.3.4 in case $M = M' = N = N' = k$ uses a product given by Yoneda composition and the Eckmann-Hilton argument, which simultaneously shows that the product is graded commutative, that is,

$$\alpha \smile \beta = (-1)^{|\alpha||\beta|} \beta \smile \alpha$$

for all $\alpha, \beta \in \mathrm{H}^*(A, k)$. See Suárez-Álvarez [**211**] for a general context for this type of argument. More generally, when $M = N = k$ and $M' = N' = B$ is an A-module algebra, we may compose the cup product with the map induced by multiplication $B \otimes B \to B$ to obtain a ring structure on $\mathrm{H}^*(A, B)$. In the next section, we will let B be the algebra A itself, under the adjoint action of A, as defined there.

Remark 9.3.7. Just as for Hochschild cohomology of finite-dimensional algebras discussed in the last chapter, for many finite-dimensional Hopf algebras A, the cohomology $\mathrm{H}^*(A, k)$ is known to be finitely generated. For example, it is finitely generated if A is a finite group algebra [**75**, **92**, **217**] and more generally a finite group scheme (noncocommutative Hopf algebra) [**79**]. Many noncocommutative Hopf algebras are also known to have finitely generated cohomology; see, for example, [**20**, **91**, **96**, **155**, **209**]. In fact, this is conjectured always to be the case, notwithstanding the existence of counterexamples such as Example 8.2.2 to an analogous statement about Hochschild cohomology. See [**79**] where the question was raised and [**73**] where there is a more general conjecture for finite tensor categories.

Example 9.3.8. Let G be a group, and let $A = kG$, the group algebra. Then $\mathrm{H}^*(kG, k) = \mathrm{Ext}_{kG}^*(k, k)$ is the *group cohomology* of G with coefficients in k, also written $\mathrm{H}^*(G, k)$, and the cup product gives it the structure of a graded commutative ring. As a small example, let p be a prime, let k be a field of characteristic p, and let $G = \mathbb{Z}/p\mathbb{Z}$. Then the group cohomology of G with coefficients in k is

$$H^*(G, k) \cong \begin{cases} k[y], & \text{if } p = 2, \\ k[y, z]/(y^2), & \text{if } p > 2, \end{cases}$$

where $|y| = 1$, $|z| = 2$. (Note that $kG \cong k[x]/(x^p)$ since $\mathrm{char}(k) = p$, which can be seen by taking $x = g - 1$. Then see Example 2.5.10.)

Let $M = M' = k$ and $N' = N$ in Lemma 9.3.4. By composing with the isomorphism $k \otimes N \cong N$, we obtain an action of $H^*(A, k)$ on $\mathrm{Ext}_A^*(N, N)$, via $- \otimes N$ followed by Yoneda composition. On the other hand, we have an action of $H^*(A, k)$ on $\mathrm{Ext}_A^*(k, N \otimes N^*)$ by Yoneda composition.

In the following statement, we apply Lemma 9.3.1 with $U = k$ and $V = W = N$.

Theorem 9.3.9. *Let N be an A-module. Then the action of $H^*(A, k)$ on $\mathrm{Ext}_A^*(N, N)$, given by $- \otimes N$ followed by Yoneda composition, corresponds to that on $\mathrm{Ext}_A^*(k, N \otimes N^*)$, given by Yoneda composition, under the isomorphism*

$$\mathrm{Ext}_A^*(N, N) \cong \mathrm{Ext}_A^*(k, N \otimes N^*).$$

Proof. Let $P_\bullet$ be a projective resolution of k, so that $P_\bullet \otimes N$ is a projective resolution of $k \otimes N \cong N$. We must check that the following diagram commutes for each m, n, where ϕ_m, ϕ_{m+n} are the isomorphisms given by Lemma 9.2.5 with $V = W = N$ and $U = P_m, P_{m+n}$, respectively, and the horizontal maps are the chain level maps corresponding to the cup product of Lemma 9.3.4:

$$\begin{array}{ccc} \mathrm{Hom}_A(P_m, k) \otimes \mathrm{Hom}_A(P_n \otimes N, N) & \longrightarrow & \mathrm{Hom}_A(P_{m+n} \otimes N, N) \\ \Big\downarrow{\scriptstyle 1 \otimes \phi_n} & & \Big\downarrow{\scriptstyle \phi_{m+n}} \\ \mathrm{Hom}_A(P_m, k) \otimes \mathrm{Hom}_A(P_n, \mathrm{Hom}_k(N, N)) & \longrightarrow & \mathrm{Hom}_A(P_{m+n}, \mathrm{Hom}_k(N, N)) \end{array}$$

Let $\zeta \in \mathrm{Ext}_A^m(k, k)$ and $\eta \in \mathrm{Ext}_A^n(N, N)$, represented by $f \in \mathrm{Hom}_A(P_m, k)$ and $g \in \mathrm{Hom}_A(P_n \otimes N, N)$, respectively. Extend f to a chain map $f_\bullet$ with $f_i \in \mathrm{Hom}_A(P_{m+i}, P_i)$. The top horizontal map takes $f \otimes g$ to $g(f_n \otimes 1)$, and applying ϕ_{m+n} we have

$$\phi_{m+n}(g(f_n \otimes 1))(x)(v) = g(f_n(x) \otimes v)$$

for all $x \in P_{m+n}$, $v \in N$. On the other hand, $(1 \otimes \phi_n)(f \otimes g) = f \otimes \phi_n(g)$, and applying the bottom horizontal map we find

$$(\phi_n(g) f_n)(x)(v) = g(f_n(x) \otimes v).$$

Therefore the diagram commutes. $\qquad\qquad\qquad\qquad\qquad\qquad\qquad\quad \square$

There is another action of $\mathrm{H}^*(A, k)$ on $\mathrm{Ext}_A^*(M, M)$, given by $M \otimes -$ followed by Yoneda composition. In case A is cocommutative, this action is the same as that given by $- \otimes M$. (More generally, this is true if A is *quasi-triangular*, that is, there are functorial isomorphisms $M \otimes N \cong N \otimes M$ for all A-modules M, N.) In general these actions will not be the same; see, for example, [**25**]. We state next a counterpart of Theorem 9.3.9 for this action under the assumption that the antipode S is bijective, and include a proof to highlight this subtle distinction. Let $\mathrm{Hom}_k'(V, W)$ denote the A-module that is $\mathrm{Hom}_k(V, W)$ as a vector space, but with action as described in Remark 9.2.3. Let $^*V = \mathrm{Hom}_k(V, k)$, with A-module structure as described in Remark 9.2.3. Then, by Remark 9.2.8, there are isomorphisms of A-modules:

$$\mathrm{Hom}_k'(U \otimes V, W) \cong \mathrm{Hom}_k'(V, \mathrm{Hom}_k'(U, W)) \cong \mathrm{Hom}_k'(V, {}^*U \otimes W).$$

It follows that

$$(9.3.10) \qquad \mathrm{Hom}_A'(U \otimes V, W) \cong \mathrm{Hom}_A'(V, {}^*U \otimes W),$$

and consequently

$$\mathrm{Ext}_A^*(U \otimes V, W) \cong \mathrm{Ext}_A^*(V, {}^*U \otimes W).$$

Theorem 9.3.11. *Assume the antipode S of A is bijective and let M be an A-module. The action of $\mathrm{H}^*(A, k)$ on $\mathrm{Ext}_A^*(M, M)$, given by $M \otimes -$ followed by Yoneda composition, corresponds to that on $\mathrm{Ext}_A^*(k, {}^*M \otimes M)$, given by Yoneda composition, under the isomorphism*

$$\mathrm{Ext}_A^*(M, M) \cong \mathrm{Ext}_A^*(k, {}^*M \otimes M).$$

Proof. Let $P_\bullet$ be a projective resolution of k, so that $M \otimes P_\bullet$ is a projective resolution of $M \otimes k \cong M$. We must check that the following diagram commutes for each m, n, where ϕ_m, ϕ_{m+n} are the isomorphisms given in (9.3.10), with $U = W = M$ and $V = P_n, P_{m+n}$, respectively, and the horizontal maps are the chain level maps corresponding to the cup product of Lemma 9.3.4:

$$
\begin{array}{ccc}
\mathrm{Hom}_A(P_m, k) \otimes \mathrm{Hom}_A(M \otimes P_n, M) & \longrightarrow & \mathrm{Hom}_A(M \otimes P_{m+n}, M) \\
\downarrow{\scriptstyle 1 \otimes \phi_n} & & \downarrow{\scriptstyle \phi_{m+n}} \\
\mathrm{Hom}_A(P_m, k) \otimes \mathrm{Hom}_A(P_n, \mathrm{Hom}_k'(M, M)) & \longrightarrow & \mathrm{Hom}_A(P_{m+n}, \mathrm{Hom}_k'(M, M))
\end{array}
$$

Let $\zeta \in \mathrm{Ext}_A^m(k, k)$ and $\eta \in \mathrm{Ext}_A^n(M, M)$, represented by $f \in \mathrm{Hom}_A(P_m, k)$ and $g \in \mathrm{Hom}_A(M \otimes P_n, M)$, respectively. Extend f to a chain map $f_\bullet$ with $f_i \in \mathrm{Hom}_A(P_{m+i}, P_i)$. The top horizontal map takes $f \otimes g$ to $g(1_M \otimes f_n)$, and applying ϕ_{m+n} we have

$$\phi_{m+n}(g(1_M \otimes f_n))(x)(v) = g(v \otimes f_n(x))$$

for all $x \in P_{m+n}$, $v \in M$. On the other hand, $(1 \otimes \phi_n)(f \otimes g) = f \otimes \phi_n(g)$, and applying the bottom horizontal map we find

$$(\phi_n(g)f_n)(x)(v) = g(v \otimes f_n(x)).$$

Therefore the diagram commutes. $\qquad\square$

As noted in Remark 9.2.8, for any A-modules M, N, there are A-module isomorphisms ${}^*(N^*) \cong N$ and $({}^*M)^* \cong M$. This provides further relationships among various actions. Set $M = N^*$, so that ${}^*M \cong N$. Applying the isomorphisms in the statements of Theorems 9.3.9 and 9.3.11, we see that

$$\mathrm{Ext}_A^*(N, N) \cong \mathrm{Ext}_A^*(k, N \otimes M) \cong \mathrm{Ext}_A^*(M, M).$$

In this way, the second described action in each part of Theorem 9.3.12 below makes sense.

Theorem 9.3.12. *Assume the antipode S of A is bijective.*

(i) *Let N be a finite-dimensional A-module. The action of $\mathrm{H}^*(A, k)$ on $\mathrm{Ext}_A^*(N, N)$, given by $- \otimes N$ followed by Yoneda composition, corresponds to the action given by $N^* \otimes -$ followed by Yoneda composition.*

(ii) *Let M be a finite-dimensional A-module. The action of $\mathrm{H}^*(A, k)$ on $\mathrm{Ext}_A^*(M, M)$, given by $M \otimes -$ followed by Yoneda composition, corresponds to the action given by $- \otimes {}^*M$ followed by Yoneda composition.*

Proof. For (i), let $M = N^*$, and so ${}^*M \cong N$, as noted above. Apply Theorems 9.3.9 and 9.3.11. The proof of (ii) is similar. $\qquad\square$

Exercise 9.3.13. Verify details in the proof of Lemma 9.3.1, in particular that there is a quasi-isomorphism between $\mathrm{Hom}_A(P_\bullet, W \otimes V^*)$ and $\mathrm{Hom}_A(P_\bullet \otimes V, W)$.

Exercise 9.3.14. Verify details in the proof of Lemma 9.3.4, in particular that the cup product $\alpha \smile \beta$ is represented by the $(m + n)$-extension $\mathbf{f} \otimes \mathbf{g}$.

Exercise 9.3.15. Let A be a Hopf algebra, and let M, N be A-modules. Show that there is a well-defined action of $\mathrm{H}^*(A, k)$ on $\mathrm{Ext}_A^*(M, N)$ given by $M \otimes -$ followed by Yoneda composition. Similarly show that there is an action given by $N \otimes -$ followed by Yoneda composition. Are these two actions the same? Compare with actions that begin with $- \otimes M$ or $- \otimes N$.

9.4. Bimodules and Hochschild cohomology

In this section, we describe connections among Hochshild cohomology, Hopf algebra cohomology, and actions on Ext spaces. We begin by finding some relationships between the Hopf algebras A and A^e and their modules. Some of these relationships originate in the following embedding of A as a subalgebra of A^e.

Lemma 9.4.1. *Let $\delta : A \to A^e$ be the function defined by*

$$\delta(a) = \sum a_1 \otimes S(a_2)$$

for all $a \in A$. Then δ is an injective algebra homomorphism.

Proof. First note that $\delta(1) = 1 \otimes 1$, the identity in A^e. Let $a, b \in A$. Since S is an algebra antihomomorphism,

$$\begin{aligned}
\delta(ab) &= \sum a_1 b_1 \otimes S(a_2 b_2) \\
&= \sum a_1 b_1 \otimes S(b_2) S(a_2) \\
&= \left(\sum a_1 \otimes S(a_2)\right)\left(\sum b_1 \otimes S(b_2)\right) = \delta(a)\delta(b),
\end{aligned}$$

multiplication in the second factor being opposite to that in A.

To see that δ is injective, compose with the k-linear function $\psi : A^e \to A$ defined by $\psi(a \otimes b) = a\varepsilon(b)$. We have, for all $a \in A$,

$$\psi\delta(a) = \psi\left(\sum a_1 \otimes S(a_2)\right) = \sum a_1 \varepsilon(S(a_2)) = \sum a_1 \varepsilon(a_2) = a,$$

that is, $\psi\delta$ is the identity map on A. This implies that δ is injective. $\square$

We will identify A with the subalgebra $\delta(A)$ of A^e. This will allow us to induce modules from A to A^e, using tensor products. Let M be an A-module, and consider A^e to be a right A-module via right multiplication by elements of $\delta(A)$. Then the vector space $A^e \otimes_A M$ is an A^e-module, the action given by left multiplication on the factor A^e. We next determine the A^e-module induced from the trivial A-module k in this way.

Lemma 9.4.2. *There is an isomorphism of A^e-modules*

$$A \cong A^e \otimes_A k,$$

where $A^e \otimes_A k$ is the A^e-module induced from the trivial A-module k via the embedding of A into A^e given by the map δ of Lemma 9.4.1.

Proof. Let $f : A \to A^e \otimes_A k$ be the function defined by $f(a) = a \otimes 1 \otimes 1$, and let $g : A^e \otimes_A k \to A$ be the function defined by $g(a \otimes b \otimes 1) = ab$ for all $a, b \in A$. We will check that f is an A^e-module homomorphism, with inverse function g.

Let $a, b, c \in A$. Then, since $c = \sum c_1 \varepsilon(c_2)$, we have

$$
\begin{aligned}
f((b \otimes c)(a)) \;=\; f(bac) \;&=\; bac \otimes 1 \otimes 1 \\
&=\; \sum bac_1 \otimes \varepsilon(c_2) \otimes 1 \\
&=\; \sum bac_1 \otimes S(c_2)c_3 \otimes 1.
\end{aligned}
$$

Now identifying A with $\delta(A) \subseteq A^e$, since the rightmost factor is in k with action of A given by ε, and the tensor product is over A, we may rewrite this as

$$
\begin{aligned}
\sum ba \otimes c_2 \otimes \varepsilon(c_1) \;&=\; \sum ba \otimes \varepsilon(c_1)c_2 \otimes 1 \\
&=\; ba \otimes c \otimes 1 \\
&=\; (b \otimes c)(a \otimes 1 \otimes 1) \;=\; (b \otimes c)f(a).
\end{aligned}
$$

Therefore f is an A^e-module homomorphism.

Now let $a, b \in A$. We have

$$
\begin{aligned}
gf(a) \;=\; g(a \otimes 1 \otimes 1) \;&=\; a, \\
\text{and } fg(a \otimes b \otimes 1) \;=\; f(ab) \;&=\; ab \otimes 1 \otimes 1 \\
&=\; \sum ab_1 \varepsilon(b_2) \otimes 1 \otimes 1 \\
&=\; \sum ab_1 \otimes \varepsilon(b_2) \otimes 1 \\
&=\; \sum ab_1 \otimes S(b_2)b_3 \otimes 1 \\
&=\; \sum a \otimes b_2 \otimes \varepsilon(b_1) \\
&=\; \sum a \otimes \varepsilon(b_1)b_2 \otimes 1 \;=\; a \otimes b \otimes 1.
\end{aligned}
$$

Therefore f and g are inverse functions. $\qquad\square$

Remark 9.4.3. There is also a right module version. Let $\delta' : A \to A^{\mathrm{op}} \otimes A$ be the function defined by

$$(9.4.4) \qquad\qquad \delta'(a) = \sum S(a_1) \otimes a_2$$

for all $a \in A$. Then δ' is an injective algebra homomorphism, by a similar proof to that of Lemma 9.4.1. There is an isomorphism of right $A^{\mathrm{op}} \otimes A$-modules (equivalently, left A^e-modules)

$$A \cong k \otimes_{\delta'(A)} (A^{\mathrm{op}} \otimes A),$$

by a proof similar to that of Lemma 9.4.2.

We will consider A to be an A-module under the left *adjoint action*, that is, for all $a, b \in A$,

$$a \cdot b = \sum a_1 b S(a_2).$$

Denote this A-module by A^{ad}. More generally, if M is any A-bimodule, denote by M^{ad} the A-module with action given by $a \cdot m = \sum a_1 m S(a_2)$ for all $a \in A$, $m \in M$.

The following theorem is due to Ginzburg and Kumar [**91**]; our proof is from [**172**, **197**]. The multiplication on $\mathrm{H}^*(A, A^{ad})$ in the theorem statement is the cup product of Lemma 9.3.4 with $M = N = k$ and $M' = N' = A^{ad}$, followed by the map induced by multiplication $A^{ad} \otimes A^{ad} \to A^{ad}$.

Theorem 9.4.5. *Let A be a Hopf algebra over k with bijective antipode S. There is an isomorphism of algebras*

$$\mathrm{HH}^*(A) \cong \mathrm{H}^*(A, A^{ad}).$$

Proof. By Lemma 9.2.9, since the antipode S is bijective, A^e is a projective right A-module. Thus we may apply the Eckmann-Shapiro Lemma (Lemma A.6.2) with A in place of B, A^e in place of A, $M = A$, and $N = k$, together with Lemma 9.4.2 to obtain an isomorphism of vector spaces,

$$\mathrm{Ext}^*_{A^e}(A, A) \cong \mathrm{Ext}^*_A(k, A^{ad}).$$

It remains to prove that cup products are preserved by this isomorphism. This follows from the proof of [**197**, Proposition 3.1], valid more generally in this context, as we explain next.

Let $P.$ denote a projective resolution of k as an A-module. Let $X. = A^e \otimes_A P.$, a projective resolution of $A^e \otimes_A k \cong A$ as an A^e-module, since A^e is flat as a right module over $A \cong \delta(A)$.

There is a chain map $\iota : P. \to X.$ of A-modules defined by $\iota(p) = (1 \otimes 1) \otimes p$ for all $p \in P_i$. Let $f \in \mathrm{Hom}_{A^e}(X_i, A)$ be a cocycle representing a cohomology class in $\mathrm{Ext}^*_{A^e}(A, A)$. The corresponding cohomology class in $\mathrm{Ext}^*_A(k, A^{ad})$ is represented by $f\iota$.

Since $k \otimes k \cong k$, the Künneth Theorem (Theorem A.5.2) implies that $P.\otimes P.$ is also a projective resolution of k as an A-module. By the Comparison Theorem (Theorem A.2.7), there is a chain map $D : P. \to P. \otimes P.$ of A-modules lifting the identity map on k. Note that D induces an isomorphism on cohomology. It also induces a chain map $D' : X. \to X. \otimes_A X.$ as follows. There is a map of chain complexes $\theta : A^e \otimes_A (P. \otimes P.) \to X. \otimes_A X.$ of A^e-modules, given by

$$\theta((a \otimes b) \otimes (p \otimes q)) = ((a \otimes 1) \otimes p) \otimes ((1 \otimes b) \otimes q).$$

Take the map from $A^e \otimes_A P.$ to $A^e \otimes_A (P. \otimes P.)$ induced by D. Let D' be the composition of this map with θ. Again D' is unique up to homotopy.

Now let $f \in \mathrm{Hom}_{A^e}(X_i, A)$ and $g \in \mathrm{Hom}_{A^e}(X_j, A)$ be cocycles. The above observations imply that the following diagram commutes:

$$
\begin{array}{ccccccc}
X_\bullet & \xrightarrow{D'} & X_\bullet \otimes_A X_\bullet & \xrightarrow{f \otimes g} & A \otimes_A A & \xrightarrow{\sim} & A \\
\iota \uparrow & & & & & & \| \\
P_\bullet & \xrightarrow{D} & P_\bullet \otimes P_\bullet & \xrightarrow{f\iota \otimes g\iota} & A \otimes A & \xrightarrow{\pi} & A
\end{array}
$$

where π is multiplication. The top row yields the product in $\mathrm{Ext}^*_{A^e}(A, A)$ and the bottom row yields the product in $\mathrm{Ext}^*_A(k, A^{ad})$. $\qquad\square$

The following consequence of Theorem 9.4.5 was proven first by Linckelmann [**141**].

Corollary 9.4.6. *Let A be a finite-dimensional commutative Hopf algebra over k. There is an isomorphism of algebras $\mathrm{HH}^*(A) \cong A \otimes \mathrm{H}^*(A, k)$.*

Proof. Since A is finite dimensional, its antipode S is bijective. Since A is commutative, it acts trivially on the A-module A^{ad}, that is, $(A^{ad})^A = A^{ad}$. Let $P_\bullet$ be a projective resolution of k as an A-module. Due to the trivial action on A, since A is finite dimensional as a vector space over k, there is an isomorphism of complexes of vector spaces, $\mathrm{Hom}_A(P_\bullet, A^{ad}) \cong \mathrm{Hom}_A(P_\bullet, k) \otimes A$. Apply the Universal Coefficients Theorem (Theorem A.5.4) to this complex, and then Theorem 9.4.5 and analysis of the cup products to see that the statement holds. $\qquad\square$

Another consequence of Theorem 9.4.5 is the following.

Corollary 9.4.7. *Let A be a Hopf algebra over k with bijective antipode. Let $I = \mathrm{Ker}(\varepsilon)$, the augmentation ideal. Then as an algebra,*

$$
\mathrm{HH}^*(A) \cong \mathrm{H}^*(A, k) \oplus \mathrm{H}^*(A, I^{ad}),
$$

a direct sum of the subalgebra $\mathrm{H}^*(A, k)$ and the ideal $\mathrm{H}^*(A, I^{ad})$.

We may thus view $\mathrm{H}^*(A, k)$ as both a quotient and a subalgebra of $\mathrm{HH}^*(A)$.

Proof. Under the left adjoint action of A on itself, the trivial module k is isomorphic to the submodule of A^{ad} given by all scalar multiples of the identity 1. In fact k is a direct summand of A^{ad}, its complement being $I = \mathrm{Ker}(\varepsilon)$. As $\mathrm{Ext}^*_A(k, -)$ is additive, the result follows. $\qquad\square$

Viewing $\mathrm{H}^*(A, k)$ as a subalgebra of $\mathrm{HH}^*(A)$, it is natural to ask what structure on $\mathrm{HH}^*(A)$ might be induced by the Gerstenhaber bracket. It is known that the bracket of two elements of $\mathrm{H}^*(A, k)$ is always 0 in case A is cocommutative [**160**] or more generally quasi-triangular [**110**].

Remark 9.4.8. There is a Tor version of Theorem 9.4.5 that is easier: for each $i \geq 0$, $\mathrm{HH}_i(A) \cong \mathrm{H}_i(A, A^{ad})$ as abelian groups. It follows that $\mathrm{H}_i(A, k)$ is a direct summand of $\mathrm{HH}_i(A)$.

There is a connection between the actions of Hopf algebra cohomology and of Hochschild cohomology on Ext spaces, as noted in [**172**].

Proposition 9.4.9. *The following diagram commutes:*

$$
\begin{array}{ccc}
\mathrm{H}^*(A, k) & & \\
{\scriptstyle A^e \otimes_A -}\Big\downarrow & \searrow {\scriptstyle -\otimes M} & \\
\mathrm{HH}^*(A) & \xrightarrow{\ -\otimes_A M\ } & \mathrm{Ext}^*_A(M, M)
\end{array}
$$

Thus the action of $\mathrm{H}^*(A, k)$ on $\mathrm{Ext}^*_A(M, M)$, given by $-\otimes M$ followed by Yoneda composition, factors through the action of $\mathrm{HH}^*(A)$ on $\mathrm{Ext}^*_A(M, M)$.

Proof. Let $0 \to k \to P_{n-1} \to \cdots \to P_0 \to k \to 0$ be an n-extension of A-modules. For any A-module X, consider the function

$$
f_X : (A^e \otimes_A X) \otimes_A M \to X \otimes M
$$

given by

$$
f_X(a \otimes b \otimes x \otimes m) = \sum a_1 x \otimes a_2 bm
$$

for all $a, b \in A$, $x \in X$, $m \in M$. By construction, the action of A on $(A^e \otimes_A X) \otimes_A M$ is on the leftmost factor of A only. It can be checked that f_X is a functorial A-module isomorphism, and thus the actions are equivalent as stated. $\qquad\square$

Remark 9.4.10. One consequence of the proposition, under some finiteness assumptions (see Remark 9.3.7), is a connection between the support variety theory of Chapter 8 and that defined via Hopf algebra cohomology: for a Hopf algebra A, the support variety of an A-module M can be defined alternatively as the maximal ideal spectrum of the quotient of $\mathrm{H}^*(A, k)$ by the annihilator of $\mathrm{Ext}^*_A(M, M)$. (In odd characteristic, $\mathrm{H}^*(A, k)$ may be replaced by its subalgebra generated by even degree elements, a commutative ring.) To compare to Definition 8.3.1 of support varieties, choose H to be the subalgebra of $\mathrm{HH}^*(A)$ generated by the even degree elements of $\mathrm{H}^*(A, k)$ (or all elements in characteristic 2) and $\mathrm{HH}^0(A) \cong Z(A)$. This choice yields a direct relationship between the varieties for modules given by Hochschild cohomology and by Hopf algebra cohomology. For more details, see the survey [**224**] and references therein.

Exercise 9.4.11. Let G be a finite group and $A = kG$. Identify the subalgebra $\delta(kG)$ of $(kG)^e$, where δ is defined in Lemma 9.4.1.

Exercise 9.4.12. Let k be a field of characteristic $p > 0$, and let G be a group of order p. Use Example 9.3.8 and Corollary 9.4.6 to describe the Hochschild cohomology $\mathrm{HH}^*(kG)$ as an algebra under cup product.

9.5. Finite group algebras

In this section, we apply Theorem 9.4.5 to the special case $A = kG$, the group algebra of a finite group G. This leads to a decomposition of Hochschild cohomology $\mathrm{HH}^*(kG)$ as a vector space direct sum of group cohomology rings of centralizer subgroups in isomorphism (9.5.3) below. Burghelea [**41**] gave an earlier related result on Hochschild homology. Techniques from group cohomology yield a description of the product on Hochschild cohomology $\mathrm{HH}^*(kG)$ in terms of products on group cohomology. We briefly introduce the needed techniques from group cohomology here. For details, see [**21, 22, 75**].

As before, we will use the notation

$$\mathrm{H}^*(G, k) = \mathrm{H}^*(kG, k) = \mathrm{Ext}^*_{kG}(k, k)$$

for the *group cohomology* of G with coefficients in k, and write

$$\mathrm{H}^*(G, M) = \mathrm{H}^*(kG, M) = \mathrm{Ext}^*_{kG}(k, M)$$

for any kG-module M.

By Theorem 9.4.5, there is an isomorphism of algebras,

$$(9.5.1) \qquad\qquad \mathrm{HH}^*(kG) \cong \mathrm{H}^*(G, kG^{ad}).$$

We interpret the kG-module kG^{ad}. Let $g, h \in G$. The action of g on the element h viewed as being in kG^{ad} is given by

$$g \cdot h = ghS(g) = ghg^{-1},$$

since $\Delta(g) = g \otimes g$ and $S(g) = g^{-1}$. Thus the kG-module kG^{ad} has vector space basis G on which elements of G act by conjugation. It then decomposes into a direct sum of kG-submodules corresponding to conjugacy classes. Let $g_1, \ldots, g_r$ be a set of conjugacy class representatives. For each i, the G-set that is the conjugacy class of g_i may be written $G \cdot g_i \cong G/C(g_i)$, where $C(g_i)$ is the centralizer in G of g_i. The kG-submodule of kG^{ad} having basis the conjugacy class of g_i may thus be written

$$(9.5.2) \qquad\qquad kG \cdot g_i \cong kG \otimes_{kC(g_i)} k,$$

that is, this module is the trivial module k for $kC(g_i)$ induced to kG. For finite group algebras, induction and coinduction yield isomorphic modules since kG is isomorphic to its dual $(kG)^*$ as a kG-module. Therefore, by the Eckmann-Shapiro Lemma (Lemma A.6.2),

$$\begin{aligned}
\mathrm{H}^*(G, kG \otimes_{kC(g_i)} k) &= \mathrm{Ext}^*_{kG}(k, kG \otimes_{kC(g_i)} k) \\
&\cong \mathrm{Ext}^*_{kC(g_i)}(k, k) = \mathrm{H}^*(C(g_i), k).
\end{aligned}$$

We may thus rewrite Hochschild cohomology $\mathrm{HH}^*(kG)$, first applying the isomorphisms (9.5.1) and (9.5.2):

$$(9.5.3) \qquad \mathrm{HH}^*(kG) \cong \bigoplus_{i=1}^{r} \mathrm{H}^*(G, kG \cdot g_i) \cong \bigoplus_{i=1}^{r} \mathrm{H}^*(C(g_i), k).$$

That is, the Hochschild cohomology of kG is isomorphic, as a graded vector space, to the direct sum of group cohomology rings for centralizer subgroups, one summand for each element in a set of conjugacy class representatives. See also [**22**, Theorem 2.11.2].

Analyzing the two applications of the Eckmann-Shapiro Lemma used in the isomorphisms (9.5.3), we see that we may write $\mathrm{HH}^*(kG)$ explicitly as follows in terms of some maps that we introduce next. We identify elements in $\mathrm{HH}^*(kG)$ with elements in $\mathrm{H}^*(G, kG^{ad})$, at the chain level, by restriction from $(kG)^e$ to kG under the map δ of Lemma 9.4.1. Our description of the product will be for such elements in $\mathrm{H}^*(G, kG^{ad})$.

For any subgroup H of G, and any kG-module M, there is the structure of a kH-module on M by restriction. Since kG is free as a kH module, restriction takes a kG-projective resolution of k to a kH-projective resolution of k, thus inducing a well-defined *restriction* map

$$\mathrm{res}^G_H : \mathrm{H}^*(G, M) \to \mathrm{H}^*(H, M).$$

Let $\pi_i : \mathrm{H}^*(C(g_i), kG \otimes_{kC(g_i)} k) \to \mathrm{H}^*(C(g_i), k)$ denote the map induced by the projection of $kG \otimes_{kC(g_i)} k$ onto the $kC(g_i)$-direct summand $k \otimes_{kC(g_i)} k \cong k$. Then the isomorphism above is given in one direction by

$$\begin{aligned}
\mathrm{H}^*(G, kG^{ad}) &\xrightarrow{\ \sim\ } \bigoplus_{i=1}^{r} \mathrm{H}^*(C(g_i), k) \\
\zeta &\mapsto (\pi_i \, \mathrm{res}^G_{C(g_i)} \zeta)_i.
\end{aligned}$$

In order to describe products, we will also need an expression for the inverse isomorphism. This will be expressed in terms of *corestriction* maps on group cohomology that we define next. The map

$$\mathrm{cores}^G_H : \mathrm{H}^*(H, M) \to \mathrm{H}^*(G, M)$$

is defined at the chain level as follows. Let P be a projective resolution of k as a kG-module, and let $f \in \mathrm{Hom}_{kH}(P_n, M)$. Then

$$\mathrm{cores}_H^G(f)(x) = \sum_{g \in [G/H]} g \cdot f(g^{-1} \cdot x)$$

for all $x \in P_n$, where $[G/H]$ denotes a set of coset representatives of H in G. Since f is a kH-module homomorphism, the values of this function do not depend on the choice of coset representatives. This map also commutes with the differential and so induces a well-defined map on cohomology, for which we use the same notation.

Now let $\iota : \mathrm{H}^*(C(g_i), k) \to \mathrm{H}^*(C(g_i), kG \otimes_{kC(g_i)} k)$ denote the map induced by embedding k into $kG \otimes_{kC(g_i)} k$ as $k \otimes_{kC(g_i)} k$. Then the desired inverse isomorphism may be described as follows:

$$\mathrm{H}^*(G, kG^{ad}) \xleftarrow{\;\sim\;} \bigoplus_{i=1}^{r} \mathrm{H}^*(C(g_i), k)$$

$$\sum_{i=1}^{r} \mathrm{cores}_{C(g_i)}^G \iota(\alpha_i) \quad \longleftarrow\!\shortmid \quad (\alpha_i)_i.$$

For notational convenience, set

$$\gamma_i(\alpha_i) = \mathrm{cores}_{C(g_i)}^G \iota(\alpha_i).$$

We need one more map on group cohomology. If H is a subgroup of G and $g \in G$, write $^gH = \{ghg^{-1} \mid h \in H\}$ for the conjugate subgroup. There is a *conjugation* map $g^* : \mathrm{H}^*(H, k) \to \mathrm{H}^*(^gH, k)$ given at the chain level by

$$g^*(f)(x) = g \cdot (f(g^{-1} \cdot x))$$

for all $x \in P_n$ and $f \in \mathrm{Hom}_{kH}(P_n, k)$.

The following is a special case of [**197**, Theorem 5.1], and gives the product on Hochschild cohomology $\mathrm{HH}^*(kG)$ in terms of the vector space direct sum (9.5.3). For the statement, we need the notion of double cosets. If H, L are subgroups of a group G, the set $H\backslash G / L$ of *double cosets* consists of the sets $HgL = \{hgl \mid h \in H, l \in L\}$ for $g \in G$. A set D of *double coset representatives* is a set consisting of one element from each double coset.

Theorem 9.5.4. *Let G be a finite group, and let $g_1, \ldots, g_r$ be a set of conjugacy class representatives. Let $\alpha \in \mathrm{H}^*(C(g_i), k)$, $\beta \in \mathrm{H}^*(C(g_j), k)$, and let $\gamma_i(\alpha)$ and $\gamma_j(\beta)$ be corresponding elements in $\mathrm{H}^*(G, kG^{ad}) \cong \mathrm{HH}^*(kG)$ under the isomorphism (9.5.3). Then*

$$\gamma_i(\alpha) \smile \gamma_j(\beta) = \sum_{x \in D} \gamma_l \big(\mathrm{cores}_{W(x)}^{C(g_k)} \big(\mathrm{res}_{W(x)}^{^y C(g_i)} y^* \alpha \smile \mathrm{res}_{W(x)}^{^{yx} C(g_j)} (yx)^* \beta \big) \big),$$

where D is a set of double coset representatives for $C(g_i)\backslash G/C(g_j)$, and the integer $l = l(x)$ and the group element $y = y(x)$ are chosen so that $g_l = ({}^y g_i)({}^{yx} g_j)$, and $W(x) = {}^y C(g_i) \cap {}^{yx} C(g_j)$.

For a proof and some examples, see [197]. The main idea of the proof is that group cohomology, together with the maps restriction, corestriction, and conjugation, comprises a Green functor [75]. The formula in the theorem is precisely the product formula for a Green functor, interpreted in this notation and setting of Hochschild cohomology.

The following corollary, due to Holm [117], is a special case of Corollary 9.4.6, and can also be proven directly. For group rings over more general rings k, see Cibils and Solotar [53].

Corollary 9.5.5. *Let G be a finite abelian group. There is an isomorphism of algebras* $\mathrm{HH}^*(kG) \cong kG \otimes \mathrm{H}^*(G, k)$.

Liu and Zhou [143] took Theorem 9.5.4 further. They described the Batalin-Vilkovisky structure of Hochschild cohomology $\mathrm{HH}^*(kG)$, viewing kG as a symmetric algebra, and thus the Gerstenhaber bracket (see Remark 4.5.3), in terms of the decomposition (9.5.3).

Exercise 9.5.6. Let k be a field of characteristic $p > 0$. Let G be an elementary abelian p-group, that is, $G \cong \mathbb{Z}/p\mathbb{Z} \times \cdots \times \mathbb{Z}/p\mathbb{Z}$. Describe $\mathrm{HH}^*(kG)$ and $\mathrm{H}^*(G, k)$.

Exercise 9.5.7. Let k be a field of characteristic 3. Let $G = S_3$, the symmetric group on three symbols. Apply isomorphism (9.5.3) to describe the graded vector space structure of Hochschild cohomology $\mathrm{HH}^*(kG)$ as a direct sum of group cohomology spaces.

9.6. Spectral sequences for Hopf algebras

In this section, we focus on a smash product algebra $R\#A$, as in Definition 9.1.6, and construct two spectral sequences relating its Hochschild cohomology with that of R and of Hopf algebra cohomology of A. The first of these spectral sequences has a more general version for Hopf Galois extensions due to Ştefan [208] and is related to a spectral sequence for Hochschild homology of group-graded algebras due to Lorenz [149]. The second is exclusively for smash products, however with the advantage of being multiplicative, and is due to Negron [164].

In Section 9.1 we introduced smash products arising from a *left* action of A on R, and it turns out that as a result we will need to work with *right* A-module cohomology here. First we need a lemma about some special types of right A-modules.

For any $(R\#A)^e$-module N, the space $\mathrm{Hom}_{R^e}(R, N)$ is a right A-module under the action

$$(9.6.1) \qquad\qquad (f \cdot a)(r) = \sum r S(a_1) f(1) a_2$$

for all $a \in A$, $f \in \mathrm{Hom}_{R^e}(R, N)$, and $r \in R$. To see this, note that an element of $\mathrm{Hom}_{R^e}(R, N)$ is determined by its value on 1, which must be an element x of N such that $rx = xr$ for all $r \in R$. A calculation shows that for any $a \in A$, the element $\sum S(a_1) x a_2$ also then has this property. This action generalizes the Miyashita-Ulbrich action from groups to Hopf algebras.

Lemma 9.6.2. *Let A be a Hopf algebra, let R be an A-module algebra, and let M be an $(R\#A)^e$-module. There is an isomorphism of right A-modules,*

$$\mathrm{Hom}_{R^e}(R, \mathrm{Hom}_k((R\#A)^e, M)) \cong \mathrm{Hom}_k(A^{\mathrm{op}} \otimes (R\#A), M).$$

The $(R\#A)^e$-module $\mathrm{Hom}_k((R\#A)^e, M)$ in the lemma is the coinduced module (see Section A.6) of M from k to $(R\#A)^e$. Similarly we may view the A-module $\mathrm{Hom}_k(A^{\mathrm{op}} \otimes (R\#A), M)$ as a coinduced module from k to $A^{\mathrm{op}} \otimes (R\#A)$, and A acts via the embedding δ' of A into $A^{\mathrm{op}} \otimes A$ given by (9.4.4).

Proof. Let $f \in \mathrm{Hom}_{R^e}(R, \mathrm{Hom}_k((R\#A)^e, M))$ and define an element $\phi(f)$ of $\mathrm{Hom}_k(A^{\mathrm{op}} \otimes (R\#A), M)$ by

$$\phi(f)(a \otimes ra') = f(1)(a \otimes ra')$$

for all $a, a' \in A$ and $r \in R$. Then ϕ is an A-module homomorphism: the function f is determined by $f(1)$, which has the property that $rf(1) = f(1)r$. Now $\mathrm{Hom}_k((R\#A)^e, M)$ is the module M coinduced from k to $(R\#A)^e$, so this property is equivalent to the property that $f(1)(xr \otimes y) = f(1)(x \otimes ry)$ for all $r \in R$ and $x, y \in R\#A$. Paying close attention to which factors have opposite multiplication, it follows that for all $a, a', a'' \in A$ and $r \in R$,

$$\begin{aligned}
(\phi(f \cdot a))(a' \otimes ra'') &= (f \cdot a)(1)(a' \otimes ra'') \\
&= \sum S(a_1) f(1) a_2 (a' \otimes ra'') \\
&= \sum f(1)(a' S(a_1) \otimes a_2 ra''),
\end{aligned}$$

and using the embedding δ' of A into $A^{\mathrm{op}} \otimes A$,

$$\begin{aligned}
(\phi(f) \cdot a)(a' \otimes ra'') &= \sum \phi(f)(a' S(a_1) \otimes a_2 ra'') \\
&= \sum f(1)(a' S(a_1) \otimes a_2 ra''),
\end{aligned}$$

and so ϕ is an A-module homomorphism.

We next show that ϕ is bijective. Let $g \in \mathrm{Hom}_k(A^{\mathrm{op}} \otimes (R\#A), M)$ and define an element $\psi(g)$ of $\mathrm{Hom}_{R^e}(R, \mathrm{Hom}_k((A\#R)^e, M))$ by

$$\psi(g)(1)(ar \otimes r'a') = g(a \otimes rr'a')$$

for all $a, a' \in A$ and $r, r' \in R$. Since $\mathrm{Hom}_k((A\#R)^e, M)$ is the coinduced module, $\psi(g)$ is indeed an R^e-homomorphism:

$$\psi(g)(1)(arr'' \otimes r'a') = \psi(g)(1)(ar \otimes r''r'a')$$

for all $a, a' \in A$ and $r, r', r'' \in R$. A calculation shows that ψ is an inverse map to ϕ. $\qquad\square$

Now we are ready to construct the advertised spectral sequences for the Hochschild cohomology of bimodules over the smash product $R\#A$.

Theorem 9.6.3. *Let A be a Hopf algebra, let R be an A-module algebra, and let M be an $(R\#A)^e$-module. There is a right action of A on $\mathrm{HH}^q(R, M)$ and a spectral sequence*

$$E_2^{p,q} = \mathrm{H}^p(A, \mathrm{HH}^q(R, M)) \implies \mathrm{HH}^{p+q}(R\#A, M).$$

We will give two proofs of the theorem. The first is due to Ştefan, and generalizes directly to crossed product algebras and Hopf Galois extensions as in [**208**]. The second proof, due to Negron [**164**], involves a *multiplicative* spectral sequence, but is exclusively for smash product algebras without a clear generalization to these other settings. We emphasize again that the Hopf algebra cohomology in the theorem statement is that of the *right A-module* $\mathrm{HH}^q(R, M)$.

Proof. Our first proof is due to Ştefan [**208**]. Let $P_{\bullet}$ be a projective resolution of k as a right A-module. Let $Q_{\bullet}$ be an injective resolution of M as an $(R\#A)^e$-module. By restricting to the subalgebra R^e of $(R\#A)^e$, since $(R\#A)^e$ is free over R^e, $Q_{\bullet}$ becomes an R^e-injective resolution of M. To see this, note that by the Nakayama relations in Lemma A.6.1,

$$\mathrm{Hom}_{R^e}(U, I) \cong \mathrm{Hom}_{(R\#A)^e}((R\#A)^e \otimes_{R^e} U, I)$$

for all R^e-modules U and $(R\#A)^e$-modules I, so $\mathrm{Hom}_{R^e}(-, I)$ is an exact functor when I is an injective $(R\#A)^e$-module, implying that the restriction of I to R^e is also injective.

We use this resolution $Q_{\bullet}$ to give the Hochschild cohomology $\mathrm{HH}^*(R, M)$ a right A-module structure as the action induced by that of equation (9.6.1). Take the right A-action on each $\mathrm{Hom}_{R^e}(R, Q_q)$ given by (9.6.1). A calculation shows that this action of A commutes with the differentials and thus induces an action on Hochschild cohomology as claimed.

Let

$$C^{p,q} = \mathrm{Hom}_A(P_p, \mathrm{Hom}_{R^e}(R, Q_q)),$$

a double complex that we may view as:

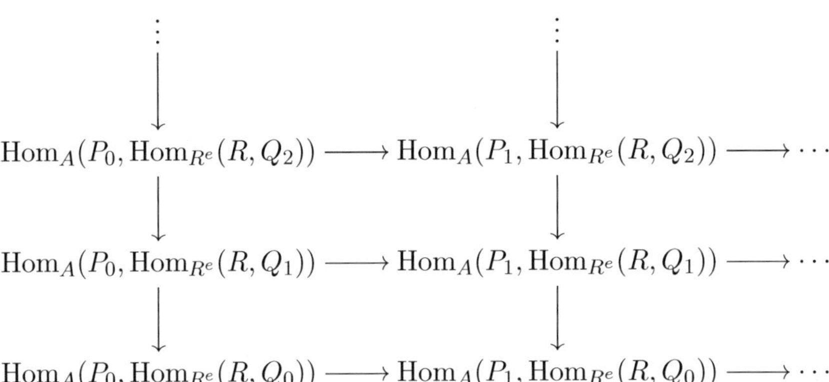

We will analyze the two first quadrant spectral sequences associated to this double complex in the notation of Section A.7. They each converge to $\mathrm{H}^*(C)$. Since each P_i is projective as a right A-module, $\mathrm{Hom}_A(P_i, -)$ is exact. Therefore, starting with vertical differentials, the cohomology in each column $\mathrm{Hom}_A(P_i, \mathrm{Hom}_{R^e}(R, Q_\bullet))$ is isomorphic to $\mathrm{Hom}_A(P_i, -)$ applied to the cohomology of $\mathrm{Hom}_{R^e}(R, Q_\bullet)$, which is $\mathrm{HH}^*(R, M)$. So we have

$$\mathrm{H}''(C) = \mathrm{Hom}_A(P_\bullet, \mathrm{HH}^\bullet(R, M)),$$

and consequently

$$\mathrm{H}'(\mathrm{H}''(C)) = \mathrm{H}^\bullet(A, \mathrm{HH}^\bullet(R, M)).$$

Therefore $E_2^{p,q}$ is as claimed.

On the other hand, starting with horizontal differentials:

$$(9.6.4) \qquad \mathrm{H}'(C) \cong \mathrm{H}^\bullet(A, \mathrm{Hom}_{R^e}(R, Q_\bullet)).$$

We claim that for each q, $\mathrm{H}^p(A, \mathrm{Hom}_{R^e}(R, Q_q)) = 0$ whenever $p > 0$. This is a special case of [**208**, Proposition 3.2]. To see this in our setting, first note that there is an injective $(R \# A)^e$-module homomorphism

$$i : Q_q \to \mathrm{Hom}_k((R \# A)^e, Q_q)$$

given by $i(m)(x \otimes y) = xmy$ for all $m \in Q_q$ and $x, y \in R \# A$. Since Q_q is an injective $(R \# A)^e$-module, this map splits, and so Q_q is a direct summand of $\mathrm{Hom}_k((R \# A)^e, Q_q)$ as an $(R \# A)^e$-module. This implies that $\mathrm{Hom}_{R^e}(R, Q_q)$ is a direct summand of $\mathrm{Hom}_{R^e}(R, \mathrm{Hom}_k((R \# A)^e, Q_q))$ as a right A-module. So to prove our claim, it suffices to show that

$$\mathrm{H}^p(A, \mathrm{Hom}_{R^e}(R, \mathrm{Hom}_k((R \# A)^e, Q_q))) = 0$$

whenever $p > 0$. Now this is the degree p cohomology of the complex

$$\mathrm{Hom}_A(P_\bullet, \mathrm{Hom}_{R^e}(R, \mathrm{Hom}_k((R \# A)^e, Q_q))),$$

and by Lemma 9.6.2, this is isomorphic to

$$\mathrm{Hom}_A(P_\bullet, \mathrm{Hom}_k(A^{\mathrm{op}} \otimes (R\#A), Q_q)).$$

In turn, this is isomorphic to $\mathrm{Hom}_k(P_\bullet \otimes_A (A^{\mathrm{op}} \otimes (R\#A)), Q_q)$ as a graded vector space. An isomorphism is given in each degree p by sending an element f of $\mathrm{Hom}_A(P_p, \mathrm{Hom}_k(A^{\mathrm{op}} \otimes (R\#A), Q_q))$ to the element $\phi(f)$ of $\mathrm{Hom}_k(P_p \otimes (R\#A), Q_q)$ defined by $\phi(f)(x \otimes y) = f(x)(y)$ for all $x \in P_p$, $y \in R\#A$. An inverse map takes $g \in \mathrm{Hom}_k(P_p \otimes_A (A^{\mathrm{op}} \otimes (R\#A)), Q_q)$ to $\psi(g)$, where $\psi(g)(x)(y) = g(x \otimes y)$. A calculation shows that for each g, the function $\psi(g)$ is a homomorphism of right A-modules, and that ϕ, ψ are indeed inverse maps. Now $\mathrm{Hom}_k(P_\bullet \otimes_A (A^{\mathrm{op}} \otimes (R\#A)), Q_q)$ has homology 0 in positive degrees, since $- \otimes_A (A^{\mathrm{op}} \otimes (R\#A))$ and $\mathrm{Hom}_k(-, Q_q)$ are exact functors, proving our claim.

As a result, the cohomology $\mathrm{H}'(C)$ given by (9.6.4) is concentrated in the leftmost column. In the qth position, by Remark 9.2.4 (see Lemma 9.2.2), it is

$$\begin{aligned}
\mathrm{H}^0(A, \mathrm{Hom}_{R^e}(R, Q_q)) &\cong \mathrm{Hom}_A(k, \mathrm{Hom}_{R^e}(R, Q_q)) \\
&\cong \mathrm{Hom}_{R^e}(R, Q_q)^A.
\end{aligned}$$

Now we claim there is an isomorphism of vector spaces

$$(9.6.5) \qquad \mathrm{Hom}_{R^e}(R, Q_q)^A \cong \mathrm{Hom}_{(R\#A)^e}(R\#A, Q_q)$$

for each q. To see this, consider an element f of $\mathrm{Hom}_{R^e}(R, Q_q)^A$, which is determined by $f(1)$. By hypothesis, $rf(1) = f(1)r$ and $af(1) = f(1)a$ for all $a \in A$. (The latter equation holds since $f(1)a = \sum a_1 S(a_2) f(1) a_3 = \sum a_1 \varepsilon(a_2) f(1) = af(1)$.) Define $\gamma(f)(ra) = (ra)f(1)$ for all $r \in R$ and $a \in A$. Then $\gamma(f)$ is an $(R\#A)^e$-module homomorphism. The map γ has inverse given by sending a function g in $\mathrm{Hom}_{(R\#A)^e}(R\#A, Q_q)$ to the function taking 1 to $g(1)$. It follows from the isomorphism (9.6.5) that $\mathrm{H}''(\mathrm{H}'(C)) = \mathrm{HH}^\bullet(R\#A, M)$, and this is the cohomology of $C^{\bullet,\bullet}$. Therefore the first spectral sequence converges to $\mathrm{HH}^\bullet(R\#A, M)$, as claimed.

Our second proof is due to Negron [**164**]. Again let $P_\bullet$ be a projective resolution of k as an A-module. Let $K_\bullet$ be a projective resolution of R as an R^e-module that is A-equivariant, that is, each K_q is an A-module for which the structure maps $R \otimes K_q \to K_q$ and $K_q \otimes R \to K_q$ are A-module homomorphisms, and this A-module structure commutes with the differentials. Existence of an A-equivariant projective R^e-module resolution of R is proven in [**164**]. It may be checked that there is a right A-module structure on $\mathrm{Hom}_{R^e}(K_q, M)$ given by

$$(f \cdot a)(x) = \sum S(a_1) f(a_2 \cdot x) a_3$$

for all $f \in \mathrm{Hom}_{R^e}(K_q, M)$, $a \in A$, and $x \in K_q$, and that

$$\mathrm{HH}^*(R\#A, M) \cong \mathrm{H}^*(\mathrm{Hom}_A(P_\bullet, \mathrm{Hom}_{R^e}(K_\bullet, M))).$$

Just as in the first proof, the bicomplex $\mathrm{Hom}_A(P_\bullet, \mathrm{Hom}_{R^e}(K_\bullet, M))$ gives rise to a spectral sequence converging to $\mathrm{HH}^*(R\#A, M)$, and the E_2 page is $\mathrm{H}^*(A, \mathrm{HH}^*(R, M))$. A calculation shows that this spectral sequence is multiplicative due to the use of a projective resolution of R as an R^e-module, as opposed to the injective resolution of M in the first proof. $\qquad\square$

In the special case of a semisimple Hopf algebra, Theorem 9.6.3 has the following consequence.

Corollary 9.6.6. *Let A be a semisimple Hopf algebra, let R be an A-module algebra, and let M be an $(R\#A)^e$-module. There is an isomorphism of graded vector spaces,*

$$\mathrm{HH}^*(R\#A, M) \cong (\mathrm{HH}^*(R, M))^A,$$

and an isomorphism of graded algebras,

$$\mathrm{HH}^*(R\#A) \cong (\mathrm{HH}^*(R, R\#A))^A.$$

Proof. Since A is semisimple, the spectral sequence of Theorem 9.6.3 collapses: the E_2 page is concentrated in the leftmost column, which is

$$\mathrm{H}^0(A, \mathrm{HH}^*(R, M)) \cong (\mathrm{HH}^*(R, M))^A.$$

The second isomorphism now follows by results of Negron [**164**] as outlined in the second proof of Theorem 9.6.3. The total complex of $P_\bullet \otimes P_\bullet$ is also a projective resolution of k as an A-module and the total complex of $K_\bullet \otimes_R K_\bullet$ is also a projective resolution of R as an R^e-module, and so the Comparison Theorem (Theorem A.2.7) implies there are chain maps $P_\bullet \to P_\bullet \otimes P_\bullet$ and $K_\bullet \to K_\bullet \otimes_R K_\bullet$ lifting the augmentation maps of the complexes. These chain maps may be used to put a differential graded algebra structure on $\mathrm{Hom}_A(P_\bullet, \mathrm{Hom}_{R^e}(K_\bullet, M))$. For details, see [**164**, Theorem 6.5]. $\qquad\square$

Exercise 9.6.7. Verify that equation (9.6.1) gives a well-defined right A-module structure to $\mathrm{Hom}_{R^e}(R, N)$.

Exercise 9.6.8. Verify the last statement in the proof of Lemma 9.6.2, that ψ is an inverse map to ϕ.

Exercise 9.6.9. Use Corollary 9.6.6 with $A = kG$ to give a different proof of the first part of Theorem 3.5.2.

Homological Algebra Background

We collect here some homological algebra terminology, notation, and results that we will use throughout the book. Proofs and more details may be found in standard homological algebra texts such as [**21**, **112**, **168**, **187**, **223**].

A.1. Complexes

Let R be a ring. We always assume that R has multiplicative identity 1 and all modules are unital modules, that is, 1 acts as the identity map. Modules will be left modules unless otherwise specified.

A *complex* $C_{\bullet}$ of R-modules, also written $(C_{\bullet}, d_{\bullet})$, is a sequence of R-modules and R-module homomorphisms, called *differentials*,

$$C_{\bullet}: \qquad \cdots \longrightarrow C_2 \xrightarrow{d_2} C_1 \xrightarrow{d_1} C_0 \xrightarrow{d_0} C_{-1} \xrightarrow{d_{-1}} C_{-2} \longrightarrow \cdots,$$

where $d_{n-1}d_n = 0$ for all $n \in \mathbb{Z}$. The *degree* (also called *dimension*) of an element x of C_n is n, and we write $|x| = n$. The differential then is considered to have degree -1 as a map. For each n, the kernel $\mathrm{Ker}(d_n)$ is the R-submodule of C_n consisting of *n-cycles*, the image $\mathrm{Im}(d_{n+1})$ is the R-submodule of C_n consisting of *n-boundaries*, and the quotient $\mathrm{H}_n(C_{\bullet}) = \mathrm{Ker}(d_n)/\mathrm{Im}(d_{n+1})$ is the *nth homology* of $C_{\bullet}$. Two n-cycles x and y are *homologous* if $x - y$ is an n-boundary. Write $\mathrm{H}_*(C_{\bullet}) = \bigoplus_{n \geq 0} \mathrm{H}_n(C_{\bullet})$, the *homology* of $C_{\bullet}$.

We will sometimes omit the subscript on $C_{\bullet}$, writing C instead, to denote the complex. We will also sometimes identify $C_{\bullet}$ with the R-module

$\bigoplus_{n\in\mathbb{Z}} C_n$ and $d_\bullet$ with the homomorphism from this R-module to itself that agrees with d_n on each C_n.

We take a *chain complex* to be a complex for which $C_n = 0$ for $n < 0$, and a *cochain complex* to be a complex for which $C_n = 0$ for $n > 0$. Some authors use these terms more generally to refer to complexes. Some complexes may be indexed differently, replacing n by $-n$ in $C_\bullet$ above, with the maps still oriented as shown so that the indexing agrees with the left to right ordering of integers on a standard number line. A cochain complex then has differential of degree $+1$, and we may choose to write the index as a superscript:

$$C^\bullet : \qquad 0 \longrightarrow C^0 \xrightarrow{d_0} C^1 \xrightarrow{d_1} C^2 \xrightarrow{d_2} \cdots .$$

In this context, elements in the kernel of d_n are *n-cocycles*, and elements in the image of d_{n-1} are *n-coboundaries*. Two n-cocycles are *cohomologous* if their difference is an n-coboundary. We write $\mathrm{H}^n(C^\bullet) = \mathrm{Ker}(d_n)/\mathrm{Im}(d_{n-1})$ and $\mathrm{H}^*(C^\bullet) = \bigoplus_{n\geq 0} \mathrm{H}^n(C^\bullet)$, and refer to this as the *cohomology* of the cochain complex $C^\bullet$. The following definitions and statements may be rephrased in terms of this other indexing choice.

We say that $C_\bullet$ is *acyclic*, or *exact*, if $\mathrm{H}_n(C_\bullet) = 0$ for all n. A *short exact sequence* is an exact complex of the form $0 \to U \to V \to W \to 0$.

Let (C, d) be a complex and $n \in \mathbb{Z}$. The *shifted* (or *translated*) *complex* $C[n]$ has

$$C[n]_i = C_{i+n}$$

and differentials $(-1)^n d$, so for example, $C[n]_0 = C_n$. (For a cochain complex, we take instead $C[n]^i = C^{i-n}$.)

Let (C, d) and (C', d') be complexes. A *chain map* $f_\bullet : C_\bullet \to C'_\bullet$ consists of an R-module homomorphism $f_n : C_n \to C'_n$ for which $f_{n-1} d_n = d'_n f_n$ for each n, that is, the following diagram commutes:

$$
\begin{array}{ccccccc}
\cdots \longrightarrow & C_1 & \xrightarrow{d_1} & C_0 & \xrightarrow{d_0} & C_{-1} & \longrightarrow \cdots \\
& \downarrow{\scriptstyle f_1} & & \downarrow{\scriptstyle f_0} & & \downarrow{\scriptstyle f_{-1}} & \\
\cdots \longrightarrow & C'_1 & \xrightarrow{d'_1} & C'_0 & \xrightarrow{d'_0} & C'_{-1} & \longrightarrow \cdots
\end{array}
$$

A chain map induces a map on homology, and is a *quasi-isomorphism* if this induced map is an isomorphism.

Two chain maps $f_\bullet, g_\bullet : C_\bullet \to C'_\bullet$ are *chain homotopic* if there are R-module homomorphisms $s_n : C_n \to C'_{n+1}$ such that

$$(\text{A.1.1}) \qquad f_n - g_n = s_{n-1} d_n + d'_{n+1} s_n$$

for all n. We call $s_\bullet$ a *homotopy* for $f_\bullet - g_\bullet$. Chain homotopy is an equivalence relation, and two chain homotopic maps induce the same maps on homology. If $g_\bullet$ is the zero map, we call $s_\bullet$ a *chain contraction* of $f_\bullet$. If there is a chain contraction of the identity map on $C_\bullet$, it is also sometimes called a *contracting homotopy*, and it follows that $C_\bullet$ is acyclic. This conclusion holds under the weaker hypothesis that there are some functions $s_n : C_n \to C'_{n+1}$ (not necessarily R-module homomorphisms) satisfying equation (A.1.1). Sometimes when R is a k-algebra, there are k-linear maps s_n satisfying this hypothesis that are not R-module homomorphisms.

Differential graded algebras. Let k be a field. A *differential graded algebra* is a complex (C, d) for which C is a graded k-algebra and the *Leibniz rule* holds: for all $x \in C_i$ and $y \in C_j$,

$$d(xy) = d(x)y + (-1)^i x d(y).$$

A *differential graded Lie algebra* is a complex (C, d) for which C is a graded Lie algebra over k and for all $x \in C_i$ and $y \in C_j$,

$$d([x, y]) = [d(x), y] + (-1)^i [x, d(y)].$$

A *differential graded coalgebra* is a complex (C, d) for which C is a graded coalgebra over k, i.e., there is a graded k-linear map $\Delta : C \to C \otimes C$ that is coassociative (i.e., $(\Delta \otimes 1)\Delta = (1 \otimes \Delta)\Delta$) and

$$(d \otimes 1 + 1 \otimes d)\Delta = \Delta d.$$

(Here we apply the sign convention (2.3.1).) A differential graded coalgebra C is *counital* if there is a k-linear map $\varepsilon : C \to k$ such that $(\varepsilon \otimes 1)\Delta = 1 = (1 \otimes \varepsilon)\Delta$.

Pushout and pullback. Let A, B, Y be R-modules, and let $\alpha : Y \to A$, $\beta : Y \to B$ be R-module homomorphisms. A *pushout* of α, β is an R-module X together with R-module homomorphisms $\phi : A \to X$, $\psi : B \to X$ such that $\phi\alpha = \psi\beta$ and for any R-module Z and R-homomorphisms $\tilde\phi : A \to Z$, $\tilde\psi : B \to Z$ for which $\tilde\phi\alpha = \tilde\psi\beta$, there is a unique R-module homomorphism $\eta : X \to Z$ such that $\tilde\phi = \eta\phi$, $\tilde\psi = \eta\psi$:

$$
\begin{array}{ccc}
Y & \xrightarrow{\ \alpha\ } & A \\
\downarrow{\scriptstyle\beta} & & \downarrow{\scriptstyle\phi} \\
B & \xrightarrow{\ \psi\ } & X
\end{array}
$$

Note that we may take

$$(A.1.2) \qquad X = A \oplus B / \{(-\alpha(y), \beta(y)) \mid y \in Y\}$$

and ϕ, ψ to be the maps induced by inclusion into $A \oplus B$.

Let A, B, X be R-modules, and let $\phi : A \to X$, $\psi : B \to X$ be R-module homomorphisms. A *pullback* of ϕ, ψ is an R-module Y together with R-module homomorphisms $\alpha : Y \to A$, $\beta : Y \to B$ such that $\phi\alpha = \psi\beta$ and for any R-module Z and R-module homomorphisms $\tilde\alpha : Z \to A$, $\tilde\beta : Z \to B$

for which $\phi\tilde\alpha = \psi\tilde\beta$, there is a unique R-module homomorphism $\eta : Z \to Y$ such that $\tilde\alpha = \alpha\eta$, $\tilde\beta = \beta\eta$. Note that we may take

$$Y = \{(a,b) \in A \oplus B \mid \phi(a) = \psi(b)\}$$

and α, β to be the maps induced by projection from $A \oplus B$.

A.2. Resolutions and dimensions

An R-module P is *projective* if for every surjective R-module homomorphism $f : U \to V$ and R-module homomorphism $g : P \to V$, there exists an R-module homomorphism $h : P \to U$ such that $fh = g$:

$$(A.2.1) \qquad \begin{array}{ccc} & & P \\ & {}^{h}\swarrow & \downarrow{}^{g} \\ U & \xrightarrow{\ f\ } V & \longrightarrow 0 \end{array}$$

There are many equivalent definitions of projective module, for example, an R-module is projective if and only if it is a direct summand of a free module (i.e., $R^{\oplus I}$ for some indexing set I). A *projective cover* of an R-module M is a projective R-module P together with a surjective R-module homomorphism $\varepsilon : P \to M$ such that for any projective R-module P' and surjective R-module homomorphism $\nu : P' \to M$, there is a surjective R-module homomorphism $\psi : P' \to P$ such that $\nu = \varepsilon\psi$.

An R-module I is *injective* if for every injective R-module homomorphism $f : V \to U$ and R-module homomorphism $g : V \to I$, there exists an R-module homomorphism $h : U \to I$ such that $hf = g$:

$$(A.2.2) \qquad \begin{array}{ccc} & & I \\ & {}^{h}\nearrow & \uparrow{}^{g} \\ U & \xleftarrow{\ f\ } V & \longleftarrow 0 \end{array}$$

An R-module F is *flat* if for every short exact sequence of right R-modules $0 \to U \to V \to W \to 0$, the induced sequence of abelian groups $0 \to U \otimes_R F \to V \otimes_R F \to W \otimes_R F \to 0$ is exact. Every projective module is flat.

Let M be an R-module. A *projective resolution* of M is a chain complex $P_\bullet$ consisting of projective R-modules P_n $(n \geq 0)$ for which $\mathrm{H}_0(P_\bullet) \cong M$ and $\mathrm{H}_n(P_\bullet) = 0$ for all $n \neq 0$. Thus $P_\bullet$ is quasi-isomorphic to the complex that is M concentrated in degree 0 and 0 elsewhere, with all maps 0:

$$\begin{array}{ccccccccc} \cdots \longrightarrow & P_2 & \longrightarrow & P_1 & \longrightarrow & P_0 & \longrightarrow & 0 & \longrightarrow \cdots \\ & \downarrow & & \downarrow & & \downarrow{}^{\varepsilon} & & \downarrow & \\ \cdots \longrightarrow & 0 & \longrightarrow & 0 & \longrightarrow & M & \longrightarrow & 0 & \longrightarrow \cdots \end{array}$$

As a consequence of the definition, the following sequence is exact:

$$(A.2.3) \qquad \cdots \xrightarrow{d_2} P_1 \xrightarrow{d_1} P_0 \xrightarrow{\varepsilon} M \longrightarrow 0.$$

We refer to the complex (A.2.3) as the *augmented complex* of $P_\bullet$. Sometimes this augmented complex is abbreviated $P_\bullet \xrightarrow{\varepsilon} M$ and referred to as the projective resolution of M, when it is clear from the context what is meant. We also sometimes refer to $P_\bullet$ as the *truncated complex* of (A.2.3).

Note that projective resolutions of R-modules always exist: every R-module M is a homomorphic image of a projective R-module, for example, the free module on a set of generators of M. One may use this fact to build projective resolutions as follows. Let P_0 be a projective R-module having M as a homomorphic image under a map ε. Let $K_1 = \operatorname{Ker}(\varepsilon)$. Then K_1 is a homomorphic image of a projective R-module P_1 via some map $\varepsilon_1 : P_1 \to K_1$. Denote by i_1 the inclusion map $i_1 : K_1 \to P_0$ and set $d_1 = i_1\varepsilon_1$. Let $K_2 = \operatorname{Ker}(d_1)$ and continue.

(A.2.4)

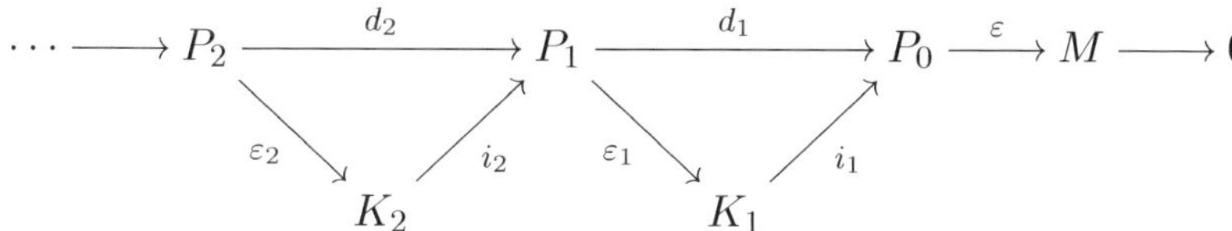

We call K_i an *ith syzygy module* of M. It depends on some choices, and Lemma A.2.6 below is a precise statement about these choices. First we state Schanuel's Lemma.

Lemma A.2.5 (Schanuel's Lemma). *Let* $0 \to K \to P \xrightarrow{\varepsilon} M \to 0$ *and* $0 \to K' \to P' \xrightarrow{\varepsilon'} M \to 0$ *be two short exact sequences of R-modules with* P, P' *projective. Then* $K \oplus P' \cong K' \oplus P$.

Schanuel's Lemma immediately implies:

Lemma A.2.6. *If K_i and K_i' are ith syzygy modules of the R-module M, then there are projective R-modules P, P' such that $K_i \oplus P \cong K_i' \oplus P'$.*

Another way to state Lemma A.2.6 is to say that K_i and K_i' are equivalent under the following equivalence relation. Two R-modules U and V are equivalent if there is an isomorphism of R-modules $U \oplus P \cong V \oplus P'$ for some projective R-modules P, P'.

The *Heller operator* Ω is defined by $\Omega(M) = K_1$ (notation from diagram (A.2.4)), understood to take values in an equivalence class of R-modules. Sometimes we write $\Omega_R = \Omega$ to emphasize the choice of ring R. This operator is often used in settings where projective direct summands do not matter, such as Theorem A.3.3 below. In some contexts, $\Omega(M)$ may instead be defined more specifically and uniquely up to isomorphism, for example by taking $P_\bullet$ to be a minimal resolution as defined below. It should be clear from the context which is meant. Sometimes this operator is iterated, and we write $\Omega^1 = \Omega$ and $\Omega^n(M) = \Omega(\Omega^{n-1}(M))$ for all $n \geq 2$.

The next theorem in particular implies a relation among projective resolutions.

Theorem A.2.7 (Comparison Theorem). *Let $(P_\bullet, d_\bullet)$ and $(Q_\bullet, d'_\bullet)$ be chain complexes of R-modules with $M = \mathrm{H}_0(P_\bullet)$, $N = \mathrm{H}_0(Q_\bullet)$, and let $\varepsilon : P_0 \to M$ and $\varepsilon' : Q_0 \to N$ be corresponding augmentation maps. Assume that the augmented complex $\cdots \to Q_1 \to Q_0 \to N \to 0$ is exact and that P_i is projective for each i. If $f : M \to N$ is an R-module homomorphism, then there is a chain map $f_\bullet : P_\bullet \to Q_\bullet$ for which $f\varepsilon = \varepsilon' f_0$, that is, the following diagram commutes:*

$$
\begin{array}{ccccccccc}
\cdots & \longrightarrow & P_2 & \longrightarrow & P_1 & \longrightarrow & P_0 & \overset{\varepsilon}{\longrightarrow} & M & \longrightarrow & 0 \\
& & \downarrow{\scriptstyle f_2} & & \downarrow{\scriptstyle f_1} & & \downarrow{\scriptstyle f_0} & & \downarrow{\scriptstyle f} & & \\
\cdots & \longrightarrow & Q_2 & \longrightarrow & Q_1 & \longrightarrow & Q_0 & \overset{\varepsilon'}{\longrightarrow} & N & \longrightarrow & 0
\end{array}
$$

The chain map $f_\bullet$ is unique up to chain homotopy.

In particular, if $P_\bullet, Q_\bullet$ are projective resolutions of M, N, respectively, the Comparison Theorem states that there is a chain map $f_\bullet : P_\bullet \to Q_\bullet$ lifting $f : M \to N$.

A projective resolution $P_\bullet$ of an R-module M is *minimal* if for every projective resolution $P'_\bullet \to M$, there is a chain map $f_\bullet : P'_\bullet \to P_\bullet$ lifting the identity map on M such that f_n is surjective for each n. Equivalently, P_0 is a projective cover of M and P_i is a projective cover of the ith syzygy module K_i for each i.

An *injective resolution* of an R-module M is a cochain complex $I_\bullet$ consisting of injective R-modules I_n for which $\mathrm{H}_0(I_\bullet) \cong M$ and $\mathrm{H}_n(I_\bullet) = 0$ for all $n \neq 0$. In other words, $M \cong \mathrm{Ker}(d_0)$ and the following sequence is exact:

$$
(A.2.8) \qquad\qquad 0 \longrightarrow M \overset{\iota}{\longrightarrow} I_0 \overset{d_0}{\longrightarrow} I_1 \overset{d_1}{\longrightarrow} \cdots ,
$$

where ι is an isomorphism from M to $\mathrm{Ker}(d_0)$ followed by inclusion into I_0. We refer to the complex (A.2.8) as the *augmented complex* of $I_\bullet$, and sometimes as the injective resolution of M when it is clear from the context what is intended. We also sometimes refer to $I_\bullet$ as the *truncated complex* of (A.2.8).

Note that injective resolutions of R-modules always exist. Baer's Theorem states that every R-module can be embedded in an injective R-module, and an injective resolution can be built in a similar fashion to that described for a projective resolution above. Let $L_1 = \operatorname{Coker}(\iota) = I_0/\operatorname{Im}(\iota)$, then embed L_1 into an injective module I_1 via $\iota_1 : L_1 \to I_1$, set $\delta_0 = \iota_1\pi_0$, where $\pi_0 : I_0 \to L_1$ is the quotient map, and so on:

(A.2.9)

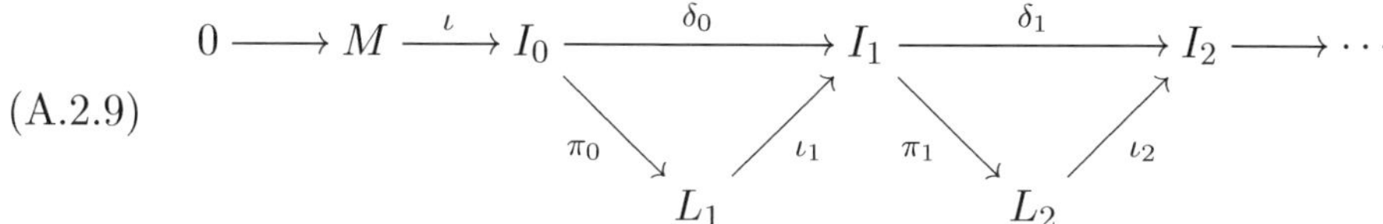

Again, the module L_1 is unique up to injective direct summands, due to a dual version of Schanuel's Lemma, which states that if $0 \to N \to I \to L \to 0$ and $0 \to N \to I' \to L' \to 0$ are exact sequences with I, I' injective, then there is an isomorphism $L \oplus I' \cong L' \oplus I$.

We define the operator Ω^{-1} by $\Omega^{-1}(M) = L_1$, understood to take values in an equivalence class of modules, and call it a *first cosyzygy module* of M. Similarly, $\Omega^{-i}(M) = L_i$ is an *ith cosyzygy module* of M.

The above notation is chosen with the following setting in mind. Assume R is a *self-injective* algebra over a field k, that is, R is injective as an R-module (under left multiplication). Then projective A-modules are also injective, and vice versa. Combining diagrams (A.2.4) and (A.2.9) we obtain:

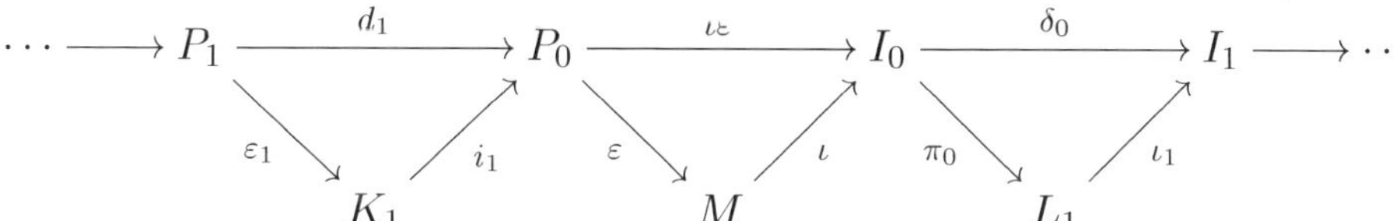

The modules in the top row are all projective and injective and may be viewed as terms in projective and injective resolutions of the modules in the bottom row. It follows that $\Omega^{-1}(\Omega(M))$ is equivalent to M. There are projective modules P, P' such that

$$\Omega^{-1}(\Omega(M)) \oplus P \cong M \oplus P'$$

in this case that R is self-injective.

Other types of resolutions may be defined similarly, for example, flat resolutions.

The *projective dimension* $\operatorname{pdim}_R(M)$ of an R-module M is the smallest integer n such that there is a projective resolution of M:

$$0 \longrightarrow P_n \longrightarrow \cdots \longrightarrow P_0 \longrightarrow M \longrightarrow 0.$$

If no such n exists we write $\operatorname{pdim}_R(M) = \infty$. We similarly define injective dimension and flat dimension. Note that $\operatorname{pdim}_R(M)$ is the smallest integer n such that an nth syzygy module of M is projective. Thus $\operatorname{pdim}_R(\Omega(M)) = \operatorname{pdim}_R(M) - 1$ if M is not projective. We also see that

$$(\text{A.2.10}) \quad \operatorname{pdim}_R(M) = \sup\{j \mid \operatorname{Ext}_R^j(M, N) \neq 0 \text{ for some } R\text{-module } N\}.$$

The *left global dimension* of R is

$$\operatorname{gldim}_l R = \sup\{\operatorname{pdim}_R(M) \mid M \text{ is a left } R\text{-module}\}.$$

We may similarly define right global dimension, gldim_r. Authors often instead use the notation gldim without a subscript in contexts where it is understood which is meant, or in case R is commutative.

An important class of examples is provided by the following theorem.

Theorem A.2.11 (Hilbert's Syzygy Theorem). *Let k be a field. Then*

$$\operatorname{gldim} k[x_1, \ldots, x_n] = n.$$

A ring R is (left) *hereditary* if every left ideal is projective, equivalently if $\operatorname{gldim}_l R \leq 1$. Thus for example, $k[x]$ is hereditary by Theorem A.2.11.

A.3. Ext and Tor

Let M and N be R-modules. Let $P_{\bullet} \xrightarrow{\varepsilon} M$ be a projective resolution of M. Applying $\operatorname{Hom}_R(-, N)$ to the sequence $\cdots \xrightarrow{d_2} P_1 \xrightarrow{d_1} P_0 \to 0$, we obtain

$$(\text{A.3.1}) \qquad 0 \longrightarrow \operatorname{Hom}_R(P_0, N) \xrightarrow{d_1^*} \operatorname{Hom}_R(P_1, N) \xrightarrow{d_2^*} \cdots,$$

where $d_i^*(f) = f d_i$ for all i and $f \in \operatorname{Hom}_R(P_{i-1}, N)$. We set $d_0^* = 0$. Note that $d_{i+1}^* d_i^* = 0$ since $d_i d_{i+1} = 0$, so the sequence (A.3.1) is a (cochain) complex of abelian groups (that is, $\mathbb{Z}$-modules). If R is commutative, it is a complex of R-modules. If R is an algebra over a field k, it is a complex of k-vector spaces. We define $\operatorname{Ext}_R^n(M, N)$ to be the cohomology of this complex:

$$\operatorname{Ext}_R^n(M, N) = \operatorname{H}^n(\operatorname{Hom}_R(P_{\bullet}, N)) = \operatorname{Ker}(d_{n+1}^*)/\operatorname{Im}(d_n^*)$$

for $n \geq 0$, and

$$\operatorname{Ext}_R^*(M, N) = \bigoplus_{n \geq 0} \operatorname{Ext}_R^n(M, N).$$

An application of the Comparison Theorem (Theorem A.2.7) shows that $\operatorname{Ext}_R^*(M, N)$ does not depend on choice of projective resolution of M. Note that

$$\operatorname{Ext}_R^0(M, N) \cong \operatorname{Hom}_R(M, N).$$

By construction, if M is itself a projective R-module, then $\operatorname{Ext}_R^n(M, N) = 0$ for all $n > 0$.

Equivalently, we may define $\mathrm{Ext}_R^n(M, N)$ via an injective resolution. Let $N \xrightarrow{\iota} I_{\bullet}$ be an injective resolution of N. Apply $\mathrm{Hom}_R(M, -)$ to the sequence $0 \to I_0 \xrightarrow{d_0} I_1 \xrightarrow{d_1} \cdots$ to obtain a sequence

$$(A.3.2) \qquad 0 \longrightarrow \mathrm{Hom}_R(M, I_0) \xrightarrow{(d_0)_*} \mathrm{Hom}_R(M, I_1) \xrightarrow{(d_1)_*} \cdots,$$

where $(d_i)_*(f) = d_i f$ for all i and $f \in \mathrm{Hom}_R(M, I_i)$. Set $(d_{-1})_* = 0$. Then $(d_{i+1})_*(d_i)_* = 0$ for all $i \geq -1$, so the sequence (A.3.2) is a (cochain) complex of abelian groups. It can be shown that

$$\mathrm{Ext}_R^n(M, N) \cong \mathrm{H}^n(\mathrm{Hom}_R(M, I_{\bullet})) = \mathrm{Ker}((d_n)_*)/\mathrm{Im}((d_{n-1})_*).$$

If N is an injective R-module, we now see that $\mathrm{Ext}_R^n(M, N) = 0$ for all $n > 0$.

For each n, the group $\mathrm{Ext}_R^n(M, N)$ has an interpretation in terms of exact sequences: an *n-extension* of M by N is an exact sequence of R-modules

$$\mathbf{f}: \qquad 0 \longrightarrow N \longrightarrow U_{n-1} \longrightarrow \cdots \longrightarrow U_1 \longrightarrow U_0 \longrightarrow M \longrightarrow 0.$$

If $\mathbf{g}$ is another n-extension of M by N, a map of extensions from $\mathbf{f}$ to $\mathbf{g}$ is a chain map which is the identity map (denoted 1) on M and on N:

$$\begin{array}{ccccccccccc}
0 & \longrightarrow & N & \longrightarrow & U_{n-1} & \longrightarrow \cdots \longrightarrow & U_1 & \longrightarrow & U_0 & \longrightarrow & M & \longrightarrow & 0 \\
& & \downarrow{\scriptstyle 1} & & \downarrow{\scriptstyle \phi_{n-1}} & & \downarrow{\scriptstyle \phi_1} & & \downarrow{\scriptstyle \phi_0} & & \downarrow{\scriptstyle 1} \\
0 & \longrightarrow & N & \longrightarrow & V_{n-1} & \longrightarrow \cdots \longrightarrow & V_1 & \longrightarrow & V_0 & \longrightarrow & M & \longrightarrow & 0
\end{array}$$

Maps of n-extensions generate an equivalence relation. There is a well-defined binary operation on equivalence classes, called the Baer sum, under which the set of equivalence classes of n-extensions is an abelian group. A representative of the additive inverse of the n-extension $\mathbf{f}$ above can be taken to be the sequence with the same modules in which the map $U_0 \to M$ is replaced by its additive inverse, and all other maps are the same. The group $\mathrm{Ext}_R^n(M, N)$ is isomorphic to the group of equivalence classes of n-extensions of M by N. We outline this one-to-one correspondence next.

Let $\mathbf{f}$ be the above n-extension of M by N. We will define an element of $\mathrm{Ext}_R^n(M, N)$ corresponding to it. Let $P_{\bullet} \to M$ be a projective resolution of M. By the Comparison Theorem (Theorem A.2.7), there is a chain map $\hat{f}_{\bullet}$:

$$\begin{array}{ccccccccccc}
P_{n+1} & \xrightarrow{d_{n+1}} & P_n & \xrightarrow{d_n} & P_{n-1} & \longrightarrow \cdots \longrightarrow & P_1 & \xrightarrow{d_1} & P_0 & \xrightarrow{\varepsilon} & M & \longrightarrow & 0 \\
\downarrow & & \downarrow{\scriptstyle \hat{f}_n} & & \downarrow{\scriptstyle \hat{f}_{n-1}} & & \downarrow{\scriptstyle \hat{f}_1} & & \downarrow{\scriptstyle \hat{f}_0} & & \downarrow{\scriptstyle 1} \\
0 & \longrightarrow & N & \longrightarrow & U_{n-1} & \longrightarrow \cdots \longrightarrow & U_1 & \longrightarrow & U_0 & \longrightarrow & M & \longrightarrow & 0
\end{array}$$

Then $\hat{f}_n \in \mathrm{Hom}_R(P_n, N)$ and $\hat{f}_n d_{n+1} = 0$, so $\hat{f}_n$ is a cocycle. Set $f = \hat{f}_n$. Since $\hat{f}_\bullet$ is unique up to chain homotopy, any two such maps represent the same element of $\mathrm{Ext}_R^n(M, N)$.

Conversely, let $f \in \mathrm{Hom}_R(P_n, N)$ for which $f d_{n+1} = 0$. We will define an n-extension $\mathbf{f}$ of M by N corresponding to f. Let X be a pushout of $P_n \xrightarrow{d_n} P_{n-1}$ and $P_n \xrightarrow{f} N$. We may take

$$X = (P_{n-1} \oplus N)/\{(-d_n(x), f(x)) \mid x \in P_n\}.$$

Then the following diagram commutes:

$$
\begin{array}{ccccccccccccc}
P_{n+1} & \xrightarrow{d_{n+1}} & P_n & \xrightarrow{d_n} & P_{n-1} & \xrightarrow{d_{n-1}} & P_{n-2} & \longrightarrow & \cdots & \longrightarrow & P_0 & \xrightarrow{\varepsilon} & M & \longrightarrow & 0 \\
\downarrow & & \downarrow{\scriptstyle f} & & \downarrow{\scriptstyle \binom{1}{0}} & & \downarrow{\scriptstyle 1} & & & & \downarrow{\scriptstyle 1} & & \downarrow{\scriptstyle 1} & & \\
0 & \longrightarrow & N & \xrightarrow{\binom{0}{1}} & X & \xrightarrow{(d_{n-1},0)} & P_{n-2} & \xrightarrow{d_{n-2}} & \cdots & \longrightarrow & P_0 & \xrightarrow{\varepsilon} & M & \longrightarrow & 0
\end{array}
$$

(Equivalently, we may replace P_n by $K_n = \mathrm{Ker}(d_{n-1})$ in the pushout diagram.) The lower sequence is an n-extension of M by N.

The following theorem is called "dimension shifting" since it allows any Ext^n group to be expressed as an Ext^1 group (shifting degree, or dimension, from n to 1). It follows from close inspection of diagrams (A.2.4) and (A.2.9).

Theorem A.3.3 (Dimension shifting). *Let M and N be R-modules. For each $i \geq 1$, let $\Omega^i M$ denote an ith syzygy module of M, and $\Omega^{-i} N$ an ith cosyzygy module of N. Then*

$$
\begin{aligned}
\mathrm{Ext}_R^n(M, N) &\cong \mathrm{Ext}_R^1(\Omega^{n-1} M, N), \\
\mathrm{Ext}_R^n(M, N) &\cong \mathrm{Ext}_R^1(M, \Omega^{1-n} N)
\end{aligned}
$$

for all $n \geq 2$.

As observed in the last section, if R is self-injective, then $\Omega^{-1}(\Omega(N))$ is equivalent to N (that is, isomorphic up to projective direct summands). Thus we have the following corollary.

Corollary A.3.4. *If R is a self-injective algebra over a field k, then*

$$\mathrm{Ext}_R^n(M, N) \cong \mathrm{Ext}_R^n(\Omega(M), \Omega(N))$$

for all R-modules M, N and $n \geq 1$.

Now suppose M is a right R-module and N is a left R-module. Let $P_\bullet \to M$ be a (right R-module) projective resolution of M. Apply $- \otimes_R N$ to obtain a sequence of $\mathbb{Z}$-modules:

$$\cdots \longrightarrow P_2 \otimes_R N \xrightarrow{d_2 \otimes 1} P_1 \otimes_R N \xrightarrow{d_1 \otimes 1} P_0 \otimes_R N \longrightarrow 0.$$

Here, as elsewhere, in order to minimize notational clutter, we suppress the subscript R on the tensor symbol $\otimes$, for maps and elements, when it is clear from the context that they involve tensor products over R. We set $d_0 = 0$. This is a chain complex and we define $\mathrm{Tor}_n^R(M, N)$ to be its homology:

$$\mathrm{Tor}_n^R(M, N) = \mathrm{H}_n(P_{\bullet} \otimes_R N) = \mathrm{Ker}(d_n \otimes 1)/\mathrm{Im}(d_{n+1} \otimes 1)$$

for $n \geq 0$, and

$$\mathrm{Tor}_*^R(M, N) = \bigoplus_{n \geq 0} \mathrm{Tor}_n^R(M, N).$$

By the Comparison Theorem (Theorem A.2.7), $\mathrm{Tor}_n^R(M, N)$ does not depend on choice of projective resolution of M. Note that

$$\mathrm{Tor}_0^R(M, N) \cong M \otimes_R N.$$

Equivalently, we may define $\mathrm{Tor}_n^R(M, N)$ via a (left R-module) projective resolution of N. Let $Q_{\bullet} \to N$ be a projective resolution of N and apply $M \otimes_R -$ to obtain a sequence

$$\cdots \longrightarrow M \otimes_R Q_2 \xrightarrow{\ 1_M \otimes d_2\ } M \otimes_R Q_1 \xrightarrow{\ 1_M \otimes d_1\ } M \otimes_R Q_0 \longrightarrow 0.$$

It can be shown that $\mathrm{Tor}_n^R(M, N) \cong \mathrm{H}_n(M \otimes_R Q_{\bullet})$. By construction then, if either M or N is flat as an R-module, then $\mathrm{Tor}_n^R(M, N) = 0$ for all $n > 0$.

A.4. Long exact sequences

The following lemma can be used to construct long exact sequences for Ext and Tor.

Lemma A.4.1 (Snake Lemma). *Let U, U', V, V', W, W' be R-modules for which there is a commuting diagram with exact rows:*

$$
\begin{array}{ccccccc}
U' & \longrightarrow & V' & \xrightarrow{\ p\ } & W' & \longrightarrow & 0 \\
& & \downarrow{\scriptstyle f} & & \downarrow{\scriptstyle g} & & \downarrow{\scriptstyle h} \\
0 & \longrightarrow & U & \xrightarrow{\ i\ } & V & \longrightarrow & W
\end{array}
$$

There is an exact sequence

$$\mathrm{Ker}(f) \to \mathrm{Ker}(g) \to \mathrm{Ker}(h) \xrightarrow{\ \partial\ } \mathrm{Coker}(f) \to \mathrm{Coker}(g) \to \mathrm{Coker}(h),$$

where $\partial(w') = i^{-1}gp^{-1}(w')$ for all $w' \in \mathrm{Ker}(h)$. If the map $U' \to V'$ is injective, then $\mathrm{Ker}(f) \to \mathrm{Ker}(g)$ is injective, and if $V \to W$ is surjective, then $\mathrm{Coker}(g) \to \mathrm{Coker}(h)$ is surjective.

By the notation $p^{-1}(w')$ in the lemma, we mean any element in the inverse image of w'. Its value under $i^{-1}g$ followed by projection to $\mathrm{Coker}(f)$ will not depend on the choice. The name Snake Lemma refers to the curved arrow in the following diagram that illustrates the statement.

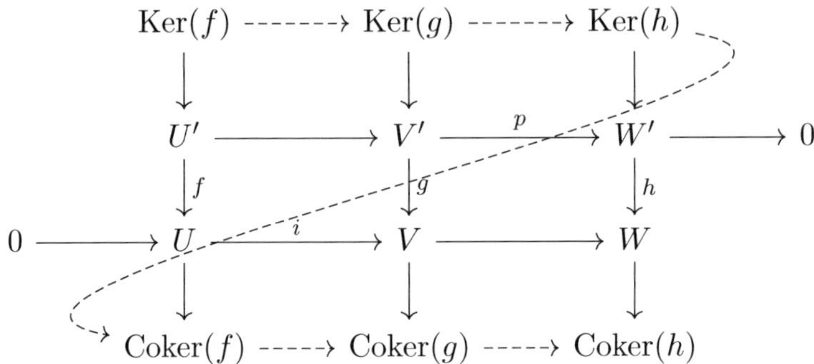

A consequence of the Snake Lemma (Lemma A.4.1) is the following theorem. A short exact sequence of complexes is a sequence

$$0 \to U_{\bullet} \xrightarrow{f_{\bullet}} V_{\bullet} \xrightarrow{g_{\bullet}} W_{\bullet} \to 0,$$

where $f_{\bullet}, g_{\bullet}$ are chain maps and for each i, f_i is injective, g_i is surjective, and $\mathrm{Im}(f_i) = \mathrm{Ker}(g_i)$.

Theorem A.4.2. *Let* $0 \to U_{\bullet} \xrightarrow{f_{\bullet}} V_{\bullet} \xrightarrow{g_{\bullet}} W_{\bullet} \to 0$ *be a short exact sequence of complexes. For each* n, *there is an abelian group homomorphism* $\partial_n :$ $\mathrm{H}_n(W_{\bullet}) \to \mathrm{H}_{n-1}(U_{\bullet})$ *such that*

$$\cdots \longrightarrow \mathrm{H}_{n+1}(W) \xrightarrow{\partial_{n+1}} \mathrm{H}_n(U) \xrightarrow{\overline{f}_n} \mathrm{H}_n(V) \xrightarrow{\overline{g}_n} \mathrm{H}_n(W) \xrightarrow{\partial_n} \cdots$$

is an exact sequence, where $\overline{f}_n, \overline{g}_n$ *denote the maps induced by* f_n, g_n.

The homomorphisms ∂_n in the theorem are called *connecting homomorphisms*.

The following lemma is called the Horseshoe Lemma due to the shape of the diagram in the statement.

Lemma A.4.3 (Horseshoe Lemma). *Let* U', U, U'' *be* R-*modules for which there is an exact sequence* $0 \to U' \to U \to U'' \to 0$, *and let* $P'_{\bullet} \to U'$,

$P''_\bullet \to U''$ be projective resolutions of U' and U'':

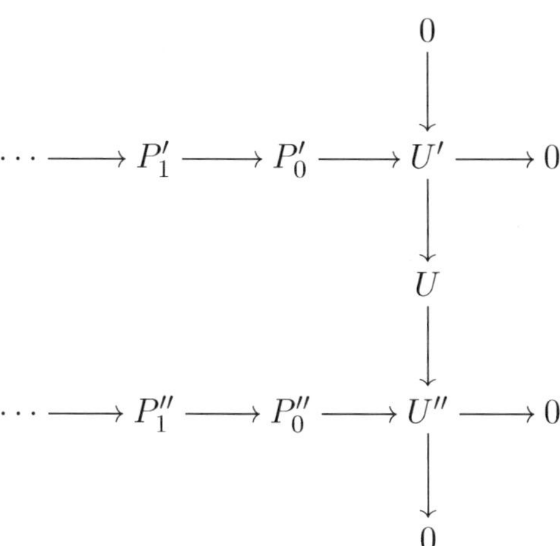

For each n, let $P_n = P'_n \oplus P''_n$. Then there are differentials d_i for which $\cdots \to P_1 \xrightarrow{d_1} P_0 \to U \to 0$ is a projective resolution of U, and the right column lifts to an exact sequence of complexes $0 \to P'_\bullet \xrightarrow{\iota_\bullet} P_\bullet \xrightarrow{\pi_\bullet} P''_\bullet \to 0$ with $\iota_\bullet, \pi_\bullet$ the standard inclusion and projection maps, respectively.

The Horseshoe Lemma (Lemma A.4.3) is used in conjunction with Theorem A.4.2 to obtain the following four long exact sequences.

Theorem A.4.4 (First long exact sequence for Ext). *Let U be an R-module, and let $0 \to V' \to V \to V'' \to 0$ be an exact sequence of R-modules. There is an exact sequence:*

$$0 \to \operatorname{Hom}_R(U, V') \longrightarrow \operatorname{Hom}_R(U, V) \longrightarrow \operatorname{Hom}_R(U, V'') \longrightarrow$$

$$\operatorname{Ext}^1_R(U, V') \longrightarrow \operatorname{Ext}^1_R(U, V) \longrightarrow \operatorname{Ext}^1_R(U, V'') \longrightarrow \operatorname{Ext}^2_R(U, V') \cdots$$

Theorem A.4.5 (Second long exact sequence for Ext). *Let V be an R-module, and let $0 \to U' \to U \to U'' \to 0$ be an exact sequence of R-modules. There is an exact sequence:*

$$0 \to \operatorname{Hom}_R(U'', V) \longrightarrow \operatorname{Hom}_R(U, V) \longrightarrow \operatorname{Hom}_R(U', V) \longrightarrow$$

$$\operatorname{Ext}^1_R(U'', V) \longrightarrow \operatorname{Ext}^1_R(U, V) \longrightarrow \operatorname{Ext}^1_R(U', V) \longrightarrow \operatorname{Ext}^2_R(U'', V) \cdots$$

Theorem A.4.6 (First long exact sequence for Tor). *Let V be a left R-module, and let $0 \to U' \to U \to U'' \to 0$ be an exact sequence of right R-modules. There is an exact sequence:*

$$\cdots \longrightarrow \operatorname{Tor}^R_2(U'', V) \longrightarrow \operatorname{Tor}^R_1(U', V) \longrightarrow \operatorname{Tor}^R_1(U, V) \longrightarrow$$

$$\operatorname{Tor}^R_1(U'', V) \longrightarrow U' \otimes_R V \longrightarrow U \otimes_R V \longrightarrow U'' \otimes_R V \longrightarrow 0$$

Theorem A.4.7 (Second long exact sequence for Tor). *Let U be a right R-module, and let $0 \to V' \to V \to V'' \to 0$ be an exact sequence of left R-modules. There is an exact sequence:*

$$\cdots \longrightarrow \operatorname{Tor}_2^R(U, V'') \longrightarrow \operatorname{Tor}_1^R(U, V') \longrightarrow \operatorname{Tor}_1^R(U, V) \longrightarrow$$

$$\operatorname{Tor}_1^R(U, V'') \longrightarrow U \otimes_R V' \longrightarrow U \otimes_R V \longrightarrow U \otimes_R V'' \longrightarrow 0$$

A.5. Double complexes

A *double complex* (or *bicomplex*) of R-modules is a set $B = \{B_{i,j}\}_{i,j \in \mathbb{Z}}$ of R-modules $B_{i,j}$ with maps

$$d_{i,j}^h : B_{i,j} \to B_{i-1,j} \quad \text{and} \quad d_{i,j}^v : B_{i,j} \to B_{i,j-1}$$

(called *horizontal* and *vertical* differentials, respectively) such that $d^h d^h = 0$, $d^v d^v = 0$, and $d^v d^h + d^h d^v = 0$. We say that B is *bounded* if for each n, there are finitely many $B_{i,j}$ with $i + j = n$ that are nonzero. The *total complex* of the double complex B is given by

$$\operatorname{Tot}(B)_n = \bigoplus_{i+j=n} B_{i,j}$$

for each n, with differential $d = d^h + d^v$. (To be more precise, we may write $\operatorname{Tot}^\oplus(B)_n = \bigoplus_{i+j=n} B_{i,j}$ and $\operatorname{Tot}^\Pi(B)_n = \prod_{i+j=n} B_{ij}$. We will generally work with bounded complexes, in which case there is no distinction.) When we refer to the homology of the double complex B, we mean the homology of its total complex.

Remark A.5.1. A double complex of R-modules may equivalently be defined as a complex of complexes, working in the abelian category (see Section A.6) whose objects are complexes of R-modules and morphisms are chain maps.

Important examples of double complexes are tensor product complexes and Hom complexes, as we define next.

Tensor product complexes. Let $(C_\bullet, d_\bullet^C)$, $(D_\bullet, d_\bullet^D)$ be complexes of right and left R-modules, respectively. Let $B_{i,j} = C_i \otimes_R D_j$ with

$$d_{i,j}^h(x \otimes y) = d_i^C(x) \otimes y \quad \text{and} \quad d_{i,j}^v(x \otimes y) = (-1)^i x \otimes d_j^D(y)$$

for all $x \in C_i$, $y \in D_j$. Written a different way, in accordance with the sign convention (2.3.1), we have

$$d_{i,j}^h = d_i^C \otimes 1_D \quad \text{and} \quad d_{i,j}^v = 1_C \otimes d_j^D.$$

Then $B_{\bullet,\bullet}$ is a double complex of $\mathbb{Z}$-modules. (If R is commutative, then B is a double complex of R-modules.) We sometimes denote $B_{\bullet,\bullet}$ by $C_\bullet \otimes_R D_\bullet$ or $C \otimes_R D$.

Theorem A.5.2 (Künneth Theorem). *Let $C_\bullet$ and $D_\bullet$ be complexes of right and left R-modules, respectively, for which C_n and $d(C_n)$ are flat R-modules for all $n \in \mathbb{Z}$. Then for all $n \in \mathbb{Z}$, there is a short exact sequence:*

$$0 \longrightarrow \bigoplus_{i+j=n} \mathrm{H}_i(C) \otimes_R \mathrm{H}_j(D) \longrightarrow \mathrm{H}_n(C \otimes_R D) \longrightarrow$$

$$\bigoplus_{i+j=n-1} \mathrm{Tor}_1^R(\mathrm{H}_i(C), \mathrm{H}_j(D)) \longrightarrow 0$$

Remark A.5.3. The hypothesis that C_n and $d(C_n)$ are flat can be replaced by the hypothesis that D_n and $d(D_n)$ are flat as left R-modules.

Viewing a module as a complex concentrated in degree 0, with differentials all 0, we obtain the following corollary.

Theorem A.5.4 (Universal Coefficients Theorem). *Let C be a complex of right R-modules in which all C_n, $d(C_n)$ are flat, and let M be a left R-module. There is a short exact sequence*

$$0 \longrightarrow \mathrm{H}_n(C) \otimes_R M \longrightarrow \mathrm{H}_n(C \otimes_R M) \longrightarrow \mathrm{Tor}_1^R(\mathrm{H}_{n-1}(C), M) \longrightarrow 0.$$

If C is quasi-isomorphic to C' and D is quasi-isomorphic to D', then $C \otimes_R D$ is quasi-isomorphic to $C' \otimes_R D'$, via tensor product maps.

Hom complexes. Let $(C_\bullet, d_\bullet^C)$, $(D_\bullet, d_\bullet^D)$ be complexes of left R-modules. Let $B_{i,j} = \mathrm{Hom}_R(C_i, D_j)$ with

$$d_{i,j}^h(f) = (-1)^{i-j+1} f d_{i+1}^C \quad \text{and} \quad d_{i,j}^v(f) = d_j^D f$$

for all $f \in \mathrm{Hom}_R(C_i, D_j)$. Then $B_{\bullet,\bullet}$ is a double complex of $\mathbb{Z}$-modules. We caution that there are other choices of sign conventions on Hom complexes in the literature. It is common instead to reindex first so that either C or D becomes a cocomplex. We sometimes denote $B_{\bullet,\bullet}$ by $\mathrm{Hom}_R(C_\bullet, D_\bullet)$ or $\mathrm{Hom}_R(C, D)$.

If C is quasi-isomorphic to C' and D is quasi-isomorphic to D', then $\mathrm{Hom}_R(C, D)$ is quasi-isomorphic to $\mathrm{Hom}_R(C', D')$.

A standard Hom complex arises as follows. Let $P_\bullet \to M$ be a projective resolution of an R-module M. Then $\mathrm{Hom}_R(P_\bullet, P_\bullet)$ is a double complex as described above. Since $P_\bullet$ is quasi-isomorphic to M as a complex concentrated in degree 0, $\mathrm{Hom}_R(P_\bullet, P_\bullet)$ is quasi-isomorphic to $\mathrm{Hom}_R(P_\bullet, M)$, whose cohomology is $\mathrm{Ext}_R^*(M, M)$. Moreover, a calculation shows that $\mathrm{Hom}_R(P_\bullet, P_\bullet)$ is a differential graded algebra under composition of functions.

Theorem A.5.5 (Acyclic Assembly Lemma). *Let B be a bounded double complex of R-modules. Then $\mathrm{Tot}(B)$ is acyclic if one of the following four conditions holds: $B_{i,j} = 0$ for all $j < 0$ and B has exact columns or exact rows, or $B_{i,j} = 0$ for all $i < 0$ and B has exact columns or exact rows.*

See [**223**, Lemma 2.7.3] for a precise statement for unbounded complexes.

A.6. Categories, functors, derived functors

A *category* $\mathcal{C}$ is a collection of *objects* $\mathrm{Obj}(\mathcal{C})$ together with a set of *morphisms* $\mathrm{Hom}_{\mathcal{C}}(A, B)$ for each pair of objects A, B of $\mathcal{C}$, including an *identity morphism* $1_A \in \mathrm{Hom}_{\mathcal{C}}(A, A)$ for each object A and a binary operation called *composition* $\circ : \mathrm{Hom}_{\mathcal{C}}(A, B) \times \mathrm{Hom}_{\mathcal{C}}(B, C) \to \mathrm{Hom}_{\mathcal{C}}(A, C)$ for every triple A, B, C of objects of $\mathcal{C}$, such that

$$(hg)f = h(gf) \quad \text{and} \quad 1_B f = f 1_A$$

for all $f \in \mathrm{Hom}_{\mathcal{C}}(A, B)$, $g \in \mathrm{Hom}_{\mathcal{C}}(B, C)$, $h \in \mathrm{Hom}_{\mathcal{C}}(C, D)$, and objects A, B, C, D of $\mathcal{C}$. (Here, as elsewhere, we have written gf in place of $g \circ f$ to denote composition.)

A morphism $f \in \mathrm{Hom}_{\mathcal{C}}(A, B)$ is an *isomorphism* if there is a morphism $g \in \mathrm{Hom}_{\mathcal{C}}(B, A)$ such that $gf = 1_A$ and $fg = 1_B$.

We work primarily with categories of left or right modules or bimodules for a ring. The morphisms are module homomorphisms, the identity morphism is the identity homomorphism, and composition is function composition. If R is a ring, we use the notation R-Mod (respectively, R-mod) to denote the categories of all left R-modules (respectively, all finitely generated left R-modules). The notation Mod-R (respectively, mod-R) denotes similar categories of right R-modules. We abbreviate $\mathrm{Hom}_{R\text{-Mod}}$ (respectively, $\mathrm{Hom}_{R\text{-mod}}$, $\mathrm{Hom}_{\text{Mod-}R}$, $\mathrm{Hom}_{\text{mod-}R}$) by Hom_R in all these cases, and it will always be clear from the context which is meant. Note that for any pair of R-modules A, B, the set $\mathrm{Hom}_R(A, B)$ is in fact an abelian group under addition of functions.

Let $\mathcal{C}$ and $\mathcal{D}$ be categories. A *functor* $F : \mathcal{C} \to \mathcal{D}$ assigns an object $F(A)$ of $\mathcal{D}$ to each object A of $\mathcal{C}$, and a morphism $F(f) \in \mathrm{Hom}_{\mathcal{D}}(F(A), F(B))$ to each morphism $f \in \mathrm{Hom}_{\mathcal{C}}(A, B)$ for each pair of objects A, B of $\mathcal{C}$ in such a way that $F(1_A) = 1_{F(A)}$ for all A and $F(g \circ f) = F(g) \circ F(f)$ for all morphisms f, g that can be composed. The identity functor $1_{\mathcal{C}} : \mathcal{C} \to \mathcal{C}$ is given by $1_{\mathcal{C}}(A) = A$ and $1_{\mathcal{C}}(f) = f$ for all objects A and morphisms f of $\mathcal{C}$. We have in fact defined a *covariant functor*, to be more precise. A *contravariant functor* similarly assigns to each object A of $\mathcal{C}$ an object $F(A)$ of $\mathcal{D}$ and to each morphism $f \in \mathrm{Hom}_{\mathcal{C}}(A, B)$ a morphism $F(f) \in \mathrm{Hom}_{\mathcal{D}}(F(B), F(A))$ such that $F(1_A) = 1_{F(A)}$ and $F(g \circ f) = F(f) \circ F(g)$ for all morphisms f, g that can be composed.

Let $\mathcal{C}$ and $\mathcal{D}$ be categories. Let $F : \mathcal{C} \to \mathcal{D}$ and $G : \mathcal{C} \to \mathcal{D}$ be functors. A *natural transformation* $\eta : F \to G$ assigns a morphism $\eta_A : F(A) \to G(A)$ to each object A of $\mathcal{C}$ in such a way that $G(f) \circ \eta_A = \eta_B \circ F(f)$ for all objects A, B of $\mathcal{C}$ and morphisms $f \in \operatorname{Hom}_{\mathcal{C}}(A, B)$, that is, the following diagram commutes:

$$
\begin{array}{ccc}
F(A) & \xrightarrow{\ F(f)\ } & F(B) \\
{\scriptstyle \eta_A}\downarrow & & \downarrow{\scriptstyle \eta_B} \\
G(A) & \xrightarrow{\ G(f)\ } & G(B)
\end{array}
$$

If η_A is an isomorphism for each object A, we say that η is a *natural isomorphism*, and write $F \cong G$. The adjective "natural" is often used in this technical sense.

Two categories $\mathcal{C}$ and $\mathcal{D}$ are *equivalent* if there are functors $F : \mathcal{C} \to D$, $G : \mathcal{D} \to \mathcal{C}$ and natural isomorphisms $FG \cong 1_{\mathcal{D}}$, $GF \cong 1_{\mathcal{C}}$.

Let R and S be rings. A functor $F : R\text{-Mod} \to S\text{-Mod}$ is *additive* if F induces homomorphisms of abelian groups $\operatorname{Hom}_R(A, B) \cong \operatorname{Hom}_S(F(A), F(B))$ for all R-modules A, B. The rings R and S are *Morita equivalent* if R-Mod and S-Mod are equivalent categories via additive functors $F : R\text{-Mod} \to S\text{-Mod}$ and $G : S\text{-Mod} \to R\text{-Mod}$. Note that the original definition of Morita equivalence requires F and G to have a particular form; this definition is equivalent.

Let $\mathcal{C}, \mathcal{D}$ be categories and let $F : \mathcal{C} \to \mathcal{D}$, $G : \mathcal{D} \to \mathcal{C}$ be functors. Suppose that there are natural isomorphisms

$$
\operatorname{Hom}_{\mathcal{D}}(F(X), Y) \xrightarrow{\ \sim\ } \operatorname{Hom}_{\mathcal{C}}(X, G(Y))
$$

for each object X of $\mathcal{C}$ and each object Y of $\mathcal{D}$. Then we say that F is a *left adjoint* to G, and that G is a *right adjoint* to F.

Examples of adjoint functors are provided by induction and coinduction of modules as follows. Let A be a ring, and let B be a subring of A. Let N be a B-module. The *induced* (also called *tensor induced*) A-module is $A \otimes_B N$ with action of A given by multiplication on the left factor A. The *coinduced* A-module is $\operatorname{Hom}_B(A, N)$ with action of A given by

$$
(a \cdot f)(a') = f(a'a)
$$

for all $a, a' \in A$ and $f \in \operatorname{Hom}_B(A, N)$.

The following lemma is a statement about adjoint functors.

Lemma A.6.1 (Nakayama relations). *Let A be a ring, and let B be a subring of A. Let M be an A-module, and let N be a B-module. Then*

$$
\begin{aligned}
\operatorname{Hom}_B(N, M) &\cong \operatorname{Hom}_A(A \otimes_B N, M), \\
\operatorname{Hom}_B(M, N) &\cong \operatorname{Hom}_A(M, \operatorname{Hom}_B(A, N)).
\end{aligned}
$$

That is, restriction from A to B has a left adjoint given by induction and a right adjoint given by coinduction.

The following lemma is a consequence of Lemma A.6.1.

Lemma A.6.2 (Eckmann-Shapiro Lemma). *Let A be a ring and let B be a subring of A such that A is projective as a right B-module. Let M be an A-module and let N be a B-module. Then*

$$\operatorname{Ext}_B^n(N, M) \;\cong\; \operatorname{Ext}_A^n(A \otimes_B N, M),$$
$$\operatorname{Ext}_B^n(M, N) \;\cong\; \operatorname{Ext}_A^n(M, \operatorname{Hom}_B(A, N)).$$

More generally we will be interested in additive functors on abelian categories, which are a generalization of categories of modules that retain enough structure for homological algebra. We define these next after some other needed definitions.

A *zero object* of a category $\mathcal{C}$ is an object A such that $|\operatorname{Hom}_\mathcal{C}(A, B)| = 1$ and $|\operatorname{Hom}_\mathcal{C}(B, A)| = 1$ for all objects B of $\mathcal{C}$ (or in other words, A is both an *initial* and a *terminal* object). We often write 0 instead of A.

Let $\{A_i\}_{i \in I}$ be a set of objects A_i of $\mathcal{C}$ indexed by some set I. A *product* $\prod_{i \in I} A_i$ is an object A, together with morphisms $\pi_i \in \operatorname{Hom}_\mathcal{C}(A, A_i)$ for all $i \in I$ satisfying the following universal property. If B is an object of $\mathcal{C}$ and $\psi_i \in \operatorname{Hom}_\mathcal{C}(B, A_i)$ for all $i \in I$, then there is a unique $\theta \in \operatorname{Hom}_\mathcal{C}(B, A)$ such that the following diagram commutes for all $i \in I$:

$$
\begin{array}{ccc}
 & & A \\
 & {\scriptstyle \theta}\nearrow & \big\downarrow {\scriptstyle \pi_i} \\
B & \xrightarrow[\psi_i]{} & A_i
\end{array}
$$

A *coproduct* $\coprod_{i \in I} A_i$ is an object A together with morphisms $\iota_i \in \operatorname{Hom}_\mathcal{C}(A_i, A)$ satisfying: if B is an object of $\mathcal{C}$ and $\phi_i \in \operatorname{Hom}_\mathcal{C}(A_i, B)$ for all $i \in I$, then there is a unique $\tau \in \operatorname{Hom}_\mathcal{C}(A, B)$ such that the following diagram commutes for all $i \in I$:

$$
\begin{array}{ccc}
 & A & \\
{\scriptstyle \iota_i}\big\uparrow & & \searrow {\scriptstyle \tau} \\
A_i & \xrightarrow[\phi_i]{} & B
\end{array}
$$

For categories of modules, product is direct product and coproduct is direct sum.

A category $\mathcal{C}$ is *additive* if $\operatorname{Hom}_\mathcal{C}(A, B)$ is an abelian group for every object A, B in $\mathcal{C}$, composition of morphisms is $\mathbb{Z}$-bilinear, and $\mathcal{C}$ has a zero object, finite products, and coproducts.

Let $\mathcal{C}$ be an additive category and $f \in \mathrm{Hom}_{\mathcal{C}}(A, B)$ for objects A, B of $\mathcal{C}$. A *kernel* of f is an object K in $\mathcal{C}$ and a morphism $j \in \mathrm{Hom}_{\mathcal{C}}(K, A)$ such that $fj = 0$, and whenever C is an object and $g \in \mathrm{Hom}_{\mathcal{C}}(C, A)$ satisfies $fg = 0$, there is a unique $\bar{g} \in \mathrm{Hom}_{\mathcal{C}}(C, K)$ such that $j\bar{g} = g$. That is, the following diagram commutes:

$$
\begin{array}{ccc}
K & \xrightarrow{\ \ j\ \ } A & \xrightarrow{\ \ f\ \ } B \\
 & \nwarrow \quad \uparrow g & \\
 & \bar{g} \quad & \\
 & C &
\end{array}
$$

A *cokernel* of f is an object D in $\mathcal{C}$ and a morphism $p \in \mathrm{Hom}_{\mathcal{C}}(B, D)$ such that $pf = 0$, and whenever C is an object and $g \in \mathrm{Hom}_{\mathcal{C}}(B, C)$ satisfies $gf = 0$, there is a unique $\bar{g} \in \mathrm{Hom}_{\mathcal{C}}(D, C)$ such that $\bar{g}p = g$. That is, the following diagram commutes:

$$
\begin{array}{ccc}
A & \xrightarrow{\ \ f\ \ } B & \xrightarrow{\ \ p\ \ } D \\
 & g \downarrow \quad \swarrow \bar{g} & \\
 & C &
\end{array}
$$

Let $\mathcal{C}$ be a category. Let A, B be objects of $\mathcal{C}$ and $f \in \mathrm{Hom}_{\mathcal{C}}(A, B)$. Then f is a *monomorphism* if whenever C is an object of $\mathcal{C}$ and $g, h \in \mathrm{Hom}_{\mathcal{C}}(C, A)$, if $fg = fh$ then $g = h$. The morphism f is an *epimorphism* if whenever C is an object of $\mathcal{C}$ and $g, h \in \mathrm{Hom}_{\mathcal{C}}(B, C)$, if $gf = hg$ then $g = h$.

A category $\mathcal{C}$ is *abelian* if it is additive, every morphism has both a kernel and a cokernel, every monomorphism is a kernel, and every epimorphism is a cokernel. Categories of R-modules for a ring R are abelian.

Let $\mathcal{C}$ be an abelian category. A sequence $A \xrightarrow{\ f\ } B \xrightarrow{\ g\ } C$ of morphisms f, g in $\mathcal{C}$ is *exact* if f is a kernel of g, and g is a cokernel of f. Projective and injective objects of $\mathcal{C}$ can be defined via diagrams (A.2.1) and (A.2.2), as well as projective and injective resolutions (which do not always exist in general). Many of the standard homological constructions and properties of the previous sections make sense in any abelian category.

Let $\mathcal{C}, \mathcal{D}$ be abelian categories. A covariant functor $F : \mathcal{C} \to \mathcal{D}$ is *left exact* (respectively, *right exact*) if for every exact sequence $0 \to A \to B \to C$ (respectively, $A \to B \to C \to 0$), the sequence

$$0 \to F(A) \to F(B) \to F(C)$$

is exact (respectively, $F(A) \to F(B) \to F(C) \to 0$ is exact). For example, $\mathrm{Hom}_R(D, -)$ is left exact. A contravariant functor $F : \mathcal{C} \to \mathcal{D}$ is *left exact* (respectively, *right exact*) if for every exact sequence $A \to B \to C \to 0$

(respectively, $0 \to A \to B \to C$), the sequence

$$0 \to F(C) \to F(B) \to F(A)$$

is exact (respectively, $F(C) \to F(B) \to F(A) \to 0$ is exact). For example, $\mathrm{Hom}_R(-, D)$ is left exact. In either case, F is *exact* if it is both left and right exact.

Let $\mathcal{C}, \mathcal{D}$ be abelian categories and $F : \mathcal{C} \to \mathcal{D}$ an additive (covariant) functor. Assume $\mathcal{C}$ *has enough projectives*, that is, assume that for each object A of $\mathcal{C}$, there is an epimorphism from a projective object in $\mathcal{C}$ to A. For each object A in $\mathcal{C}$ choose a projective resolution $P_{\bullet}$ of A, which exists since $\mathcal{C}$ has enough projectives. Apply the functor F:

$$F(P_{\bullet}) : \qquad \cdots \longrightarrow F(P_2) \xrightarrow{F(d_2)} F(P_1) \xrightarrow{F(d_1)} F(P_0) \longrightarrow 0.$$

Then $F(P_{\bullet})$ is a complex and we define the *left derived functor* of F to be $L_{\bullet}F$, where $L_n F(A) = \mathrm{H}_n(F(P_{\bullet}))$. Note that if F is right exact, then $L_0 F(A) \cong F(A)$. (The adjective "left" here indicates the objects are on the left with 0 at the end.) A typical example is: let R be a ring, $\mathcal{C} = \mathrm{Mod}\text{-}R$, $\mathcal{D} = \mathbb{Z}\text{-Mod}$, B an object of R-Mod, and F the functor $- \otimes_R B$; then

$$L_n F(A) = \mathrm{Tor}_n^R(A, B).$$

Assume $\mathcal{D}$ *has enough injectives*, that is, assume that for each object A of $\mathcal{C}$, there is a monomorphism from A to an injective object. For each object B in $\mathcal{C}$ choose an injective resolution $I_{\bullet}$ of B. Apply the functor F:

$$F(I_{\bullet}) : \qquad 0 \longrightarrow F(I_0) \longrightarrow F(I_1) \longrightarrow F(I_2) \longrightarrow \cdots .$$

Then $F(I_{\bullet})$ is a complex and we define the *right derived functor* of F to be $R^{\bullet}F$, where $R^n F(B) = \mathrm{H}^n(F(I_{\bullet}))$. Note that if F is left exact, then $R^0 F(B) \cong F(B)$. (The adjective "right" here indicates the objects are on the right with 0 at the beginning.) A typical example is: let R be a ring, $\mathcal{C} = R$-Mod, $\mathcal{D} = \mathbb{Z}$-Mod, A an object of $\mathcal{C}$, and $F(B) = \mathrm{Hom}_R(A, B)$; then

$$R^n F(B) = \mathrm{Ext}_R^n(A, B).$$

Similarly one defines derived functors of contravariant functors. For a right derived functor, use a projective resolution, and for a left derived functor, use an injective resolution.

A.7. Spectral sequences

We will define cohomology spectral sequences here; homology spectral sequences are similar but with arrows reversed.

Definition A.7.1. A *cohomology spectral sequence* in an abelian category $\mathcal{C}$ is a set $\{E_r^{pq} \mid p, q, r \in \mathbb{Z},\ r \geq 0\}$ of objects in $\mathcal{C}$, together with morphisms

$$d_r^{pq} : E_r^{pq} \to E_r^{p+r,q-r+1}$$

for which $d_r^2 = 0$ and $E_{r+1} \cong \mathrm{H}^{\bullet}(E_r)$, that is, for all p, q, r,

$$E_{r+1}^{p,q} \cong \mathrm{Ker}(d_r^{pq})/\mathrm{Im}(d_r^{p-r,q+r-1}).$$

For each r, the set E_r of objects E_r^{pq} together with the morphisms d_r^{pq} is the *rth page* of the spectral sequence. A page can be visualized in a plane, such as pages E_0, E_1, E_2 below.

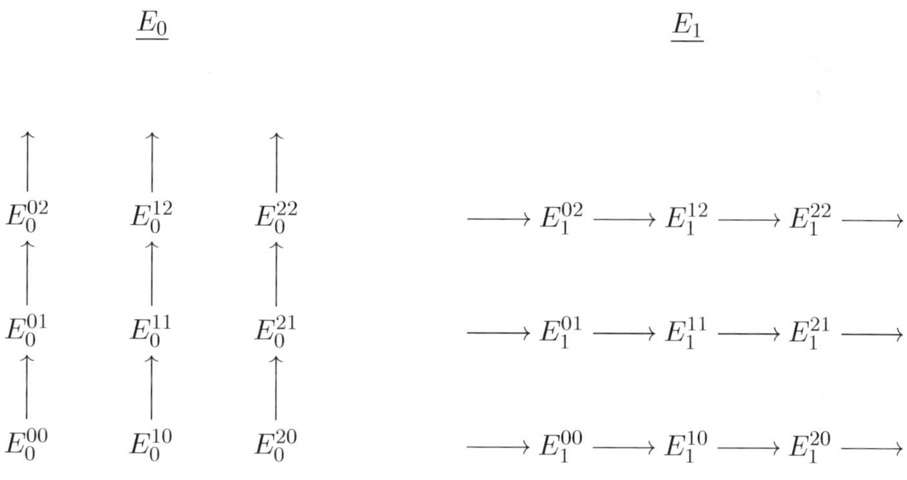

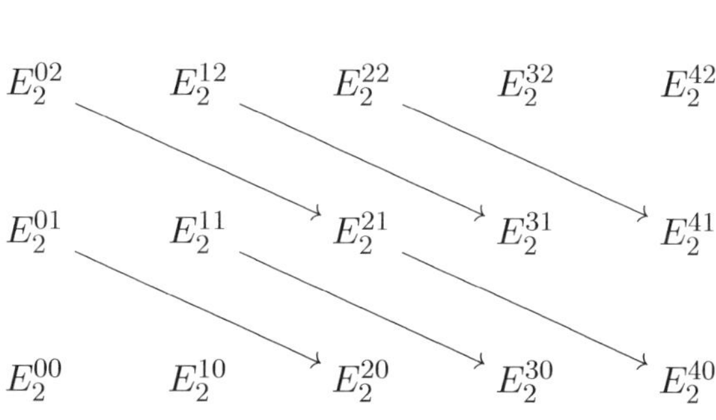

Definition A.7.2. A spectral sequence (E, d) is *bounded* if for each $n \in \mathbb{Z}$ there are finitely many nonzero E_0^{pq} with $p + q = n$.

By its definition, in case (E, d) is bounded, for each pair p, q, there is an r_0 (depending on p, q) such that $E_r^{pq} \cong E_{r_0}^{pq}$ for all $r \geq r_0$. In this case, we write

$$E_\infty^{pq} = E_{r_0}^{pq}.$$

Definition A.7.3. A bounded spectral sequence (E, d) *converges* if there is a family $\{H^n \mid n \in \mathbb{Z}\}$ of objects of $\mathcal{C}$, each having a finite filtration

$$0 = F^t H^n \subseteq \cdots \subseteq F^{p+1} H^n \subseteq F^p H^n \subseteq F^{p-1} H^n \subseteq \cdots \subseteq F^s H^n = H^n,$$

and isomorphisms $E_\infty^{pq} \cong F^p H^{p+q} / F^{p+1} H^{p+q}$ for all p, q.

In this book we use some spectral sequences associated to double complexes. We will define these spectral sequences next. In this context, common notation for a double complex is $B = (B^{pq}, d', d'')$, where d' and d'' are the horizontal and vertical differentials. For each n, write $B^n = \mathrm{Tot}(B)_n = \bigoplus_{p+q=n} B^{p,q}$, a complex with $d = d' + d''$. The notation B will then sometimes refer to this complex, when no confusion will arise.

For each p, n, let

$$(A.7.4) \qquad\qquad F^p B^n = \bigoplus_{p' \geq p} B^{p', n-p'}.$$

That is, we truncate the double complex at the pth column, replacing all objects to the left of this column by 0, and then sum over diagonal lines whose indices sum to a fixed value n.

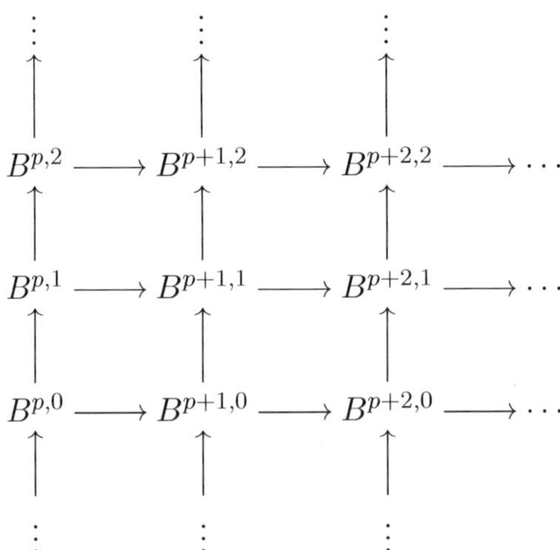

If B is bounded, then for each n, this yields a finite filtration of B^n. If B is a first quadrant double complex (that is, $B^{pq} = 0$ whenever $p < 0$ or $q < 0$), then $F^0 B^n = B^n$ and $F^p B^n = 0$ for all $p > n$.

For each p, q, r, let

$$C_r^{pq} = \{x \in F^p B^{p+q} \mid d(x) \in F^{p+r} B^{p+q+1}\}.$$

In particular, $C_0^{pq} = F^p B^{p+q}$ by definition. Also by definition, if $x \in C_r^{pq}$, then $d(x)$ has component 0 within the band $p \leq p' \leq p + r$, in order that $d(x)$ be in $F^{p+r} B^{p+q+1}$ as specified. Let $E_0^{pq} = C_0^{pq}$ and

$$(A.7.5) \qquad E_r^{pq} = \frac{C_r^{pq} + F^{p+1} B^{p+q}}{d(C_{r-1}^{p-r+1,q+r-2}) + F^{p+1} B^{p+q}}$$

for each $r > 0$ and $p, q \in \mathbb{Z}$. By the definitions there are induced morphisms

$$d_r^{pq} : E_r^{pq} \to E_r^{p+r,q-r+1}$$

for which $d_r^2 = 0$. By the definitions,

$$\mathrm{H}^*(E_r) \cong E_{r+1}.$$

Note that $E_1^{p,q}$ is often written $\mathrm{H}''(B)^{p,q}$, that is, the cohomology of B with vertical differentials only, and E_2^{pq} is often written $\mathrm{H}' \mathrm{H}''(B)^{p,q}$, the cohomology of $\mathrm{H}''(B)^{pq}$ with respect to the differential induced by horizontal differentials on B only.

We have assumed B is bounded, and so for each p, q, there is an r_0 for which the differentials $d_{r_0}^{pq}$ as well as $d_{r_0}^{p-r_0,q+r_0-1}$ (that is, those starting and ending at position p, q) are zero maps. So $E_\infty^{pq} = E_{r_0}^{pq}$.

Let $p, q \in \mathbb{Z}$, $n = p + q$, and let $x \in B^{p,q}$ be a cocycle such that $x \notin F^{p+1} B^n$. Then x determines an element of E_r^{pq} for all $r \geq 1$ and d_r is 0 on the corresponding element of E_∞^{pq}. This describes a morphism from $F^p \mathrm{H}^{p+q}(B)$, which is the image of $\mathrm{H}^{p+q}(F^p B)$ in $\mathrm{H}^{p+q}(B)$, to E_∞^{pq}. Moreover, this is an epimorphism since $E_\infty^{pq} = E_{r_0}^{pq}$ for some r_0. The kernel of the epimorphism is $F^{p+1} \mathrm{H}^{p+q}(B)$ by the definitions. As a consequence, $\mathrm{H}^*(B)$ is filtered with filtration given by $F^p \mathrm{H}^*(B)$ and

$$F^p \mathrm{H}^n(B) / F^{p+1} \mathrm{H}^n(B) \cong E_\infty^{pq}(B)$$

for fixed $p + q = n$. That is, E_r converges to $\mathrm{H}^*(B)$.

Theorem A.7.6. *Let B be a bounded bicomplex. Give B the filtration (A.7.4) and let E be the corresponding spectral sequence given by (A.7.5). Then E_r converges to $\mathrm{H}^*(B)$.*

The statement that E_r converges to $\mathrm{H}^*(B)$ is often written

$$E_r \implies \mathrm{H}^*(B).$$

Note that we could have chosen to filter the complex instead by truncating rows, resulting in another spectral sequence. The E_1 page is then denoted $\mathrm{H}'(B)$, and the E_2 page $\mathrm{H}'' \mathrm{H}'(B)$. Comparison of these two spectral sequences can be useful.

The spectral sequence E is *multiplicative* if E_0 has a bigraded product

$$E_0^{p,q} \times E_0^{p',q'} \to E_0^{p+p',q+q'}$$

satisfying the Leibniz rule:

$$d(xy) = d(x)y + (-1)^{p+q}xd(y)$$

for $x \in E_0^{p,q}, y \in E_0^{p',q'}$. It follows that for all r, E_r has a bigraded product such that the Leibniz rule holds. In the context of Theorem A.7.6, if E is multiplicative, then it converges to the associated graded algebra of $\mathrm{H}^*(B)$.

Bibliography

[1] H. Abbaspour, *On algebraic structures of the Hochschild complex*, Free loop spaces in geometry and topology, IRMA Lect. Math. Theor. Phys., vol. 24, Eur. Math. Soc., Zürich, 2015, pp. 165–222. MR3445567

[2] A. Adem and R. J. Milgram, *Cohomology of finite groups*, Grundlehren der Mathematischen Wissenschaften [Fundamental Principles of Mathematical Sciences], vol. 309, Springer-Verlag, Berlin, 1994. MR1317096

[3] C. Aholt, *Equivalence of the Ext-algebra structures of an R-module*, Bachelor Thesis, 2008.

[4] J. Alev, M. A. Farinati, T. Lambre, and A. L. Solotar, *Homologie des invariants d'une algèbre de Weyl sous l'action d'un groupe fini* (French, with English and French summaries), J. Algebra **232** (2000), no. 2, 564–577, DOI 10.1006/jabr.2000.8406. MR1792746

[5] M. P. Allocca, L_∞-*algebra representation theory*, Ph.D. Thesis, 2010.

[6] J. L. Alperin, *Local representation theory: Modular representations as an introduction to the local representation theory of finite groups*, Cambridge Studies in Advanced Mathematics, vol. 11, Cambridge University Press, Cambridge, 1986. MR860771

[7] C. Amiot, *Preprojective algebras and Calabi-Yau duality*, Mathematisches Forschungsinstitut Oberwolfach, 2014. Oberwolfach report on joint work with O. Iyama, S. Oppermann, and I. Reiten.

[8] D. J. Anick, *On the homology of associative algebras*, Trans. Amer. Math. Soc. **296** (1986), no. 2, 641–659, DOI 10.2307/2000383. MR846601

[9] M. A. Armenta and B. Keller, *Derived invariance of the cap product in Hochschild theory* (English, with English and French summaries), C. R. Math. Acad. Sci. Paris **355** (2017), no. 12, 1205–1207, DOI 10.1016/j.crma.2017.11.005. MR3730498

[10] I. Assem, D. Simson, and A. Skowroński, *Elements of the representation theory of associative algebras 1: Techniques of representation theory*, London Mathematical Society Student Texts, vol. 65, Cambridge University Press, 2006.

[11] M. Auslander, I. Reiten, and S. O. Smalø, *Representation theory of Artin algebras*, Cambridge Studies in Advanced Mathematics, vol. 36, Cambridge University Press, Cambridge, 1995. MR1314422

[12] L. L. Avramov and S. Iyengar, *Finite generation of Hochschild homology algebras*, Invent. Math. **140** (2000), no. 1, 143–170, DOI 10.1007/s002220000051. MR1779800

[13] L. L. Avramov and S. Iyengar, *Gaps in Hochschild cohomology imply smooth-ness for commutative algebras*, Math. Res. Lett. **12** (2005), no. 5-6, 789–804, DOI 10.4310/MRL.2005.v12.n6.a1. MR2189239

[14] L. L. Avramov and M. Vigué-Poirrier, *Hochschild homology criteria for smoothness*, Internat. Math. Res. Notices **1** (1992), 17–25, DOI 10.1155/S1073792892000035. MR1149001

[15] BACH, *A Hochschild homology criterium for the smoothness of an algebra*, Comment. Math. Helv. **69** (1994), no. 2, 163–168. MR1282365

[16] B. Bakalov and A. Kirillov Jr., *Lectures on tensor categories and modular functors*, University Lecture Series, vol. 21, American Mathematical Society, Providence, RI, 2001. MR1797619

[17] M. J. Bardzell, A. C. Locateli, and E. N. Marcos, *On the Hochschild cohomology of truncated cycle algebras*, Comm. Algebra **28** (2000), no. 3, 1615–1639, DOI 10.1080/00927870008826917. MR1742678

[18] M. J. Bardzell, *The alternating syzygy behavior of monomial algebras*, J. Algebra **188** (1997), no. 1, 69–89, DOI 10.1006/jabr.1996.6813. MR1432347

[19] M. Barr, *Harrison homology, Hochschild homology and triples*, J. Algebra **8** (1968), 314–323, DOI 10.1016/0021-8693(68)90062-8. MR220799

[20] C. P. Bendel, D. K. Nakano, B. J. Parshall, and C. Pillen, *Cohomology for quantum groups via the geometry of the nullcone*, Mem. Amer. Math. Soc. **229** (2014), no. 1077, x+93. MR3204911

[21] D. J. Benson, *Representations and cohomology. I: Basic representation theory of finite groups and associative algebras*, Cambridge Studies in Advanced Mathematics, vol. 30, Cambridge University Press, Cambridge, 1991. MR1110581

[22] D. J. Benson, *Representations and cohomology. II: Cohomology of groups and modules*, Cambridge Studies in Advanced Mathematics, vol. 31, Cambridge University Press, Cambridge, 1991. MR1156302

[23] D. J. Benson, J. F. Carlson, and J. Rickard, *Thick subcategories of the stable module category*, Fund. Math. **153** (1997), no. 1, 59–80. MR1450996

[24] D. J. Benson and K. Erdmann, *Hochschild cohomology of Hecke algebras*, J. Algebra **336** (2011), 391–394, DOI 10.1016/j.jalgebra.2011.03.022. MR2802551

[25] D. Benson and S. Witherspoon, *Examples of support varieties for Hopf algebras with noncommutative tensor products*, Arch. Math. (Basel) **102** (2014), no. 6, 513–520, DOI 10.1007/s00013-014-0659-8. MR3227473

[26] R. Berger, *Dimension de Hochschild des algèbres graduées* (French, with English and French summaries), C. R. Math. Acad. Sci. Paris **341** (2005), no. 10, 597–600, DOI 10.1016/j.crma.2005.09.039. MR2179797

[27] F. Bergeron and N. Bergeron, *Orthogonal idempotents in the descent algebra of B_n and applications*, J. Pure Appl. Algebra **79** (1992), no. 2, 109–129, DOI 10.1016/0022-4049(92)90153-7. MR1163285

[28] P. A. Bergh and K. Erdmann, *Homology and cohomology of quantum complete intersections*, Algebra Number Theory **2** (2008), no. 5, 501–522, DOI 10.2140/ant.2008.2.501. MR2429451

[29] P. A. Bergh and D. A. Jorgensen, *Tate-Hochschild homology and cohomology of Frobenius algebras*, J. Noncommut. Geom. **7** (2013), no. 4, 907–937, DOI 10.4171/JNCG/139. MR3148613

[30] P. A. Bergh and S. Oppermann, *Cohomology of twisted tensor products*, J. Algebra **320** (2008), no. 8, 3327–3338, DOI 10.1016/j.jalgebra.2008.08.005. MR2450729

[31] P. A. Bergh and Ø. Solberg, *Relative support varieties*, Q. J. Math. **61** (2010), no. 2, 171–182, DOI 10.1093/qmath/han032. MR2646083

[32] M. Bordemann and A. Makhlouf, *Formality and deformations of universal enveloping algebras*, Internat. J. Theoret. Phys. **47** (2008), no. 2, 311–332, DOI 10.1007/s10773-007-9435-x. MR2396767

[33] A. Braverman and D. Gaitsgory, *Poincaré-Birkhoff-Witt theorem for quadratic algebras of Koszul type*, J. Algebra **181** (1996), no. 2, 315–328, DOI 10.1006/jabr.1996.0122. MR1383469

[34] B. Briggs and V. Gélinas, *The A_∞-centre of the Yoneda algebra and the characteristic action of Hochschild cohomology on the derived category*, 2017. arXiv:1702.00721.

[35] K. S. Brown, *Cohomology of groups*, Graduate Texts in Mathematics, vol. 87, Springer-Verlag, New York-Berlin, 1982. MR672956

[36] R.-O. Buchweitz, *Finite representation type and periodic Hochschild (co-)homology*, Trends in the representation theory of finite-dimensional algebras (Seattle, WA, 1997), Contemp. Math., vol. 229, Amer. Math. Soc., Providence, RI, 1998, pp. 81–109, DOI 10.1090/conm/229/03311. MR1676212

[37] R.-O. Buchweitz, *Morita contexts, idempotents, and Hochschild cohomology—with applications to invariant rings*, Commutative algebra (Grenoble/Lyon, 2001), Contemp. Math., vol. 331, Amer. Math. Soc., Providence, RI, 2003, pp. 25–53, DOI 10.1090/conm/331/05901. MR2011764

[38] R.-O. Buchweitz, E. L. Green, D. Madsen, and Ø. Solberg, *Finite Hochschild cohomology without finite global dimension*, Math. Res. Lett. **12** (2005), no. 5-6, 805–816, DOI 10.4310/MRL.2005.v12.n6.a2. MR2189240

[39] R.-O. Buchweitz, E. L. Green, N. Snashall, and Ø. Solberg, *Multiplicative structures for Koszul algebras*, Q. J. Math. **59** (2008), no. 4, 441–454, DOI 10.1093/qmath/ham056. MR2461267

[40] R.-O. Buchweitz and S. Liu, *Hochschild cohomology and representation-finite algebras*, Proc. London Math. Soc. (3) **88** (2004), no. 2, 355–380, DOI 10.1112/S0024611503014394. MR2032511

[41] D. Burghelea, *The cyclic homology of the group rings*, Comment. Math. Helv. **60** (1985), no. 3, 354–365, DOI 10.1007/BF02567420. MR814144

[42] M. C. R. Butler and A. D. King, *Minimal resolutions of algebras*, J. Algebra **212** (1999), no. 1, 323–362, DOI 10.1006/jabr.1998.7599. MR1670674

[43] A. Čap, H. Schichl, and J. Vanžura, *On twisted tensor products of algebras*, Comm. Algebra **23** (1995), no. 12, 4701–4735, DOI 10.1080/00927879508825496. MR1352565

[44] J. F. Carlson, *Periodic modules over modular group algebras*, J. London Math. Soc. (2) **15** (1977), no. 3, 431–436, DOI 10.1112/jlms/s2-15.3.431. MR0472985

[45] J. F. Carlson, *The varieties and the cohomology ring of a module*, J. Algebra **85** (1983), no. 1, 104–143, DOI 10.1016/0021-8693(83)90121-7. MR723070

[46] J. F. Carlson, *The variety of an indecomposable module is connected*, Invent. Math. **77** (1984), no. 2, 291–299, DOI 10.1007/BF01388448. MR752822

[47] J. F. Carlson, L. Townsley, L. Valeri-Elizondo, and M. Zhang, *Cohomology rings of finite groups*, Algebra and Applications, vol. 3, Kluwer Academic Publishers, Dordrecht, 2003. With an appendix: Calculations of cohomology rings of groups of order dividing 64 by Carlson, Valeri-Elizondo and Zhang. MR2028960

[48] H. Cartan and S. Eilenberg, *Homological algebra*, Princeton University Press, Princeton, N. J., 1956. MR0077480

[49] S. Chouhy and A. Solotar, *Projective resolutions of associative algebras and ambiguities*, J. Algebra **432** (2015), 22–61, DOI 10.1016/j.jalgebra.2015.02.019. MR3334140

[50] C. Cibils, *On the Hochschild cohomology of finite-dimensional algebras*, Comm. Algebra **16** (1988), no. 3, 645–649, DOI 10.1080/00927878808823591. MR917337

[51] C. Cibils, *Hochschild cohomology algebra of radical square zero algebras*, Algebras and modules, II (Geiranger, 1996), CMS Conf. Proc., vol. 24, Amer. Math. Soc., Providence, RI, 1998, pp. 93–101. MR1648617

[52] C. Cibils, E. Marcos, M. J. Redondo, and A. Solotar, *Cohomology of split algebras and of trivial extensions*, Glasg. Math. J. **45** (2003), no. 1, 21–40, DOI 10.1017/S0017089502008959. MR1972690

[53] C. Cibils and A. Solotar, *Hochschild cohomology algebra of abelian groups*, Arch. Math. (Basel) **68** (1997), no. 1, 17–21, DOI 10.1007/PL00000389. MR1421841

[54] A. B. Conner, *A(infinity)-structures, generalized Koszul properties, and combinatorial topology*, ProQuest LLC, Ann Arbor, MI, 2011. Thesis (Ph.D.)–University of Oregon. MR2942011

[55] J. Cuntz and D. Quillen, *Algebra extensions and nonsingularity*, J. Amer. Math. Soc. **8** (1995), no. 2, 251–289, DOI 10.2307/2152819. MR1303029

[56] R. K. Dennis and K. Igusa, *Hochschild homology and the second obstruction for pseudoisotopy*, Algebraic K-theory, Part I (Oberwolfach, 1980), Lecture Notes in Math., vol. 966, Springer, Berlin-New York, 1982, pp. 7–58. MR689365

[57] Y. Doi, *Homological coalgebra*, J. Math. Soc. Japan **33** (1981), no. 1, 31–50.

[58] V. Dolgushev, D. Tamarkin, and B. Tsygan, *The homotopy Gerstenhaber algebra of Hochschild cochains of a regular algebra is formal*, J. Noncommut. Geom. **1** (2007), no. 1, 1–25, DOI 10.4171/JNCG/1. MR2294189

[59] S. Eilenberg and S. MacLane, *Group extensions and homology*, Ann. of Math. (2) **43** (1942), 757–831, DOI 10.2307/1968966. MR0007108

[60] S. Eilenberg and S. Mac Lane, *Cohomology theory in groups*, Bull. Amer. Math. Soc. **50** (1944), no. 1, 53.

[61] S. Eilenberg and S. MacLane, *Cohomology theory in abstract groups. I*, Ann. of Math. (2) **48** (1947), 51–78, DOI 10.2307/1969215. MR0019092

[62] D. Eisenbud, *Homological algebra on a complete intersection, with an application to group representations*, Trans. Amer. Math. Soc. **260** (1980), no. 1, 35–64, DOI 10.2307/1999875. MR570778

[63] K. Erdmann, *Nilpotent elements in Hochschild cohomology*, Geometric and topological aspects of the representation theory of finite groups, Springer Proc. Math. Stat., vol. 242, Springer, Cham, 2018, pp. 51–66. MR3901156

[64] K. Erdmann and M. Hellstrøm-Finnsen, *Hochschild cohomology of some quantum complete intersections*, J. Algebra Appl. **17** (2018), no. 11, 1850215, 22, DOI 10.1142/S0219498818502158. MR3879091

[65] K. Erdmann, M. Holloway, R. Taillefer, N. Snashall, and Ø. Solberg, *Support varieties for selfinjective algebras*, K-Theory **33** (2004), no. 1, 67–87, DOI 10.1007/s10977-004-0838-7. MR2199789

[66] K. Erdmann and T. Holm, *Twisted bimodules and Hochschild cohomology for self-injective algebras of class A_n*, Forum Math. **11** (1999), no. 2, 177–201, DOI 10.1515/form.1999.002. MR1680594

[67] K. Erdmann, T. Holm, and N. Snashall, *Twisted bimodules and Hochschild cohomology for self-injective algebras of class A_n. II*, Algebr. Represent. Theory **5** (2002), no. 5, 457–482, DOI 10.1023/A:1020551906728. MR1935856

[68] K. Erdmann and N. Snashall, *On Hochschild cohomology of preprojective algebras. I, II*, J. Algebra **205** (1998), no. 2, 391–412, 413–434, DOI 10.1006/jabr.1998.7547. MR1632808

[69] K. Erdmann and N. Snashall, *Preprojective algebras of Dynkin type, periodicity and the second Hochschild cohomology*, Algebras and modules, II (Geiranger, 1996), CMS Conf. Proc., vol. 24, Amer. Math. Soc., Providence, RI, 1998, pp. 183–193. MR1648626

[70] K. Erdmann and Ø. Solberg, *Radical cube zero selfinjective algebras of finite complexity*, J. Pure Appl. Algebra **215** (2011), no. 7, 1747–1768, DOI 10.1016/j.jpaa.2010.10.010. MR2771643

[71] P. Etingof, S. Gelaki, D. Nikshych, and V. Ostrik, *Tensor categories*, Mathematical Surveys and Monographs, vol. 205, American Mathematical Society, Providence, RI, 2015. MR3242743

[72] P. Etingof and D. Kazhdan, *Quantization of Lie bialgebras. I*, Selecta Math. (N.S.) **2** (1996), no. 1, 1–41, DOI 10.1007/BF01587938. MR1403351

[73] P. Etingof and V. Ostrik, *Finite tensor categories*, Mosc. Math. J. **4** (2004), no. 3, 627–654, 782–783.

[74] C.-H. Eu and T. Schedler, *Calabi-Yau Frobenius algebras*, J. Algebra **321** (2009), no. 3, 774–815, DOI 10.1016/j.jalgebra.2008.11.003. MR2488552

[75] L. Evens, *The cohomology ring of a finite group*, Trans. Amer. Math. Soc. **101** (1961), 224–239, DOI 10.2307/1993372. MR137742

[76] L. Evens, *The cohomology of groups*, Oxford Mathematical Monographs, The Clarendon Press, Oxford University Press, New York, 1991. Oxford Science Publications. MR1144017

[77] M. Farinati, *Hochschild duality, localization, and smash products*, J. Algebra **284** (2005), no. 1, 415–434, DOI 10.1016/j.jalgebra.2004.09.009. MR2115022

[78] J. Feldvoss and S. Witherspoon, *Support varieties and representation type of self-injective algebras*, Homology Homotopy Appl. **13** (2011), no. 2, 197–215, DOI 10.4310/HHA.2011.v13.n2.a13. MR2854335

[79] E. M. Friedlander and A. Suslin, *Cohomology of finite group schemes over a field*, Invent. Math. **127** (1997), no. 2, 209–270, DOI 10.1007/s002220050119. MR1427618

[80] A. M. Garsia, *Combinatorics of the free Lie algebra and the symmetric group*, Analysis, et cetera, Academic Press, Boston, MA, 1990, pp. 309–382. MR1039352

[81] E. Gawell and Q. R. Xantcha, *Centers of partly (anti-)commutative quiver algebras and finite generation of the Hochschild cohomology ring*, Manuscripta Math. **150** (2016), no. 3-4, 383–406, DOI 10.1007/s00229-015-0816-9. MR3514736

[82] M. Gerstenhaber, *The cohomology structure of an associative ring*, Ann. of Math. (2) **78** (1963), 267–288, DOI 10.2307/1970343. MR0161898

[83] M. Gerstenhaber and S. D. Schack, *Relative Hochschild cohomology, rigid algebras, and the Bockstein*, J. Pure Appl. Algebra **43** (1986), no. 1, 53–74, DOI 10.1016/0022-4049(86)90004-6. MR862872

[84] M. Gerstenhaber and S. D. Schack, *A Hodge-type decomposition for commutative algebra cohomology*, J. Pure Appl. Algebra **48** (1987), no. 3, 229–247, DOI 10.1016/0022-4049(87)90112-5. MR917209

[85] M. Gerstenhaber and S. D. Schack, *Algebraic cohomology and deformation theory*, Deformation theory of algebras and structures and applications (Il Ciocco, 1986), NATO Adv. Sci. Inst. Ser. C Math. Phys. Sci., vol. 247, Kluwer Acad. Publ., Dordrecht, 1988, pp. 11–264, DOI 10.1007/978-94-009-3057-5_2. MR981619

[86] M. Gerstenhaber and S. D. Schack, *The cohomology of presheaves of algebras. I. Presheaves over a partially ordered set*, Trans. Amer. Math. Soc. **310** (1988), no. 1, 135–165, DOI 10.2307/2001114. MR965749

[87] A. Giaquinto, *Topics in algebraic deformation theory*, Higher structures in geometry and physics, Progr. Math., vol. 287, Birkhäuser/Springer, New York, 2011, pp. 1–24, DOI 10.1007/978-0-8176-4735-3_1. MR2762537

[88] V. Ginzburg, *Lectures on noncommutative geometry*, 2005. arXiv:0506603.

[89] V. Ginzburg, *Calabi-Yau algebras*, 2006. arXiv:0612139.

[90] V. Ginzburg and D. Kaledin, *Poisson deformations of symplectic quotient singularities*, Adv. Math. **186** (2004), no. 1, 1–57, DOI 10.1016/j.aim.2003.07.006. MR2065506

[91] V. Ginzburg and S. Kumar, *Cohomology of quantum groups at roots of unity*, Duke Math. J. **69** (1993), no. 1, 179–198, DOI 10.1215/S0012-7094-93-06909-8. MR1201697

[92] E. Golod, *The cohomology ring of a finite p-group* (Russian), Dokl. Akad. Nauk SSSR **125** (1959), 703–706. MR0104720

[93] K. R. Goodearl and R. B. Warfield Jr., *An introduction to noncommutative Noetherian rings*, London Mathematical Society Student Texts, vol. 16, Cambridge University Press, Cambridge, 1989. MR1020298

[94] J. Goodman and U. Krähmer, *Untwisting a twisted Calabi-Yau algebra*, J. Algebra **406** (2014), 272–289, DOI 10.1016/j.jalgebra.2014.02.018. MR3188338

[95] N. S. Gopalakrishnan and R. Sridharan, *Homological dimension of Ore-extensions*, Pacific J. Math. **19** (1966), 67–75. MR200324

[96] I. G. Gordon, *Cohomology of quantized function algebras at roots of unity*, Proc. London Math. Soc. (3) **80** (2000), no. 2, 337–359, DOI 10.1112/S002461150001217X. MR1734320

[97] E. L. Green and N. Snashall, *The Hochschild cohomology ring modulo nilpotence of a stacked monomial algebra*, Colloq. Math. **105** (2006), no. 2, 233–258, DOI 10.4064/cm105-2-6. MR2237910

[98] E. L. Green, N. Snashall, and Ø. Solberg, *The Hochschild cohomology ring of a selfinjective algebra of finite representation type*, Proc. Amer. Math. Soc. **131** (2003), no. 11, 3387–3393, DOI 10.1090/S0002-9939-03-06912-0. MR1990627

[99] E. L. Green, N. Snashall, and Ø. Solberg, *The Hochschild cohomology ring modulo nilpotence of a monomial algebra*, J. Algebra Appl. **5** (2006), no. 2, 153–192, DOI 10.1142/S0219498806001648. MR2223465

[100] E. L. Green and Ø. Solberg, *An algorithmic approach to resolutions*, J. Symbolic Comput. **42** (2007), no. 11-12, 1012–1033, DOI 10.1016/j.jsc.2007.05.002. MR2368070

[101] L. Grimley, *Brackets on Hochschild cohomology of noncommutative algebras*, Ph.D. Thesis, 2016.

[102] L. Grimley, V. C. Nguyen, and S. Witherspoon, *Gerstenhaber brackets on Hochschild cohomology of twisted tensor products*, J. Noncommut. Geom. **11** (2017), no. 4, 1351–1379, DOI 10.4171/JNCG/11-4-4. MR3743225

[103] J. A. Guccione and J. J. Guccione, *Hochschild homology of twisted tensor products*, K-Theory **18** (1999), no. 4, 363–400, DOI 10.1023/A:1007890230081. MR1738899

[104] G. Halbout, *Globalization of Tamarkin's formality theorem*, Lett. Math. Phys. **71** (2005), no. 1, 39–48, DOI 10.1007/s11005-004-5926-3. MR2136736

[105] D. Happel, *Hochschild cohomology of finite-dimensional algebras*, Séminaire d'Algèbre Paul Dubreil et Marie-Paul Malliavin, 39ème Année (Paris, 1987/1988), Lecture Notes in Math., vol. 1404, Springer, Berlin, 1989, pp. 108–126, DOI 10.1007/BFb0084073. MR1035222

[106] T. J. Harding and F. J. Bloore, *Isomorphism of de Rham cohomology and relative Hochschild cohomology of differential operators*, Ann. Inst. H. Poincaré Phys. Théor. **58** (1993), no. 4, 433–452. MR1241705

[107] D. K. Harrison, *Commutative algebras and cohomology*, Trans. Amer. Math. Soc. **104** (1962), 191–204, DOI 10.2307/1993575. MR142607

[108] R. Hermann, *Exact sequences, Hochschild cohomology, and the Lie module structure over the M-relative center*, J. Algebra **454** (2016), 29–69, DOI 10.1016/j.jalgebra.2016.01.021. MR3473419

[109] R. Hermann, *Homological epimorphisms, recollements and Hochschild cohomology—with a conjecture by Snashall-Solberg in view*, Adv. Math. **299** (2016), 687–759, DOI 10.1016/j.aim.2016.05.022. MR3519480

[110] R. Hermann, *Monoidal categories and the Gerstenhaber bracket in Hochschild cohomology*, Mem. Amer. Math. Soc. **243** (2016), no. 1151, v+146, DOI 10.1090/memo/1151. MR3518219

[111] E. Herscovich, *Using torsion theory to compute the algebraic structure of Hochschild (co)homology*, Homology Homotopy Appl. **20** (2018), no. 1, 117–139, DOI 10.4310/HHA.2018.v20.n1.a8. MR3775352

[112] P. J. Hilton and U. Stammbach, *A course in homological algebra*, Graduate Texts in Mathematics, vol. 4, Springer-Verlag, New York-Berlin, 1971. MR0346025

[113] V. Hinich, *Tamarkin's proof of Kontsevich formality theorem*, Forum Math. **15** (2003), no. 4, 591–614, DOI 10.1515/form.2003.032. MR1978336

[114] G. Hochschild, *On the cohomology groups of an associative algebra*, Ann. of Math. (2) **46** (1945), 58–67, DOI 10.2307/1969145. MR0011076

[115] G. Hochschild, *Relative homological algebra*, Trans. Amer. Math. Soc. **82** (1956), 246–269, DOI 10.2307/1992988. MR80654

[116] G. Hochschild, B. Kostant, and A. Rosenberg, *Differential forms on regular affine algebras*, Trans. Amer. Math. Soc. **102** (1962), 383–408, DOI 10.2307/1993614. MR142598

[117] T. Holm, *The Hochschild cohomology ring of a modular group algebra: the commutative case*, Comm. Algebra **24** (1996), no. 6, 1957–1969, DOI 10.1080/00927879608825682. MR1386022

[118] T. Holm, *Hochschild cohomology rings of algebras $k[X]/(f)$*, Beiträge Algebra Geom. **41** (2000), no. 1, 291–301. MR1745598

[119] J. Huebschmann, *On the construction of A_∞-structures*, Georgian Math. J. **17** (2010), no. 1, 161–202.

[120] J. Huebschmann and J. Stasheff, *Formal solution of the master equation via HPT and deformation theory*, Forum Math. **14** (2002), no. 6, 847–868, DOI 10.1515/form.2002.037. MR1932522

[121] T. W. Hungerford, *Algebra*, Springer-Verlag, 1974.

[122] B. E. Johnson, *Cohomology in Banach algebras*, American Mathematical Society, Providence, R.I., 1972. Memoirs of the American Mathematical Society, No. 127. MR0374934

[123] T. V. Kadeishvili, *The algebraic structure in the homology of an $A(\infty)$-algebra* (Russian, with English and Georgian summaries), Soobshch. Akad. Nauk Gruzin. SSR **108** (1982), no. 2, 249–252 (1983). MR720689

[124] H. Kajiura and J. Stasheff, *Homotopy algebras inspired by classical open-closed string field theory*, Comm. Math. Phys. **263** (2006), no. 3, 553–581, DOI 10.1007/s00220-006-1539-2. MR2211816

[125] C. Kassel, *Quantum groups*, Graduate Texts in Mathematics, vol. 155, Springer-Verlag, New York, 1995. MR1321145

[126] B. Keller, *A-infinity algebras in representation theory*, Representations of algebra. Vol. I, II, Beijing Norm. Univ. Press, Beijing, 2002, pp. 74–86. MR2067371

[127] B. Keller, *Derived invariance of higher structures on the Hochschild complex*, 2003. http://www.math.jussieu.fr/~keller/publ/dih.pdf.

[128] B. Keller, *Hochschild cohomology and derived Picard groups*, J. Pure Appl. Algebra **190** (2004), no. 1-3, 177–196, DOI 10.1016/j.jpaa.2003.10.030. MR2043327

[129] B. Keller, *Deformation quantization after Kontsevich and Tamarkin*, Déformation, quantification, théorie de Lie, Panor. Synthèses, vol. 20, Soc. Math. France, Paris, 2005, pp. 19–62. MR2274224

[130] B. Keller, *On differential graded categories*, International Congress of Mathematicians. Vol. II, Eur. Math. Soc., Zürich, 2006, pp. 151–190. MR2275593

[131] Y. Kobayashi, *Gröbner bases on path algebras and the Hochschild cohomology algebras*, Sci. Math. Jpn. **64** (2006), no. 1, 103–129. MR2242347

[132] M. Kontsevich, *Homological algebra of mirror symmetry*, Proceedings of the International Congress of Mathematicians, Vol. 1, 2 (Zürich, 1994), Birkhäuser, Basel, 1995, pp. 120–139. MR1403918

[133] M. Kontsevich, *Deformation quantization of Poisson manifolds*, Lett. Math. Phys. **66** (2003), no. 3, 157–216, DOI 10.1023/B:MATH.0000027508.00421.bf. MR2062626

[134] M. Kontsevich and Y. Soibelman, *Deformations of algebras over operads and the Deligne conjecture*, Conférence Moshé Flato 1999, Vol. I (Dijon), Math. Phys. Stud., vol. 21, Kluwer Acad. Publ., Dordrecht, 2000, pp. 255–307. MR1805894

[135] N. Kowalzig and U. Krähmer, *Batalin-Vilkovisky structures on* Ext *and* Tor, J. Reine Angew. Math. **697** (2014), 159–219.

[136] U. Krähmer, *Poincaré duality in Hochschild (co)homology*, New techniques in Hopf algebras and graded ring theory, Contactforum, 2007, pp. 117–125.

[137] U. Krähmer, *Notes on Koszul algebras*, 2011. Semantic Scholar pdf, semanticscholar.org.

[138] T. Lada and M. Markl, *Strongly homotopy Lie algebras*, Comm. Algebra **23** (1995), no. 6, 2147–2161, DOI 10.1080/00927879508825335. MR1327129

[139] T. Lambre, G. Zhou, and A. Zimmermann, *The Hochschild cohomology ring of a Frobenius algebra with semisimple Nakayama automorphism is a Batalin-Vilkovisky algebra*, J. Algebra **446** (2016), 103–131, DOI 10.1016/j.jalgebra.2015.09.018. MR3421088

[140] J. Le and G. Zhou, *On the Hochschild cohomology ring of tensor products of algebras*, J. Pure Appl. Algebra **218** (2014), no. 8, 1463–1477, DOI 10.1016/j.jpaa.2013.11.029. MR3175033

[141] M. Linckelmann, *On the Hochschild cohomology of commutative Hopf algebras*, Arch. Math. (Basel) **75** (2000), no. 6, 410–412, DOI 10.1007/s000130050523. MR1799425

[142] M. Linckelmann, *Finite generation of Hochschild cohomology of Hecke algebras of finite classical type in characteristic zero*, Bull. Lond. Math. Soc. **43** (2011), no. 5, 871–885, DOI 10.1112/blms/bdr024. MR2854558

[143] Y. Liu and G. Zhou, *The Batalin-Vilkovisky structure over the Hochschild cohomology ring of a group algebra*, J. Noncommut. Geom. **10** (2016), no. 3, 811–858, DOI 10.4171/JNCG/249. MR3554837

[144] A. C. Locateli, *Hochschild cohomology of truncated quiver algebras*, Comm. Algebra **27** (1999), no. 2, 645–664, DOI 10.1080/00927879908826454. MR1671958

[145] J.-L. Loday, *Opérations sur l'homologie cyclique des algèbres commutatives* (French), Invent. Math. **96** (1989), no. 1, 205–230, DOI 10.1007/BF01393976. MR981743

[146] J.-L. Loday, *Cyclic homology*, 2nd ed., Grundlehren der Mathematischen Wissenschaften [Fundamental Principles of Mathematical Sciences], vol. 301, Springer-Verlag, Berlin, 1998. Appendix E by María O. Ronco; Chapter 13 by the author in collaboration with Teimuraz Pirashvili. MR1600246

[147] S. A. Lopes and A. Solotar, *Lie structure on the Hochschild cohomology of a family of subalgebras of the Weyl algebra*, 2019. arXiv:1903.01226.

[148] J. López-Peña, *Factorization structures. A Cartesian product for noncommutative geometry*, Ph.D. Thesis, 2007.

[149] M. Lorenz, *On the homology of graded algebras*, Comm. Algebra **20** (1992), no. 2, 489–507, DOI 10.1080/00927879208824353. MR1146311

[150] W. Lowen and M. Van den Bergh, *Hochschild cohomology of abelian categories and ringed spaces*, Adv. Math. **198** (2005), no. 1, 172–221, DOI 10.1016/j.aim.2004.11.010. MR2183254

[151] S. MacLane, *Homology*, 1st ed., Die Grundlehren der mathematischen Wissenschaften, vol. 114, Springer-Verlag, Berlin-New York, 1967. MR0349792

[152] M. Markl, *Deformation theory of algebras and their diagrams*, CBMS Regional Conference Series in Mathematics, vol. 116, Published for the Conference Board of the Mathematical Sciences, Washington, DC; by the American Mathematical Society, Providence, RI, 2012. MR2931635

[153] M. Markl, S. Shnider, and J. Stasheff, *Operads in algebra, topology and physics*, Mathematical Surveys and Monographs, vol. 96, American Mathematical Society, Providence, RI, 2002. MR1898414

[154] R. Martínez-Villa, *Introduction to Koszul algebras*, Rev. Un. Mat. Argentina **48** (2007), no. 2, 67–95 (2008). MR2388890

[155] M. Mastnak, J. Pevtsova, P. Schauenburg, and S. Witherspoon, *Cohomology of finite-dimensional pointed Hopf algebras*, Proc. Lond. Math. Soc. (3) **100** (2010), no. 2, 377–404, DOI 10.1112/plms/pdp030. MR2595743

[156] H. Matsumura, *Commutative ring theory*, Cambridge Studies in Advanced Mathematics, vol. 8, Cambridge University Press, Cambridge, 1986. Translated from the Japanese by M. Reid. MR879273

[157] J. E. McClure and J. H. Smith, *A solution of Deligne's Hochschild cohomology conjecture*, Recent progress in homotopy theory (Baltimore, MD, 2000), Contemp. Math., vol. 293, Amer. Math. Soc., Providence, RI, 2002, pp. 153–193, DOI 10.1090/conm/293/04948. MR1890736

[158] J. C. McConnell and J. C. Robson, *Noncommutative Noetherian rings*, Pure and Applied Mathematics (New York), John Wiley & Sons, Ltd., Chichester, 1987. With the cooperation of L. W. Small; A Wiley-Interscience Publication. MR934572

[159] J. Meinel, V. C. Nguyen, B. Pauwels, M. J. Redondo, and A. Solotar, *The Gerstenhaber structure on the Hochschild cohomology of a class of special biserial algebras*, 2018. arXiv:1803.10310.

[160] L. Menichi, *Connes-Moscovici characteristic map is a Lie algebra morphism*, J. Algebra **331** (2011), 311–337, DOI 10.1016/j.jalgebra.2010.12.025. MR2774661

[161] B. Mitchell, *Rings with several objects*, Advances in Math. **8** (1972), 1–161, DOI 10.1016/0001-8708(72)90002-3. MR0294454

[162] S. Montgomery, *Hopf algebras and their actions on rings*, CBMS Regional Conference Series in Mathematics, vol. 82, Published for the Conference Board of the Mathematical Sciences, Washington, DC; by the American Mathematical Society, Providence, RI, 1993. MR1243637

[163] I. Mori and S. P. Smith, *The classification of 3-Calabi-Yau algebras with 3 generators and 3 quadratic relations*, Math. Z. **287** (2017), no. 1–2, 215–241.

[164] C. Negron, *Spectral sequences for the cohomology rings of a smash product*, J. Algebra **433** (2015), 73–106, DOI 10.1016/j.jalgebra.2015.02.021. MR3341818

[165] C. Negron and S. Witherspoon, *An alternate approach to the Lie bracket on Hochschild cohomology*, Homology Homotopy Appl. **18** (2016), no. 1, 265–285, DOI 10.4310/HHA.2016.v18.n1.a14. MR3498646

[166] C. Negron and S. Witherspoon, *The Gerstenhaber bracket as a Schouten bracket for polynomial rings extended by finite groups*, Proc. Lond. Math. Soc. (3) **115** (2017), no. 6, 1149–1169, DOI 10.1112/plms.12057. MR3741848

[167] S. Oppermann, *Hochschild cohomology and homology of quantum complete intersections*, Algebra Number Theory **4** (2010), no. 7, 821–838, DOI 10.2140/ant.2010.4.821. MR2776874

[168] M. S. Osborne, *Basic homological algebra*, Graduate Texts in Mathematics, vol. 196, Springer-Verlag, New York, 2000. MR1757274

[169] S. Pan and G. Zhou, *Stable equivalences of Morita type and stable Hochschild cohomology rings*, Arch. Math. (Basel) **94** (2010), no. 6, 511–518, DOI 10.1007/s00013-010-0137-x. MR2653667

[170] A. Parker and N. Snashall, *A family of Koszul self-injective algebras with finite Hochschild cohomology*, J. Pure Appl. Algebra **216** (2012), no. 5, 1245–1252, DOI 10.1016/j.jpaa.2011.12.011. MR2875343

[171] M. Penkava and A. Schwarz, A_∞ *algebras and the cohomology of moduli spaces*, Lie groups and Lie algebras: E. B. Dynkin's Seminar, Amer. Math. Soc. Transl. Ser. 2, vol. 169, Amer. Math. Soc., Providence, RI, 1995, pp. 91–107, DOI 10.1090/trans2/169/07. MR1364455

[172] J. Pevtsova and S. Witherspoon, *Varieties for modules of quantum elementary abelian groups*, Algebr. Represent. Theory **12** (2009), no. 6, 567–595, DOI 10.1007/s10468-008-9100-y. MR2563183

[173] T. Pirashvili and F. Waldhausen, *Mac Lane homology and topological Hochschild homology*, J. Pure Appl. Algebra **82** (1992), no. 1, 81–98, DOI 10.1016/0022-4049(92)90012-5. MR1181095

[174] H. Poincaré, *Analysis situs*, J. de l'École Polytechnique **1** (1895), no. 2, 1–123.

[175] A. Polishchuk and L. Positselski, *Quadratic algebras*, University Lecture Series, vol. 37, American Mathematical Society, Providence, RI, 2005. MR2177131

[176] L. E. Positselski, *Nonhomogeneous quadratic duality and curvature* (Russian, with Russian summary), Funktsional. Anal. i Prilozhen. **27** (1993), no. 3, 57–66, 96, DOI 10.1007/BF01087537; English transl., Funct. Anal. Appl. **27** (1993), no. 3, 197–204. MR1250981

[177] S. B. Priddy, *Koszul resolutions*, Trans. Amer. Math. Soc. **152** (1970), 39–60, DOI 10.2307/1995637. MR265437

[178] C. Psaroudakis, Ø. Skartsæterhagen, and Ø. Solberg, *Gorenstein categories, singular equivalences and finite generation of cohomology rings in recollements*, Trans. Amer. Math. Soc. Ser. B **1** (2014), 45–95, DOI 10.1090/S2330-0000-2014-00004-6. MR3274657

[179] D. Quillen, *On the (co-) homology of commutative rings*, Applications of Categorical Algebra (Proc. Sympos. Pure Math., Vol. XVII, New York, 1968), Amer. Math. Soc., Providence, R.I., 1970, pp. 65–87. MR0257068

[180] D. Quillen, *Algebra cochains and cyclic cohomology*, Inst. Hautes Études Sci. Publ. Math. **68** (1988), 139–174 (1989). MR1001452

[181] D. Quillen, *Cyclic cohomology and algebra extensions*, K-Theory **3** (1989), no. 3, 205–246, DOI 10.1007/BF00533370. MR1040400

[182] D. E. Radford, *Hopf algebras*, Series on Knots and Everything, vol. 49, World Scientific Publishing Co. Pte. Ltd., Hackensack, NJ, 2012. MR2894855

[183] M. J. Redondo and L. Román, *Comparison morphisms between two projective resolutions of monomial algebras*, Rev. Un. Mat. Argentina **59** (2018), no. 1, 1–31. MR3825761

[184] V. S. Retakh, *Homotopy properties of categories of extensions* (Russian), Uspekhi Mat. Nauk **41** (1986), no. 6(252), 179–180. MR890505

[185] J. Rickard, *Derived equivalences as derived functors*, J. London Math. Soc. (2) **43** (1991), no. 1, 37–48, DOI 10.1112/jlms/s2-43.1.37. MR1099084

[186] M. Ronco, *On the Hochschild homology decompositions*, Comm. Algebra **21** (1993), no. 12, 4699–4712, DOI 10.1080/00927879308824824. MR1242856

[187] J. J. Rotman, *An introduction to homological algebra*, 2nd ed., Universitext, Springer, New York, 2009. MR2455920

[188] K. Sanada, *On the Hochschild cohomology of crossed products*, Comm. Algebra **21** (1993), no. 8, 2727–2748, DOI 10.1080/00927879308824703. MR1222741

[189] T. Schedler, *Deformations of algebras in noncommutative geometry*, Noncommutative algebraic geometry, Math. Sci. Res. Inst. Publ., vol. 64, Cambridge Univ. Press, New York, 2016, pp. 71–165. MR3618473

[190] W. F. Schelter, *Smooth algebras*, J. Algebra **103** (1986), no. 2, 677–685, DOI 10.1016/0021-8693(86)90160-2. MR864437

[191] M. Schlessinger and J. Stasheff, *The Lie algebra structure of tangent cohomology and deformation theory*, J. Pure Appl. Algebra **38** (1985), no. 2-3, 313–322, DOI 10.1016/0022-4049(85)90019-2. MR814187

[192] H.-J. Schneider, *Lectures on Hopf algebras*, Universidad Nacional de Córdoba, 2006. http://www.famaf.unc.edu.ar/series/pdf/pdfBMat/BMat31.pdf.

[193] F. Schuhmacher, *Deformation of L_∞-algebras*, 2004. arXiv:math.QA/0405485.

[194] S. Schwede, *An exact sequence interpretation of the Lie bracket in Hochschild cohomology*, J. Reine Angew. Math. **498** (1998), 153–172, DOI 10.1515/crll.1998.048. MR1629858

[195] A. V. Shepler and S. Witherspoon, *Poincaré-Birkhoff-Witt theorems*, Commutative algebra and noncommutative algebraic geometry. Vol. I, Math. Sci. Res. Inst. Publ., vol. 67, Cambridge Univ. Press, New York, 2015, pp. 259–290. MR3525474

[196] A. Shepler and S. Witherspoon, *Resolutions for twisted tensor products*, Pacific J. Math. **298** (2019), no. 2, 445–469, DOI 10.2140/pjm.2019.298.445. MR3936025

[197] S. F. Siegel and S. J. Witherspoon, *The Hochschild cohomology ring of a group algebra*, Proc. London Math. Soc. (3) **79** (1999), no. 1, 131–157, DOI 10.1112/S0024611599011958. MR1687539

[198] A. M. Sinclair and R. R. Smith, *Hochschild cohomology of von Neumann algebras*, London Mathematical Society Lecture Note Series, vol. 203, Cambridge University Press, Cambridge, 1995. MR1336825

[199] E. Sköldberg, *A contracting homotopy for Bardzell's resolution*, Math. Proc. R. Ir. Acad. **108** (2008), no. 2, 111–117, DOI 10.3318/PRIA.2008.108.2.111. MR2475805

[200] N. Snashall, *Support varieties and the Hochschild cohomology ring modulo nilpotence*, Proceedings of the 41st Symposium on Ring Theory and Representation Theory, Symp. Ring Theory Represent. Theory Organ. Comm., Tsukuba, 2009, pp. 68–82. MR2512413

[201] N. Snashall and Ø. Solberg, *Support varieties and Hochschild cohomology rings*, Proc. London Math. Soc. (3) **88** (2004), no. 3, 705–732, DOI 10.1112/S002461150301459X. MR2044054

[202] N. Snashall and R. Taillefer, *The Hochschild cohomology ring of a class of special biserial algebras*, J. Algebra Appl. **9** (2010), no. 1, 73–122, DOI 10.1142/S0219498810003781. MR2642814

[203] Ø. Solberg, *Support varieties for modules and complexes*, Trends in representation theory of algebras and related topics, Contemp. Math., vol. 406, Amer. Math. Soc., Providence, RI, 2006, pp. 239–270, DOI 10.1090/conm/406/07659. MR2258047

[204] R. Sridharan, *Filtered algebras and representations of Lie algebras*, Trans. Amer. Math. Soc. **100** (1961), 530–550, DOI 10.2307/1993527. MR130900

[205] M. D. Staic, *Secondary Hochschild cohomology*, Algebr. Represent. Theory **19** (2016), no. 1, 47–56, DOI 10.1007/s10468-015-9561-8. MR3465889

[206] J. D. Stasheff, *Homotopy associativity of H-spaces. I, II*, Trans. Amer. Math. Soc. 108 (1963), 275-292; ibid. **108** (1963), 293–312, DOI 10.1090/s0002-9947-1963-0158400-5. MR0158400

[207] J. Stasheff, *The intrinsic bracket on the deformation complex of an associative algebra*, J. Pure Appl. Algebra **89** (1993), no. 1-2, 231–235, DOI 10.1016/0022-4049(93)90096-C. MR1239562

[208] D. Ştefan, *Hochschild cohomology on Hopf Galois extensions*, J. Pure Appl. Algebra **103** (1995), no. 2, 221–233, DOI 10.1016/0022-4049(95)00101-2. MR1358765

[209] D. Ştefan and C. Vay, *The cohomology ring of the 12-dimensional Fomin-Kirillov algebra*, Adv. Math. **291** (2016), 584–620, DOI 10.1016/j.aim.2016.01.001. MR3459024

[210] M. Suárez-Álvarez, *Algebra structure on the Hochschild cohomology of the ring of invariants of a Weyl algebra under a finite group*, J. Algebra **248** (2002), no. 1, 291–306, DOI 10.1006/jabr.2001.9013. MR1879019

[211] M. Suárez-Álvarez, *The Hilton-Heckmann argument for the anti-commutativity of cup products*, Proc. Amer. Math. Soc. **132** (2004), no. 8, 2241–2246, DOI 10.1090/S0002-9939-04-07409-X. MR2052399

[212] M. Suárez-Álvarez, *A little bit of extra functoriality for Ext and the computation of the Gerstenhaber bracket*, J. Pure Appl. Algebra **221** (2017), no. 8, 1981–1998, DOI 10.1016/j.jpaa.2016.10.015. MR3623179

[213] R. G. Swan, *Hochschild cohomology of quasiprojective schemes*, J. Pure Appl. Algebra **110** (1996), no. 1, 57–80, DOI 10.1016/0022-4049(95)00091-7. MR1390671

[214] D. Tamarkin, *Another proof of M. Kontsevich formality theorem*, 1998. arXiv:math.QA/9803025.

[215] T. Tradler, *The Batalin Vilkovisky algebra on Hochschild cohomology induced by infinity inner products* (English, with English and French summaries), Ann. Inst. Fourier (Grenoble) **58** (2008), no. 7, 2351–2379. MR2498354

[216] M. Van den Bergh, *A relation between Hochschild homology and cohomology for Gorenstein rings*, Proc. Amer. Math. Soc. **126** (1998), no. 5, 1345–1348, DOI 10.1090/S0002-9939-98-04210-5. Erratum, Proc. Amer. Math. Soc. 130 (2002), no. 9, 2809–2810. MR1443171; MR1900889

[217] B. B. Venkov, *Cohomology algebras for some classifying spaces* (Russian), Dokl. Akad. Nauk SSSR **127** (1959), 943–944. MR0108788

[218] M. Vigué-Poirrier, *Décompositions de l'homologie cyclique des algèbres différentielles graduées commutatives* (French, with English summary), *K*-Theory **4** (1991), no. 5, 399–410, DOI 10.1007/BF00533212. MR1116926

[219] Y. V. Volkov, *BV differential on Hochschild cohomology of Frobenius algebras*, J. Pure Appl. Algebra **220** (2016), no. 10, 3384–3402, DOI 10.1016/j.jpaa.2016.04.005. MR3497967

[220] Y. Volkov, *Gerstenhaber bracket on the Hochschild cohomology via an arbitrary resolution*, Proc. Edinb. Math. Soc. (2) **62** (2019), no. 3, 817–836, DOI 10.1017/s0013091518000901. MR3974969

[221] A. A. Voronov, *Homotopy Gerstenhaber algebras*, Conférence Moshé Flato 1999, Vol. II (Dijon), Math. Phys. Stud., vol. 22, Kluwer Acad. Publ., Dordrecht, 2000, pp. 307–331. MR1805923

[222] A. A. Voronov and M. Gerstenkhaber, *Higher-order operations on the Hochschild complex* (Russian, with Russian summary), Funktsional. Anal. i Prilozhen. **29** (1995), no. 1, 1–6, 96, DOI 10.1007/BF01077036; English transl., Funct. Anal. Appl. **29** (1995), no. 1, 1–5. MR1328534

[223] C. A. Weibel, *An introduction to homological algebra*, Cambridge Studies in Advanced Mathematics, vol. 38, Cambridge University Press, Cambridge, 1994. MR1269324

[224] S. Witherspoon, *Varieties for modules of finite dimensional Hopf algebras*, Geometric and topological aspects of the representation theory of finite groups, Springer Proc. Math. Stat., vol. 242, Springer, Cham, 2018, pp. 481–495. MR3901173

[225] F. Xu, *Hochschild and ordinary cohomology rings of small categories*, Adv. Math. **219** (2008), no. 6, 1872–1893, DOI 10.1016/j.aim.2008.07.014. MR2455628

Index

Selected Published Titles in This Series

For a complete list of titles in this series, visit the
AMS Bookstore at **www.ams.org/bookstore/gsmseries/**.